T0328919

NET ZERO ENERGY BUILDINGS (NZEB)

NET ZERO ENERGY BUILDINGS (NZEB)

Concepts, Frameworks and Roadmap for Project Analysis and Implementation

SHADY ATTIA
Canadian International Development Agency, Hull, QC, Canada

Butterworth-Heinemann is an imprint of Elsevier
The Boulevard, Langford Lane, Kidlington, Oxford OX5 1GB, United Kingdom
50 Hampshire Street, 5th Floor, Cambridge, MA 02139, United States

British Library Cataloguing-in-Publication Data
A catalogue record for this book is available from the British Library

Library of Congress Cataloging-in-Publication Data
A catalog record for this book is available from the Library of Congress

ISBN: 978-0-12-812461-1

For Information on all Butterworth-Heinemann publications
visit our website at https://www.elsevier.com/books-and-journals

Publisher: Matthew Deans
Acquisition Editor: Ken McCombs
Editorial Project Manager: Andrae Akeh
Production Project Manager: Anitha Sivaraj
Cover Designer: Liesbeth De Jong

Typeset by MPS Limited, Chennai, India

Dedication

This book is dedicated to my father Prof. Dr. Eng. Galal Attia, professor of metallurgical and material engineering at the Faculty of Petroleum and Mining Engineering, Suez University in Egypt. The generous support and valuable insights of my family made my life and career journey in Germany, Egypt, the Netherlands, the United States, Switzerland, and Belgium possible.

Contents

Foreword

I am glad to introduce Shady Attia's book on *Net Zero Energy Buildings* (NZEB). NZEB are crucial and there is an urgent need to cut emissions from the buildings sector as a means to mitigate climate change while providing healthy and energy efficient buildings.

I first met Shady in 2012 during the Eurotech Energy Efficient Buildings and Communities project annual meeting in EPFL, The Swiss Federal Institute of Technology of Lausanne. At that time, he was working on developing a modeling and simulation methodology for designing robust NZEB with high indoor environmental quality. The project aimed at the integration of current building energy systems and optimization techniques in a design method that considers uncertainties and quantifies NZEB's robustness.

Shady Attia is a Leadership in Energy & Environmental Design (LEED) accredited professorial and faculty member of the United States Green Buildings Council. This gave him the opportunity to certify more than 12 LEED green building over the past 15 years. His experience as a sustainability and energy efficiency consultant gave him the opportunity to reach insights on green building performance and assessment. Also, he is an assistant professor at Liege University in Belgium and leads a young research group working on sustainability assessments of the built environment and evidence-based design decision support.

One reason I think this book is useful for building professionals is that it offers a roadmap for engaging with energy efficiency in high-performance buildings projects. The book combines a solid grounding in core concepts of this subject area—such as energy efficiency and healthy buildings—with a wider context that includes the technical, socio-cultural, and environmental dimensions. Further, the book provides a framework that can help designers during the early design process.

Net Zero Energy Buildings (NZEB): Concepts, Frameworks and Roadmap for Project Analysis and Implementation provides readers with the elements they need to understand, combine, and contextualize design decisions on NZEB. The book is based on lessons learnt from NZEB design, construction, and operation that are integrated to bring the most relevant topics, such as multidisciplinary, climate sensitivity, comfort requirements, carbon footprints, construction quality, and evidence-based design. Chapters introduce the context of high-performance buildings, present overviews of NZEB, discusses the performance thresholds for efficient buildings and cover materials, micro-grid and smart grids, construction quality, performance monitoring, post occupancy evaluation, and more.

I look forward to the development and implementation of these principles and lessons learnt from NZEB projects on the path ahead.

Bjarne Olesen
ASHRAE's President for the 2017–2018 term,
Professor at Danish Technical University,
Director of the International Center for Indoor Environment and Energy, Copenhagen, Denmark
February 15, 2018

Preface

Between 2008 and 2013, I was lucky to participate as a member in the International Energy Agency Solar Heating and Cooling Task 40: "Towards net-zero solar energy buildings." I got the opportunity to collaborate with the finest researchers in the world working in the field of high performance buildings. This project involved leading international experts and as a result of my participation, several technical reports and articles (in journals and books) were published. I chaired bi-annual experts' meetings, facilitating research efforts to identify and refine design approaches and tools to support industry adoption of innovative demand/supply technologies for *Net Zero Energy Buildings* (NZEB). My PhD resulted in developing a tool to support the integration of NZEB technologies and architecture during the early design phase.

I got the idea to write this book in 2011 when I decided to launch a series of training sessions on NZEB design. I wanted building designers, operators, and owners who wanted to learn about NZEB to attend my training. Owners, designers, and building operators suffer from flawed decision-making and decision-making stress to plan, design, construct, and operate NZEB. I dedicated time every year to present one or two workshops. This was before I joined the EPFL (École polytechnique fédérale de Lausanne) in Switzerland in 2012 and Liege University Belgium in 2014. I identified a hunger for information about energy-neutral buildings and how we should design and operate them. Therefore, I decided to write this book to share my experience as an architect, building engineer, and green buildings certification expert.

My early career background as an architectural engineer provided me with exposure to building design and construction. I learnt that modeling approaches, including building performance simulation, used during early design stages of NZEB had serious built-in problems. Approaching NZEB from a modeling or simulation path is too theoretical and calculation driven. However, my later career expertise in building engineering, and in particular in building commissioning and post-occupancy evaluation, allowed me to investigate NZEB from the other side of the spectrum—after they are constructed. Unfortunately, NZEB do not work from Day 1 as expected. NZEB require a careful follow up during operation to reach optimal operation conditions and adapt to users' behavior while addressing the rebound and prebound effects. In this book, we learn from real operating NZEB, where user behavior changes building performance. Based on analyzed case studies, this book was written taking into consideration the important variation of the performance of NZEB due to occupant behavior, indoor environmental quality requirement, climate, systems, and appliances.

Currently, NZEB emerge as an urgent necessity in the construction sector to achieve energy transition. When people start to see the benefits of NZEB they naturally find it easier adopt this concept. The unique selling point of Zero Energy Buildings is that they combine the energy demand side and the energy supply side in one facility, in a very cost-effective way. Businesses and corporations are disrupting the market with products and technologies that will enable NZEB to be affordable, smarter, more connected, and grid resilient. The accelerating pace of affordable technological innovations—including photovoltaics, heat pumps, batteries, integrated HVAC systems, controls, lighting technologies, smart grids, and automation systems—will make manufacturers and businesses play a major role in increasing NZEB market uptake. The NZEB concept is strong enough to make the architecture, engineering, and construction industry collaborate to deliver ultra-efficient and energy-neutral healthy buildings. In order to design and operate NZEB we need to be holistic in our approach. NZEB are not just about energy efficiency or technology; they are about occupant's well-being and a performance-based project delivery process. NZEB should allow users to connect to the outside and control their environment in a flexible way to increase productivity and satisfaction.

Also, we have a real opportunity to fight climate change and decarbonize our energy grids through NZEB. After Fukushima accident of 2011 was declared manmade, the focus shifted worldwide to out-phase nuclear plants and handle nuclear waste disposal. In parallel, the Oil and Gas sector is losing thousands of jobs and facing immense environmental challenges related to earthquakes, spills, emissions and pollution. Therefore, cannot afford to lose the climate change battle. Integrating embodied energy calculations for material choices and land use changes in the calculation of CO_2 emissions associated with NZEB construction and operation is crucial. We need to certify NZEB and increase the awareness and skills of thousands of professionals through training and events. We need the energy industry to enable price signals, load management, and storage for the future generation of Zero Energy Buildings. Countries must strengthen their governance and leadership to adopt NZEB concept and effectively increase the market uptake. This should include structuring and enforcing the industrial and regulatory infrastructure for NZEB and renewable energy grids. With the decreasing price of renewable energy systems and decarbonization of energy grids, NZEB can transform the world's built environment. Therefore, in this book, I explain how to achieve 100% renewable, grid-friendly, and resilient Zero Energy Buildings.

Finally, I invite you to read this book as a reference guide. I enjoyed writing this book and invested time to design its structure and select its content. My moto that kept me writing was: *"I read, I investigate, and I share."* I designed the book structure so that beginners, with *hypocognition,* can read it from the beginning to the end. *Hypocognition readers,* who are unable to recognize the key concepts and ideas of NZEB, are invited to read the book from beginning until end. On the other hand, *hypercognition readers,* who have excessive understanding of NZEB, are invited to read the book from the end to the beginning.

I hope that by publishing the results of research in the field of Zero Energy Buildings will advance your capabilities.

Shady Attia

Shady Attia
University of Liège, Liège, Belgium

Acknowledgments

The work for this research could not have been done without the support of, and discussions with, Liesbeth de Jong. Her generous guidance and valuable graphic design have made this book possible.

This book was prepared with the support of many people, including members of the IEA SHC Task40/ECBCS Annex 52 team.

Special thanks is dedicated to architect Conrad Lutz (Green Offices), architect Michel Post (Iewan Social Housing, NREL team including Paul Torcellini and Shanti Pless as well as Kumpen's architect Marcel Barattucci). Also, special thanks is due to Dr. Stephen Wittkopf, Mr. Kariem Moustafa, and Mrs. Monika Sigg for facilitating the access to the Butten Multifamily house.

The author also thanks the following individuals and organizations for offering their insights and perspectives on this work:

Prof. Bjarne Olesen, ASHRAE's President for the 2017–18 term and Director of the International Center for Indoor Environment and Energy for his foreword.

Architecture et Climat research lab at UCLouvain, and in particular Prof. Andre De Herde and Prof Geoffrey van Moeseke.

Josef Ayoub, Natural Resources Canada and the IEA SHC Task40/ECBCS Annex 52 team including Prof. Andreas Athienitis, Dr. Salvatore Carlucci, Dr. Mohamed Hamdy, and Dr. Liam O'Brien.

André Wolff and Andrae Akeh, from Elsevier publishers, and all the reviewers including the book production team who provided valuable feedback during the book's development.

Finally, my gratitude is addressed to the University of Liège (ULg), The Faculty of Applied Science (FSA), and the Department of Urban and Environmental (UEE) Engineering, for the quality of its members' research and knowledge and the institutions' infrastructures.

CHAPTER

1

Introduction to NZEB and Market Accelerators

ABBREVIATIONS

AEC	Architectural, Engineering and Construction
ASC	active systems controls
ASHRAE	American society for heating and refrigeration engineers
BAC	building automation and control
BIM	building information modeling
BMS	Building Managements Systems
DER	distributed energy resources
DOE	US Department of Energy
EPBD	energy performance of buildings directive
EPC	energy performance certificate
EC	European Commission
EUI	energy use intensity
HVAC-R	heating, ventilation, air conditioning and refrigeration
IAQ	indoor air quality
IEQ	indoor environmental quality
KPI	key performance indicators
LCA	life cycle analysis
NAHB	National Association of Housing Builders
NAR	National Association of REALTORS
nZEB	nearly Zero Energy Buildings
NZEB	Net Zero Energy Buildings
PM	particular matter
PEB	positive energy buildings
POE	postoccupancy evaluation
REHVA	Federation of European Heating, Ventilation and Air Conditioning Associations
RET	renewable energy technologies
RMI	Rocky Mountain Institute
SME	small and middle-size enterprise
US	United States
WGBC	World Green Building Council

Net Zero Energy Buildings (NZEB)
DOI: https://doi.org/10.1016/B978-0-12-812461-1.00001-0

1 INTRODUCTION

Net Zero Energy Buildings (NZEB) are going to be the next big frontier for innovation and competition in the world's real estate market and can be rapidly scaled across Europe and North America. The Architectural, Engineering, and Construction (AEC) industry is under increasing pressure to deliver cost-effective, robust, ultra-low energy buildings at a fast speed. Embracing energy efficiency in high-performance buildings implies in accepting a new paradigm in the construction cycle. All stakeholders (architects, engineers, contractors, and owners) must agree beforehand on the performance levels.

NZEB are a significant part of energy efficiency strategies worldwide. As buildings represent about 30%–40% of the final energy use worldwide, the reduction of their energy demand is the key for a sustainable future. Besides being updated with energy-efficient technologies, every stakeholder should have a solid understanding of the fundamentals of the area and be familiar with the roadmap of such projects. Therefore, this book provides AEC professionals and researchers with fundamental concepts and analytic frameworks to make decisions in the context of energy-efficient and healthy buildings.

NZEB are ultra-low energy buildings that meet their energy needs annually from renewable sources, produced onsite or near-by. Examples and models for NZEB already exist throughout the United States, Europe, Asia, and Australia. NZEB definitions and concepts vary across Europe (Attia et al., 2017a). The strength of NZEB is based on combining the supply and demand sides of energy for the same building. Also, NZEB is a promising concept for Smart Cities. NZEB empower building owners and tenants and makes them seek ultra-efficiency and generate energy onsite. At the same time, NZEB are complex and require greater attention for planning, design, construction, and operation. They require systematic performance metrics and monitoring. If design teams are not experienced, this can result in greater costs. Therefore, we wrote this book to guide decision makers and inform them about the methods and approaches to achieve NZEB construction in a speedy and cost-effective way.

Several authorities and organizations including the European Commission (EC), World Green Building Council, American Society for Heating and Refrigeration Engineers (ASHRAE), the Federation of European Heating, Ventilation and Air Conditioning Associations (REHVA), US Department of Energy (DOE), and several national ministries strongly believe that states and local governments should work together to enable NZEB for new construction in their jurisdictions to scale large development and make a concerted effort to facilitate enabling policy design and stakeholder engagement to support this global project. Such effort would not only promote innovation and development in the real estate sector, but would also help leverage the abundant solar potential in states, encourage local job creation, and increase employment overall. This is a win-win market-based solution with a business case for states, local governments, developers, and homebuyers across the world.

Local governments and professionals in the AEC industry need to understand the fundamentals of energy efficiency and performance-based buildings before engaging in related projects. They need to position themselves towards future technological solutions and best practices in the building sector. There are major barriers among professionals to manage, design, and construct NZEB: Mainly the lack of understanding about the *trias energetica* concept and performance-driven design approach benefits, and methodological and systematic advice on how to implement high-performance buildings (see Chapter 2, Section 3). This book offers a roadmap to make decisions, analyze the technological solutions, and related tools currently available. We strongly believe that states and local governments should work together to enable new construction in their jurisdictions to scale new NZEB development and make a concerted effort to facilitate enabling policy design and stakeholder engagement to support the mechanism. The worldwide cases and examples presented in this book help translate the concepts to practice, also taking into consideration a variety of climatic and societal conditions. The book provides models to increase comfort, safety, energy neutrality, and affordability to scale NZEB to single and multiple families and, at the same time, commercial buildings in the market over time. This book shows how to promote innovation and development of NZEB and help leverage the abundant solar potential in urban areas, encourage more local job creation, and increase employment overall. Cutting-edge energy efficiency technologies and construction techniques are presented with a business case for local governments, developers, and homebuyers.

The book offers a roadmap for engaging in energy efficiency in high-performance buildings projects. It combines a solid grounding in core concepts of this subject area, such as viewing energy efficiency with a wider context to include the technical, sociocultural, and environmental dimensions. We explore comprehensive building modeling techniques as a tool to achieve NZEB on a large scale. The book provides logical frameworks and different roadmaps to analyze projects in the context of environmental change. Also, the book presents worldwide examples and cases for different climates and societies. We focus on lessons learned and shared best practices. The target audience is meant to be professionals in the AEC industries—building engineers and professionals. In particular, eight types of readers would benefit from reading this book, namely:

1. Developers and building owners
2. Architects
3. Engineers and building consultants
4. Contractors
5. Heating, Ventilation, Air Conditioning and Refrigeration (HAVC-R) and Building Automation and Control (BAC) professionals and manufactures
6. Property managers and building operators
7. Government (policy makers) and public funds officials
8. Graduate students and academics

The book addresses NZEB for industrial countries following high-tech approaches, as well as NZEB for nonindustrial countries following low-tech approaches. In the following sections, we present the accelerating factor that will make NZEB a main stream win-win market-based solution with a business case for nations, local governments, developers, and homebuyers. We also explore and summarize the structure of the book by providing a summary for each chapter in Section 5.

2 CLIMATE CHANGE AND GHG EMISSIONS

The AEC industry is under increasing pressure to deliver sustainable and neutral buildings. The global ecosystem of nature, in which we all live, cannot support present rates of economic and population growth (Meadows et al., 1972). After the great disaster in East Japan in 2011, the need for self-sufficient buildings and assured energy security has become crucial. There is a new generation of standards, certification schemes, and codes to create sustainable buildings (Attia, 2018). There is a rapid development and increasing market share of rapidly-built, affordable, all-electric NZEB. Homes are being turned around health, comfort, and energy performance standards at breakneck speed. NZEB are remarkably leading the way in utilizing local materials and local skills with a smart, methodological, and lean approach to design, construction, and operation. Both policy makers and the AEC industry are confronted with a range of challenges and opportunities when it comes to reducing energy consumption, improving indoor environmental quality (IEQ) and increasing the use of renewables.

2.1 Greenhouse Gas Emissions Reduction

Greenhouse gas (GHG) emissions have an impact on economy, society, and environment. The never-ending stream of carbon pollution break records for carbon dioxide in the atmosphere. Carbon dioxide exceeded the 410 parts per million threshold in many cities across the world. The cause of the steady increase of carbon dioxide is human activities that emit carbon pollution. Those human activities are altering the basic chemistry of our atmosphere and destabilizing the climate (Berry et al., 2017). Since the industrial revolution, human activities have added more carbon dioxide than plants can take up resulting in climate variability and change. As a consequence, temperatures have risen by roughly 1°C, sea levels have increased, and several other impacts have been observed worldwide. Large wildlife distinction, heat waves, and extreme rainfall are all examples of how temperature increase can further destabilize the climate. Based on the Paris agreement and Marrakesh Agreement, countries are committed to reduce the GHG emissions and adhere to emission reduction targets. As shown in Fig. 1.1, there is a global race to curb GHG emission so that the temperature rise remains below 2°C and ideally below 1.5°C. Rocky Mountain Institute's Positive Disruption Report describes limiting global temperature rise to well below 2°C analysis and describes scenarios for rapid transitions in energy, agriculture, and land use that could limit the global average temperature increase to 1.5–2°C above preindustrial levels.

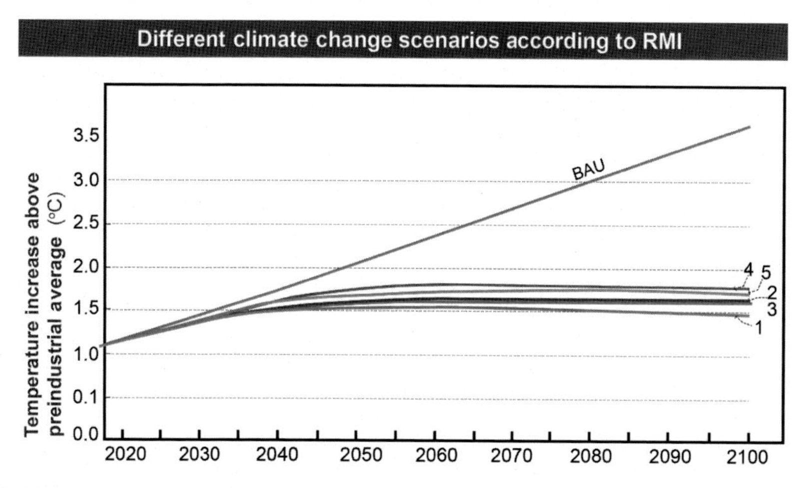

FIGURE 1.1 Global average temperature change above the preindustrial era under different scenarios (Abramczyk et al., 2017).

However, GHG emissions are increasing. The World GHG emissions resulting from burning fossil fuels increased in 2017. Based on the Global Carbon Project study, the world has not reached the peak of GHG emissions. Global Carbon Emissions are rising again after three flat years (Le Quéré and Peters, 2017). Altogether, human-caused emissions in 2017 are projected to reach 41 billion tons of carbon dioxide. The future trajectory of global emissions now depends largely on what happens in China, as well as in developing nations in Southeast Asia, Middle East, and elsewhere around the world where emissions are still increasing quickly in the background. Thus, the AEC industry has the potential to reduce GHG emissions in line with the climate target as set out in the Paris Climate Change Agreement (COP21) in December 2015. Clear and consistent policy targets play an important role in driving NZEB in the construction sector.

Also, emissions of all air pollutants are decreasing air quality (WHO, 2016). For example, in Europe ambient concentrations of particulate matter and ozone in the air are the worst they have been in decades. As shown in Fig. 1.2, this might be due to meteorological variability and the growing long-distance transportation of pollutants. However, fine particulate matter with a diameter size below 2.5 μm (PM 2.5) is now generally recognized to be the main threat to human health from air pollution. So outdoor air is the largest indoor air pollution source if air is not properly filtered and purified before taken indoors—and we need buildings to face this challenge.

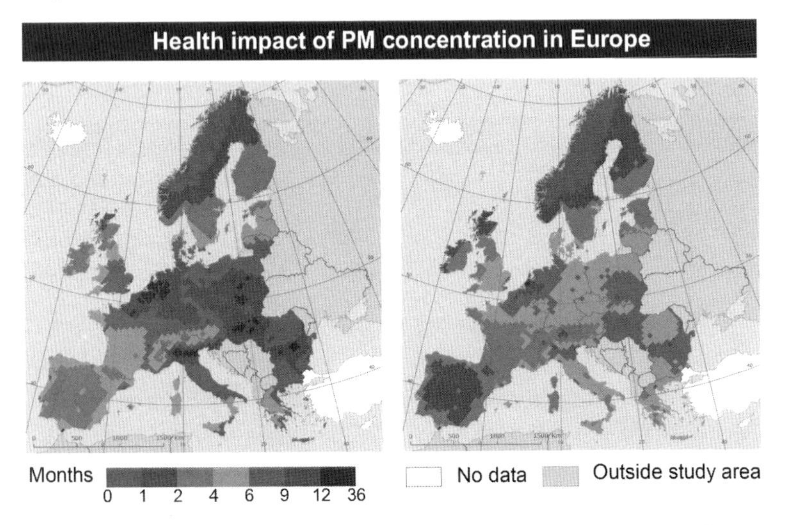

FIGURE 1.2 Estimated losses in life expectancy attributable to exposure to fine particulate matter PM 2.5 from anthropogenic emissions for 2000 (left) and 2020 (right). Source: *EEA, 2007. Air pollution in Europe, available from: https://www.eea.europa.eu/publications/eea_report_2007_2, accessed November2017 (EEA, 2007).*

2.2 Net Zero Energy Buildings as a Solution

Based on the facts above, nations must address the GHG and sequestration opportunities associated with the building sector. By 2050, all major business sectors must operate in a state of zero carbon emissions. The rapid advances in key technologies including solar power, battery storage, heat recovery systems, heat pumps, BAC, performance dashboards, LED lighting, and electric vehicles put the AEC industry at the cusp of rapid change. The electrification revolution and flexibility of the grid and its smart management are speeding structural transition and change. The World Green Building Council calls for a dramatic and ambitious transformation towards a completely zero carbon built environment, through the dual goals of:

- All new buildings must operate at net zero carbon from 2030.
- 100% of buildings must operate at net zero carbon by 2050.

Therefore, high-performance buildings including NZEB and positive energy buildings (PEB) are a means to fight climate change. The energy needs for space cooling, space heating, domestic hot water, and ventilation need to be calculated and operated to improve health, achieve comfort, and neutralize energy consumption. The concept of building energy-neutral buildings that produce the energy they consume to meet the users' needs is one of the key solutions to reach carbon reduction targets. By employing high-energy efficiency measures to decrease energy demand as much as possible, the rest of the energy mandate using renewable sources and managing energy systems must be covered. Another advantage of NZEB is their performance-based approach that ensures IEQ, energy efficiency, and energy generation at the same time. NZEB have been proven to be a reliable innovative solution. NZEB minimize their impact on the environment, while being strong enough to withstand storms and climatic challenges. They are part of the healthy buildings movement to incorporate users and health at the heart of their development and operation process. Scaling this concept on the urban level to benefit from distributed energy resources (DERs)—such as roof-top solar, heat pumps electricity batteries and smart micro grids—collectively has been already proven (see Chapter 9: Construction Quality and Cost) as a strong business case. Through concerted action of three core groups of actors—business, government and nongovernmental organizations—the impacts of climate change can be avoided and a number of societal and economic benefits will be brought about.

3 SMARTNESS AND GRID MODERNIZATION

We are on the verge of a third industrial revolution where digitalization, automation, and energy flexibility will lead the market transition.

Digital transformation and automation has cascaded though major global industries, including the construction sector. The digitation of the industry, productivity, and cost-effectiveness are tightly aligned. The digitalization and automation of the construction sector will achieve a critical mass within the construction ecosystem. Next, modernizing the grid to make it "smarter" and more resilient through the use of cutting-edge technologies, equipment, and controls that communicate and work together to deliver electricity more reliably and efficiently, is transforming the construction industry. The demand for flexible and energy-neutral buildings, and easier integration of local electric controls with building management systems (BMS), are leading to an overall optimization of building energy efficiency. In the following sections, we will discuss the impact of this revolution on the NZEB market uptake.

3.1 Digitalization and Automation

We are in a new digital age. Owners are seeking lower costs, better outcomes and designs, while construction firms are seeking competitive differentiation and higher profits. Our new digitally driven age is driving the construction sector to advance its supply chain and fabrication process. Suppliers are selling owners and contractors productivity as opposed to products and services. The data-driven automation of equipment and inventory control is increasing productivity and coordination. Building information modeling (BIM) modular prefabrication and 3D printing are changing traditional project delivery process and manufacturing. Offsite assembly and prefabrication saves labor costs, improves quality, and consistency (Bonington, 2017). Printing building components onsite, on demand, and based on virtual models has been an effective method that will gradually increase in the market. Onsite jobs, mobile devices, robots, virtual reality, and drones are increasing work productivity. The automation of materials streamlining and logistics supply, together with digital tagging and tracking, contribute to improving the project delivery process and increase collaboration and project revenues. Digitalization and automation is directly linked to the energy transition. This wave of digitalization and automation will be profound on the AEC industry, leading to streamlining and modernizing the construction process. It reduces the cost of NZEB construction and makes the project delivery process more efficient.

Building automation for heating, ventilation, air conditioning, lighting, and other systems through a BMS or BAC, is an available market solution. The totalized control buildings or centralized control units of indoor climate solutions, ventilation products, or intelligent control systems, together with model predictive control, are becoming a fast-growing market reality. Almost no legislation or standard is addressing smart buildings control for NZEB. Smart NZEB are a cornerstone of a decarbonized green economy. The increased integration of DERs, renewables, storage, and the growing peak demand for electricity is driving the market to increased flexibility, demand-response capabilities, and consumer empowerment. BACs complement smart buildings. Building automation enables the participation of demand response, encourages immediate use or storage, and enables synergies between smart buildings and electromobility. With the proliferation of self-monitoring building components and systems and their ability to communicate with the grid through climate, comfort, or price signals, building will become smart and users will be part of the larger energy markets.

3.2 Smartness and Grid Modernization

In a net zero energy world, operation of both buildings and the electric grid need to be more flexible and responsive. Smart sensing, smart controls, data analytics, machine learning, and energy mix diversification are all concepts and technologies that will change how we operate our buildings. The building industry is forced to shift its focus from single buildings to collective building agglomerations in relation to the grid. The design of a low-energy community while meeting the constraints of the burdened electric grid is challenging. Finding the optimal balance between energy efficiency, district thermal energy, and onsite PV to achieve zero-energy districts and communities is the next challenge. The deployment of infrastructure for electromobility is accelerating. There are no metrics that define building-level grid interaction or rate building-grid interaction quality. However, grid operators and energy companies are forced to integrate renewable energy onto the grid. The diversification of national energy mix and the accelerating pace of introducing electric vehicles in cities are forcing the construction sector to transform buildings into smart buildings. The use of sophisticated machine-learning algorithms to model energy, thermal cooling/heating, and power requirements for buildings, as well as benefit from weather forecasts is guiding the building industry towards integrated, connected, and efficient appliances and systems. The Internet of Things encompass being able to sense occupants' outdoor and indoor conditions to provide comfortable and healthy environment at an individual level. There is a valorization of building-grid interactions that will force the AEC industry to realize smart and high-performance buildings. We believe that the reduction of GHG emissions of national energy mixes will increase the demand for smart NZEB to reduce peak demand to

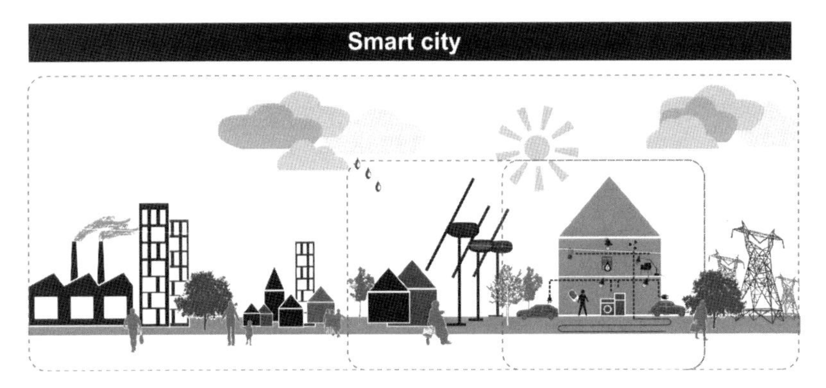

FIGURE 1.3 We are heading towards an ecosystem of interconnected smart cities, smart systems, and smart buildings.

balance self-generation and back-feed. Grid smartness is becoming a reality that depends strongly on climate within an interconnected network of smart cities; smart systems, and smart buildings (see Fig. 1.3). To keep energy companies and grid management cost-effective and increase the resilience of energy grid, future building must become smart and net zero.

4 MARKET SIZE AND PROPERTY VALUE MATTERS

NZEB are catalysts in the market transformation of the built environment. Momentum in pursuing NZEB is growing in the United States, Europe, and Asia. An increasing number of buildings achieved this concept, raising confidence that net zero goals are realistic given current building technologies and design approaches (Ayoub et al., 2017). NZEB are designed for people's security, safety, comfort, wellbeing, and productivity. They provide a fair and high-quality building solution for consumers. In the following section, we will discuss two influential factors which indicate an increasing demand for NZEB, and their increasing market share.

4.1 Market Demand for NZEB

Several online databases were developed in order to facilitate quick, easy, and tailor-made access to national and comparative international indicators on NZEB activities. However, it is difficult to estimate the market share and demand on NZEB. The year 2015 saw US$15billion of investment in zero energy buildings (ZEB) even though relevant policies are in their infancy globally, with US$12.5 billion of this investment in energy efficiency and US$2.5 billion in renewable energy (IEA, 2016). In the United States, zero energy residential markets are growing and industry leaders are already tapping into this US$1.3 trillion market. On the EU-level, the online nearly Zero Energy Buildings (nZEB) tracker shows that the European Union currently dominates the market for ZEB construction at more than US$14 billion (Navigant Research, 2015). With a potential incremental market opportunity of over US$33billion by 2037, NZEB residences offer favorable outcomes to homeowners and real estate developers, while also benefitting the economy and the environment (Lalit and Petersen, 2017). This reflects the enactment of a net zero energy building policy framework that prompted the recent growth and is expected to stimulate further growth through to 2020.

Also marketers explain, a fast-growing segment, mostly millennials, is already targeted through active searches for NZEB (Lalit and Petersen, 2017). A growing segment of homeowners in Europe and North America are concerned about carbon dioxide emissions associated with building use. NZEB provide a reasonable return of investment and make sense for operational energy cost optimization. Therefore, we can imagine that NZEB will provide a genuine experience for future homebuyers to generate energy onsite and benefit from ultra-energy efficiency measures. This shall grant a long-term loyalty to early adopters and environment conscious segments that are already increasing year after year. The key message here is that the AEC industry needs to brand NZEB to increase their market share due to their societal, economic, and environmental value.

4.2 Cost Cutting and Property Value

Scaling NZEB to district scale implementation can accelerate the market uptake and reduce the project's cost. We learned from case studies that operating buildings collectively on a cluster, community, or district level can be more efficient and cost-effective (see Fig. 1.4). Achieving NZEB on a district-level and coupling them to micro grids is promising and has the potential to maximize optimal management interactions. Implementing NZEB on a district scale, creates a stronger connection between the building's energy users and renewable energy system capacities on a larger scale. The increase of demand-supply and the direct interaction between different buildings on the same project site can lead to locally centralized energy generation. This allows for leverage of synergies such as thermal load electric supply aggregation. The use of central heating or cooling systems becomes cost-effective and boost energy efficiency, aggregating buildings in a district which can facilitate loads management and make the micro grids more robust and

FIGURE 1.4 The Sustainable City on the outskirts of Dubai (Google Earth, 2018).

reliable. As a consequence, developers and tenants can reap the benefits of those synergies, reduce their carbon footprint, save money, and decrease upfront investment.

The New Buildings Institute (NBI, 2014) report also found that net zero energy can be achieved within the cost range of other buildings of the same type. When total construction costs for these buildings are analyzed against control groups, NZEB are comparable to conventional buildings (see Chapter 9: Construction Quality and Cost). Although their cost can be lower (or neutral) compared with conventional buildings, NZEB are not taking off as fast as might have been expected. The design process is different and results in buildings that are different in function, aesthetics, and comfort (BEI, 2014). Different does not mean worse, but it does mean change. Also, people fear taking loans and cost-optimality studies do not assure many consumers. People, particularly those investing millions of dollars, tend to shy away from experimentation. As comfort and well-being grows with new design features, NZEB are poised to grow in popularity. Rocky Mountain Institute's (RMI's) latest insight brief, *R-PACE: A Game-Changer for NZE Homes*, shows that providing access to financial resources can lead to benefits of a cost-saving, high-performance smart home (Lalit and Petersen, 2017). Also in Europe, on an annual basis, NZE homes are likely to provide homeowners with more annual cost savings than the amount they pay to compensate their annual additional initial cost installment. In fact, there is a need for a comprehensive set of policies to finance and provide an incentive system. In other words, NZEB are a great investment as it ultimately costs the users less than an average building year-on-year with no incremental up-front cost. Moreover, the intangible benefits of the superior performance of an NZEB makes the business case even more compelling.

5 BOOK CHAPTERS

After exploring the market accelerators, it is now time to introduce this book's chapters. In this section, we summarize the book structure to provide an overall overview of the content. Every chapter is based on best practices and case studies that we share with readers. We also include, in each chapter, a final discussion and set of lessons learned in relation to the chapter's title.

5.1 Chapter 2: The Evolution of NZEB Definition

The definition of NZEB has evolved since 2010. In 2010, the IEA and EU set their definitions within a vision of the 2020 Horizon. However, there are many new proposals to extend the 2010 definition. The concepts of autonomy, carbon neutrality, and circularity are a concern to set a new definition and performance targets for NZEB until 2050. In this chapter, readers will be introduced to global and national definitions of NZEB with an overview for the coming years. The chapter presents the main concepts and design principles for NZEB and the different approaches available in the market to achieve NZEB status. A specialized discussion on the high-tech and low-tech approaches to reach NZEB is also presented.

5.2 Chapter 3: NZEB Performance Indicators and Thresholds

In this chapter, the concepts of energy use intensity, rating systems and certification, building performance robustness, and minimum performance threshold will be elaborated on. Another unique topic that is discussed in this chapter will be the correlation between the performance-based design and prescriptive design for NZEB. This chapter will explain the issues related to performance indicators for NZEB with a specific focus on:

- Maximum values of energy demand.
- Overarching values for primary energy consumption kWh.
- Carbon Emissions.
- Indoor Environmental Quality.
- Heating and Cooling Balance.

- Renewable Energy Share and Storage.
- Occupancy Density.
- Cost of NZEB.

Finally, the chapter provides a worldwide review of the status of NZEB in six countries and provides suggestions for NZEB performance threshold in selected countries.

5.3 Chapter 4: Integrative Project Delivery and Team Roles

In this chapter, we elaborate on the integrated design process with a focus on the iterative and circular nature of design of NZEB versus the linear traditional design process. With the help of process maps for NZEB case studies we demonstrate to the reader how to understand and avoid common problems during the design and project delivery process. We discuss the importance of integrative process design, integrated teams, and collaboration as well as how to prepare NZEB performance-based contracts. Additionally, we discuss the use of building performance modeling and BIM technology as a means to assess the design alternatives before construction begins.

5.4 Chapter 5: Occupants Well-Being and Indoor Environmental Quality

The selection of a thermal comfort model for establishing indoor optimal hygrothermal conditions for NZEB has a major impact on energy consumption. The objective of this chapter is to compare the influence of using different thermal comfort models for NZEB to guide and inform decision-making. Similarly, the impact of IEQ and occupant behavior on expected energy savings will be elaborated on in this chapter. We explore latest the ISO, REHVA, and ASHRAE standards and certification systems (i.e., WELL, LEED and BREAAM) that deal with the broadest sense of IEQ in high-performance buildings. The overall aim of the chapter is to demonstrate the strong relation between health, occupants' well-being, and achieving energy targets. Through illustrations and flow charts, readers will be informed about the correlation between the comfort/IEQ requirements and energy consumption of NZEB.

5.5 Chapter 6: Materials and Environmental Impact Assessment

By tracing the environmental impact of operational energy and embodied energy of several case studies, this chapter will explore the influence of choices for building materials for NZEB. With the mandatory performance requirements of nZEB by 2020 in the EU, we cannot remain operating under the efficiency or neutrality paradigm. Therefore, this chapter will demonstrate that how selecting the right materials play a role in mitigating the effects of climate change and help building designers create a positive impact of the built environment on their surroundings. The following topics will be elaborated in this chapter:

1. Environmental Performance.
2. Embodied Energy.
3. Environmental assessment.
4. Materials Passport.

We also provide the results of life cycle analyses (LCA) for several NZEB.

5.6 Chapter 7: Energy Systems and Loads Operation

Active systems and building services are essential for NZEB to achieve comfort during extreme climatic conditions. With the global phenomena of urban densification, the built environment is associated with air pollution, noise, excessive heat, humidity, and dust. Therefore, it is crucial to guarantee high-quality indoor and outdoor comfort associated with, and proportional to, occupant activities in the buildings. At the same time, NZEB are buildings with little heating demand and variable needs. NZEB are very sensitive to occupants' needs and less sensitive to outdoor conditions. Therefore, the sizing, integration, and installation of flexible systems with high COP for domestic hot water, mechanical cooling, free cooling, ventilation, and IEQ control must become a part of any NZEB integrated concept—even if the extreme conditions occur only occasionally and during short periods of the year. In this chapter, readers will gain practical skills in HVAC design consideration and renewable energy systems, as well as how to increase the coverage rates of renewable energy on-site (PV). The following topics will be elaborated on in direct relation to NZEB design and operation:

1. Building services integration,
2. HAVC systems, energy generation, and energy storage.
3. Mechanical ventilation with heat recovery.
4. Thermally active buildings systems (TABS).
5. Plug loads and electric lighting.
6. Building automation and control.

We also elaborate on the importance of demand control systems (i.e., temperature presence, CO_2, VOC, and humidity) and significant energy savings of mechanic heat recovery ventilation systems. One of the learned synergies of properly designed and operated mechanical heat recovery ventilation systems is that they are essential elements of NZEB, because they lead to overall improvement of IAQ and a reduction of reported health-related problems.

5.7 Chapter 8: Smart-Decarbonized Energy Grids and NZEB up Scaling

The challenge to maintain grid stability will become larger with the increasing penetration of renewable energies. Therefore, energy flexibility and costs in NZEB will play an important role in facilitating energy systems based entirely on renewable energy sources. Smart grids are necessary to control energy consumption to match actual energy generation from various energy sources—i.e., solar and wind power. The aim of this chapter is to demonstrate how energy flexibility in buildings can provide generating capacity for energy grids, and to identify critical aspects and possible solutions to manage such flexibility. This knowledge is important in order to incorporate the energy flexibility of buildings into future smart energy systems and to better accommodate renewable sources in energy systems. It is also important when developing the business case for using flexible NZEB within future systems to increase grid stability and to potentially reduce costly upgrades of energy distribution grids. Smart buildings, smart grids, and decarbonized power grids are at the cross-section of this chapter.

5.8 Chapter 9: Construction Quality and Cost

The engineering and construction sector has been slow to adopt new technologies, and has certainly never undergone a major transformation. However, technological advances are now revolutionizing almost all points in the life-cycle of a built asset, from conceptualization to demolition. In this context, NZEB are high-tech buildings that require construction quality and advanced building components. Without highly trained labor, high-performance materials, building components, and quality-driven construction practice, NZEB will not function. This chapter will elaborate on factors that influence the cost of NZEB in relation to construction quality, digitalization, standardization, and the role of prefabrication in achieving affordable and high quality NZEB. Construction best practices and aspects related to capacity building, education and training of professionals and builders are elaborated on too. Finally, we address the important role of inspection, cost control, commissioning, project management, and procurement as strategical tools to assure quality, compliance, and certification during construction.

5.9 Chapter 10: Occupant Behavior and Performance Assurance

As performance-driven buildings, NZEB operate within an energy balance objective. Within this context monitoring, it is essential to validate the building performance and include real feedback evaluating the environmental performance aspects of buildings. With the proliferation of microprocessor-based meters and data loggers, this chapter will illustrate the importance of monitoring and provide some practical principles to apply mentoring techniques to NZEB. The chapter provides some guidance on the appropriate Postoccupancy Evaluation (POE) techniques to promote good practice and to offer reference material to support practicing building assessors who want to evaluate occupant satisfaction, occupant behavior, and postoccupancy building performance. In this chapter, we focus on the role of occupant engagement, occupant control modes, and adaptation measures that can influence of overall performance of NZEB during occupancy. The influence of occupant density, and workplace or living place culture, are addressed, as well as the challenge to deal with the fluctuation of occupancy presence and modes of use in the indoor environment.

Moreover, we share lessons learned related to occupant appropriation and performance assurance measures in NZEB, and realistic estimating savings. Monitoring, energy performance certification, continuous commissioning, soft-landing, and performance visualization are discussed in this chapter in direct relation to occupant behavior.

5.10 Chapter 11: NZEB Case studies and Lessons Learned

This chapter presents five in-depth case studies of NZEB that generate as much energy as they consume over the course of a year. The case studies represent different project types, sizes, and climate zones include large office buildings, a municipality building, an industrial building, a school, and a multifamily autonomous housing block. Each case study describes the project goals, design principles, performance characteristics, and lessons learned. Based on measurement, verification, and postoccupancy evaluations, we present results that include technical insight, hands-on experience, and valuable evidence on NZEB′ operation. This chapter presents comparable information regarding the performance, cost, and occupant behavior to avoid when designing and operating a NZEB. The case studies form the foundation for other chapters providing unique insights into common best practices and mitigation strategies that are critical for any owner or designer looking to get to the net zero energy target in their operations.

5.11 Chapter 12: NZEB Road Map and Tools

Implementing and up-scaling NZEB require planed actions. National and Federal governments, and state and local authorities are beginning to move towards zero energy building targets. In this final chapter, we present two roadmaps for NZEB implementation. The roadmaps present strategical guidance for policy makers in industrial and nonindustrial countries. Both roadmaps depict a high-tech and low-tech approach to bring NZEB to local markets. The aim of these roadmaps is to guide government officials and authorities on the strategical level to set- short, medium-, and long-term targets to plan the NZEB market uptake. Also in this chapter, we present operational guidance for property owners, project managers, and building professionals for NZEB projects' implementation. We identify the important role building modeling can play to design, construct, and operate NZEB. Lastly, we discuss the means to get to NZEB status and the best practices to increase the market uptake and lower barriers. By presenting roadmaps, action plans, and tools we offer support and guidance for a broad group of market stakeholders.

6 LESSONS LEARNED #1

NZEB are becoming the main stream quality in the AEC industry. The benefits of NZEB are far-reaching, but the most benefit resides in NZEB' ability to balance energy loads by providing net-neutral or positive energy to the local grid. With electrification, we might soon shift the focus from energy management to power management. The markets are prepared to increase their uptake and governments should accelerate this process through active price signals and a comprehensive set of polices to finance and create incentive systems. The growth of green and smart technologies is accelerating the market uptake of NZEB and their interaction with smart grids and appliances. Fighting climate change and reducing GHG emissions is a great enabler allowing for the realization of large scale projects. Almost 75% of the building stock in industrial countries is inefficient. However, we cannot improve existing buildings and accelerate the pace of deep renovation without mastering the design, construction, and operation of newly constructed NZEB for all building typologies in all different climates. We need to prepare the energy and infrastructure landscape to host these smart, flexible, and interactive high-performance buildings.

One of the important enablers of NZEB is that the market demand is increasing annually by millennials and others. NZEB are proven to be cost-effective, healthy, and energy efficient. Rating systems such as LEED, BREAAM, DGNB, WELL, ARC, CASBEE, Green Star, Living Building Challenge BELS, and other certification schemes are the best way to integrate energy efficiency and health aspects for user-centered NZEB. Ultimately, the realization on district scale allowing for up-scaling NZEB to reduce their cost in a way that benefits not only the building, but also the larger community and grid, is the aim of this book. Branding NZEB as part of the solution to GHG reduction and climate change is extremely important. More demonstration projects need to be promoted to exemplify the benefits, viability, and integrated approach of NZEB. NZEB provide an opportunity to deliver an integral approach for high performance, healthy, and energy-efficient buildings. This book provides an effective opportunity for sharing knowledge about NZEB best practice and lessons learned—and we invite you to read it and reap the benefits.

References

Abramczyk, M., Campbell, M., Chitkara, A., Diawara, M., Lerch, A., Newcomb, J., 2017. Positive Disruption: Limiting Global Temperature Rise to Well Below 2°C. Rocky Mountain Institute. Available from: http://www.rmi.org/insights/reports/positive_disruption_limiting_global_temperature_rise.

Attia, S., 2018. Regenerative and Positive Impact Architecture: Learning from Case Studies. Springer International Publishing, London, UK, ISBN: 978-3-319-66717-1.

Attia, S., Eleftheriou, P., Xeni, F., Morlot, R., Ménézo, C., Kostopoulos, V., et al., 2017a. Overview of challenges of residential nearly Zero Energy Buildings (nZEB) in Southern Europe. Energy and Buildings 155, 439–458.

Ayoub, J., Aelenei, L., Aelenei, D., Scognamiglio, A. (Eds.), 2017. Solution Sets for Net Zero Energy Buildings: Feedback from 30 Buildings Worldwide. John Wiley & Sons, Hoboken, NJ.

BEI, 2014. Net zero energy market update. Available from: http://www.buildingefficiencyinitiative.org/articles/net-zero-energy-market-update (accessed November 2017).

Berry, P., Sánchez-Arcilla Conejo, A., Betts, R., Harrison, P.A., 2017. High-end climate change in Europe: impacts, vulnerability and adaptation.

Bonington, P., 2017. Construction's digital breakthrough, AEC-ST. Available from: https://www.aecst.com/blog/constructions-digital-breakthrough/ (accessed November 2017).

EEA, 2007. Air pollution in Europe. Available from: https://www.eea.europa.eu/publications/eea_report_2007_2 (accessed November 2017).

Google Earth V 7.1.5.1557, March 01, 2018. The sustainable city, Dubai 25° 02′N, 55° 27′E, Eye alt 15 meter. DigitalGlobe 2018. Available from: http://www.earth.google.com (accessed 01.03.18).

IEA, 2016. Energy efficiency market report. Available from: https://www.iea.org/eemr16/files/medium-term-energy-efficiency-2016_WEB.PDF (accessed November 2017).

Lalit, R., Petersen, A., 2017. R-PACE: A Game-Changer for Net-Zero Energy Homes. Rocky Mountain Institute (RMI), Boulder, CO.

Le Quéré, C., Peters, G., 2017. Global Carbon Project. Available form: http://www.globalcarbonproject.org/carbonbudget/17/files/International_FutureEarth_GCPBudget2017.pdf (accessed November 2017).

Meadows, D.H., Meadows, D.L., Randers, J., Behrens, W.W., 1972. The limits to growth. New York 102, 27.

Navigant Research, 2015. Energy service company market overview: expanding ESCO

NBI, 2014. Getting to zero status update. Available from: https://newbuildings.org/resource/2014-getting-zero-status-update/ (accessed November 2017).

World Health Organization, 2016. Air pollution levels rising in many of the world's poorest cities.

Further Reading

Opportunities in the United States and Europe, Navigant Research, Boulder. Available from: www.navigantresearch.com/research/energy-service-company-market-overview (accessed November 2017).

CHAPTER

2

Evolution of Definitions and Approaches

ABBREVIATIONS

AEC Architectural, Engineering, and Construction
ACH Air Change per Hour
DHW domestic hot water
ECM energy conservation measure
EE energy efficiency
EPBD Energy Performance Building Directive
EPD environmental product declaration
EU European Union
EUI Energy Use Intensity
GHG greenhouse gases
HVAC Heating, Ventilation, and Air Conditioning
IEA International Energy Agency
IEQ indoor environmental quality
LCA life cycle assessments
LCZEB Life Cycle Zero Energy Buildings
PE primary energy
PEB Positive Energy Buildings
PEB Plus Energy Buildings
PH Passive House
PHI Passive House Institute
PHIUS Passive House Institute US
PHPP Passive House Planning Package
PV photovoltaic
NZEB Net Zero Energy Buildings
nZEB nearly Zero Energy Buildings
RES renewable energy systems
RPE renewable primary energy
US DOE United States Department of Energy
US EPA United States Environmental Protection Agency
US United States
ZERH Zero Energy Ready Home

Net Zero Energy Buildings (NZEB)
DOI: https://doi.org/10.1016/B978-0-12-812461-1.00002-2

1 INTRODUCTION

Definitions of Net Zero Energy Buildings (NZEB) vary by entity, organization, and country—from NZEB to nearly Zero Energy Building (nZEB) to Net Zero Carbon Building (NZCB). They all share a common goal: To reduce or neutralize the environmental impact of buildings. Therefore, we should keep in mind that there is no single definition of NZEB but there are several different definitions or frameworks depending on the climatic, economic, or political conditions of any country. The other common goal of NZEB is that they all share the founding principles of design. Definitions are very useful to translate the design principles and fundamentals of high-performance buildings' design into a consistent framework and concept. In this chapter, we aim to understand definitions that are moving toward net zero and to analyze building energy performance principles using the Energy Use Intensity (EUI) indicator. The chapter aims to provide a scope on the evolution of NZEB worldwide and the importance of this concept to accelerate the transformation of industries toward a greener economy. Without a clear and consensus-based national NZEB definition, we cannot achieve environmental targets to reduce greenhouse gas (GHG) emissions from buildings. Definitions are essential to benchmark NZEB performance and be able to push building codes while training designers and workers and perform appropriate monitoring for different building types. In this chapter, we will discuss the historical evolution of NZEB definitions and present the fundamental principles, basics, and boundary conditions for their design. Also, we will explore the key ambitious benchmarks, design concepts, and standards in relation to climate and the type of building function, EUI, and occupancy expectations. Finally, we will discuss the different approaches and policies that are implemented in different countries to make sure NZEB become market stream. The last section of this chapter, will discuss the guidelines for setting a definition that can lead to performance robustness of NZEB, providing insights on the consensus-based nature of any successful definition. The formulation of NZEB definitions should reflect that policy on issues related to the national energy mix, grid flexibility, which industry implication, and know-how infrastructure for the architectural, engineering, and construction industry will assist to bring those buildings to the market and achieve a deep transformation.

2 DEFINITIONS OF NZEB

Setting and framing NZEB definitions makes it easier to understand and recognize, and to build on the basic principles and concepts of their design. Setting definitions assists to develop subconcepts and translate them into building performance indicators and thresholds. This is not some kind of intellectual luxury, but a basic entry to achieve mature cognition and informed decision-making. There are many reasons why we should trace the history of high-performance or performance-based buildings including NZEB. It is crucial to learn from the past and build on previous findings to avoid repetition and the struggle to identify definitions that are adaptive to their context and users' needs. Several research programs and implementation projects have been funded in the past 50 years in relation with performance-based buildings worldwide. However, a well-established characterization and lessons learned from the approaches, methodologies, and definitions of those buildings are largely missing. Therefore, in the following section we will explore the evolution of NZEB definitions and provide a series of key terminologies associated with these types of high-performance buildings.

2.1 History of NZEB

The definition of NZEB emerged strongly in recent years as a new concept for sustainable and energy-neutral development in the built environment. The importance of this concept lies in its capacity to solve part of our current environmental crisis by limiting our dependence on fossil fuels and replacing this with renewable energy through a decentralized approach. Historical definitions of zero energy are mainly based on balancing the building operation energy use including heating, cooling, ventilation, lighting, plug loads, etc. (see Fig. 2.1).

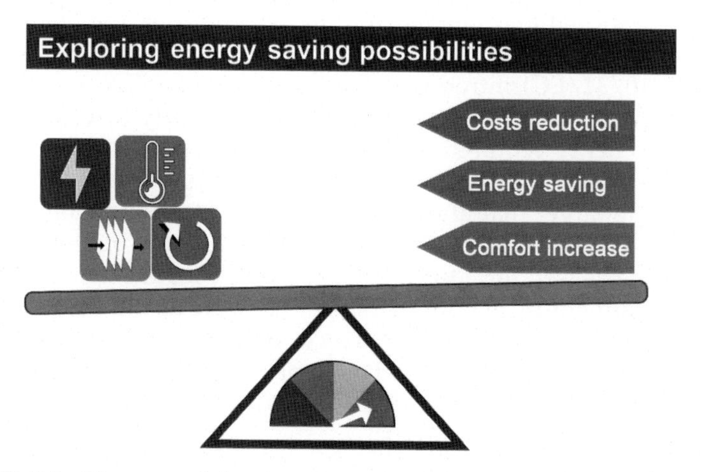

FIGURE 2.1 The energy balance concept for NZEB is mainly calculated for end-use energy or primary energy.

The earliest attempts toward zero energy buildings can be traced in North America (Steinmüller, 1979; Butti and Perlin, 1980; Attia, 2016). Original writing and research, done in North America in 1960s, 1970s, and 1980s, are proof that the concept emerged first in the United States and Canada and was voiced during the time of the establishment of United States Environmental Protection Agency (US EPA) as a consequence of the oil crises and oil embargo (Besant et al., 1979; Mazria, 1979; Butti and Perlin, 1980; Tabb, 1984, Balcomb, 1992; Deviren and Tabb, 2014). By the end of the 1970s, the idea moved to Europe and Asia with the earliest examples including the Philips Experimental House Project in 1974 (Bruno and Steinmüller, 1977; Bruno et al., 1979) and the conceived house by Van Korsgaard in Denmark (1977), Saitoh in Japan (1985), and Rostvik in Norway (1988) (Korsgaard, 1976; Esbensen and Korsgaard, 1977; Scognamiglio and Røstvik, 2013). In 1988 the renowned physicist William Shurcliff declared the super insulated building concept as mature (Nisson and Dutt, 1985; Shurcliff, 1988). By the mid-1990s, the Germans refined the concept and the Passive House Institute (PHI) was established as a fundamental step to the evolution of the NZEB concept. In parallel, the British came up with the Zero Carbon concept and Bill Dunster implemented his BedZED project. After the new Millennia, the term "net zero energy" was introduced and articulated in 2008 by the members of Task 40: The Net Zero Energy Solar Buildings project of the International Energy Agency (IEA, 2017). The "net zero energy" term presents the annual energy balance between consumed and generated energy for a building. Therefore, "net zero energy" has significant consequences for the classical production consumption balance, grid interaction, and energy mix. As a consequence, the European Union (EU) suggested a much more formal and attainable term by proposing the "nearly zero energy" in 2010. The Energy Performance of Buildings Directive (EPBD, 2010/31/EC) introduced the definition of nZEB as a building with very high energy performance where the nearly zero or very low amount of energy required should be extensively covered by renewable sources produced on-site or nearby (EPBD, 2010). This was followed by the target that after December 31, 2020, all new buildings in the 28 EU member states would be nZEB, while for public buildings the deadline is set for December 31, 2018. Meanwhile, the United States Department of Energy (US DOE) recognized the Passive House Institute US (PHIUS) work and signed an agreement with PHIUS to promote the Zero Energy Ready Home (ZERH) in the US market (Klingenberg et al., 2016). Chapter 3 provides an overview on the state of the art of NZEB in 6 countries.

In parallel, the term Plus Energy Buildings or Positive Energy Buildings (PEB) was informally coined by Rolf Disch in 1994. With the European 2020 target PEBs, that produce more energy from renewable energy sources over the course of a year, are increasing slowly worldwide. The French government is the first European country that has set an official definition and target for PEBs through the Building Positive Energy (BEPOS) mandate that will be effective starting 2018 (Attia et al., 2016, 2017). Chapter 3, Net Zero Energy Buildings Performance Indicators and Thresholds provides an overview on the state of the art of NZEB in 6 countries.

2.2 Terminology and Definitions

Ambiguity and vague terminology and definitions in the green building practice reduce the likelihood of building professionals' adherence. It leads to inconsistent interpretations and, as a result, to inappropriate performance variation and construction errors. When a topic grows in significance so do the terminology and definitions associated with it, as well as the risk of misunderstanding. Successful resolution of ambiguity and vagueness of terms and metrics for NZEB require an understanding of its meaning and characteristics. In order to present NZEB definitions we will explore the different terminology and address the energy use in buildings during its life cycle and the connection with the energy grid and energy stations in association with different energy sources and the associated carbon emissions. Here, we articulate and describe the key terminology and definitions associated with NZBE:

Performance-based building design: An approach to design buildings to meet certain measurable or predictable performance requirements. The performance requirement can be for a single criterion or multi-objective criteria. Performance-based building design requires predefined performance thresholds and performance metrics.

Benchmark: Benchmarking is the practice of comparing the measured performance of a building to itself (its building typologies, or established norms) with the goal of informing and motivating performance improvement.

Balance metric: The balance metric is a unitary measuring entity expressed to quantify the operational energy, end-use energy, embodied energy, primary energy, or carbon emissions of a NZEB.

Performance threshold: Performance thresholds are the maximum acceptable limit or value for specific metrics a project team can use to assess an individual performance criteria. There are upper- or lower-limit parameter values for a performance indicator.

Energy Use Intensity (EUI): EUI expresses a building's energy use as a function of its size or number of users. EUI is expressed as energy per square meter (foot) per year (kWh/m^2 per annum or kBTU/ft^2 per annum) or as energy per capita per year (see Fig. 2.2).

Delivered energy (imported): Energy imported by the building using the EUI index.

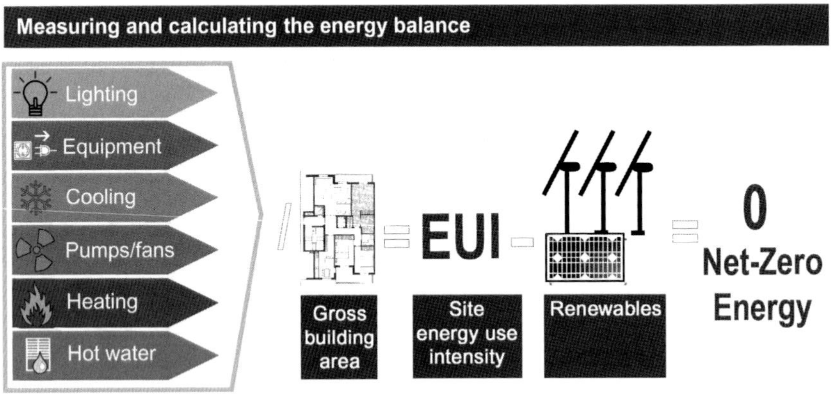

Measuring and calculating the energy balance

Lighting
Equipment
Cooling
Pumps/fans
Heating
Hot water

Gross building area

EUI

Site energy use intensity

Renewables

0

Net-Zero Energy

FIGURE 2.2 Graphical equation for the energy use breakdown and energy use intensity for net zero energy balance calculation.

Exported energy: Energy flowing from the building to the grids specified by each energy carrier using the EUI index.
Generated energy: Energy generated on-site specified using the EUI indicator.
Generation consumption balance: The percentage of supply versus demand over a period of time.
Balance boundary: Determines which energy uses are included in the balance (heating, cooling, lighting, appliances, domestic hot water (DHW), servers, etc.).
Operational energy: The amount of energy that is consumed by a building to satisfy the building energy demand at the operational stage of the building to maintain the building's functions and occupants' activities.
Embodied energy: Embodied energy is the total energy required for the extraction, processing, manufacture, and delivery of building materials to the building site. Embodied energy is measured as the quantity of nonrenewable energy per unit of building material, component, or system. It is expressed in mega joules (MJ) or gigajoules (GJ) per unit weight (kg or tonne) or area (m^2).
Primary energy use: Direct use of energy at the source, or supply to users without transformation of crude energy. The primary energy is the energy that has not been subjected to any conversion or transformation process.
Primary energy conversion factors: A conversion factor converts the balance metric into another balance metric and may include indirect effects such as transmission and delivery losses. Conversion factors reflect political and national preferences in relation to the energy mix and energy policy and do not always depend on scientific

considerations. For example, the use of waste or biofuels including pellets can reduce the factor to 0.5 or 0.6, as opposed to 2.5 for electricity depending on the energy policy and energy mix in that particular country.

Energy mix: The range of energy sources of a region, either renewable or nonrenewable from which secondary energy for direct use—usually electricity—is produced.

Exergy: Exergy analysis and efficiency provide a very effective tool to improve the energy use for buildings. Reducing the exergy losses in either a system or a process and increasing its exergy efficiency means to use energy in a more rational way. High exergy overall efficiencies mean exploiting all the available exergy content and using energy in the most rational way.

The balance and energy or energy neutrality represents the core concepts of NZEB (Marszal et al., 2011; Sartori et al., 2012). There are several informal and formal definitions of NZEB. The most important definitions are summarized below:

Nearly Zero Energy Buildings

A nZEB produces 30% or more of its required energy through the use of on-site renewable energy.

Net Zero Site Energy

A net site zero energy building produces at least as much energy as it uses annually, when accounted for at the site. The measurement time range is annual.

Net Zero Source Energy (Primary Energy)

A net zero source energy building produces at least as much energy as it uses annually, when accounted for at the source. Source energy refers to the primary energy required to generate and deliver the energy to the site. To estimate a building's total source energy, imported and exported energy is multiplied by the appropriate site to source conversion factor (Fig. 2.3).

Primary energy and primary energy indicators are calculated from delivered and exported energy as

$$PE = \frac{PE_{nren}}{A_{net}} \tag{2.1}$$

$$PE_{nren} = \sum_i \left(E_{deliv,i}, f_{deliv,nren,i}\right) - \sum_i \left(E_{exp,i}, f_{exp,nren,i}\right) \tag{2.2}$$

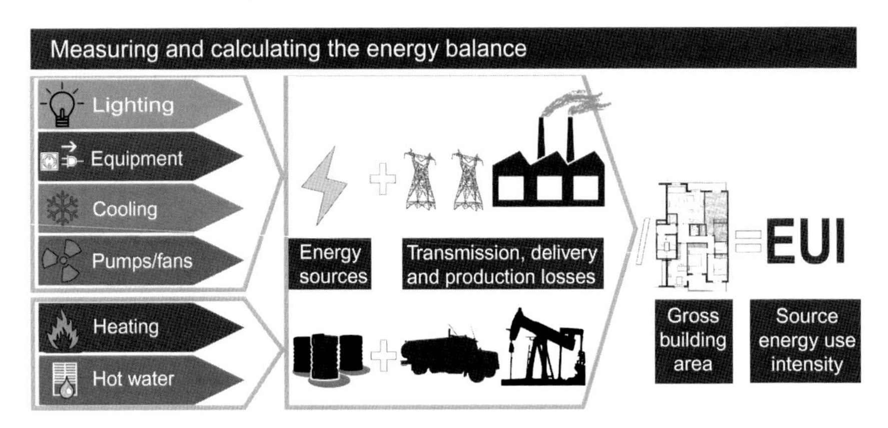

FIGURE 2.3 Graphical equation for the primary energy use breakdown and energy use intensity for net zero source energy balance calculation.

where PE is the primary energy indicator (kWh/m^2 per annum); A_{net} is the net used floor area (m^2) calculated according to national definition; PE_{nren} is the nonrenewable primary energy (kWh/annum); $E_{deliv,i}$ is the delivered energy on site (kWh/annum) for energy carrier i; $f_{deliv,nren,i}$ is the nonrenewable primary energy factor for the delivered energy carrier i; $E_{exp,i}$ is the exported energy on site (kWh/annum) for energy carrier i; $f_{exp,nren,i}$ is the nonrenewable primary energy factor for the delivered energy compensated by the exported energy for energy carrier i; which is equal to the factor of the delivered energy, if not defined nationally in different way.

Consequently, the carbon emissions of energy use can be calculated based on the delivered and exported energy.

$$PE_{CO_2} = \frac{\text{Emission CO}_2}{A_{net}} = \frac{\sum_i \left(E_{deliv,i} CE_{deliv,i}\right) - \sum_i \left(E_{exp,i} CE_{exp,i}\right)}{A_{net}} \tag{2.3}$$

where PE_{CO_2} is the CO_2 emission indicator (kgCO_2/m^2 per annum); Emission CO_2 is CO_2 emission (kgCO_2/annum); A_{net} is the net used floor area (m^2) calculated according to national definition; $E_{deliv,i}$ is the delivered energy on site (kWh/annum) for energy carrier i; $CE_{deliv,i}$ is the CO_2 emission coefficient (kgCO_2/kWh) for the delivered energy carrier i; $E_{exp,i}$ is the exported energy on site (kWh/annum) for energy carrier i; $CE_{exp,i}$ is the CO_2 emission coefficient (kgCO_2/kWh) for the exported energy carrier *I, depending on national definition.*

Net Zero Energy Cost

In a net zero energy cost building, the amount of money the owner pays the power plants or utility for energy services and energy used over a year is at least equal to the amount the energy the company pays the building tenant or owner for the energy the building exports to the grid over a year.

Life Cycle Zero Energy Buildings

For this definition, the zero-energy building is extended to include the embodied energy of the building and its components together with the annual energy use. A Life Cycle Zero Energy Buildings (LCZEB) is a building where the primary energy used during operation and the energy embodied within its constituent materials and systems, including energy generating ones, over the life of the building are equal to, or less than, the energy produced by its renewable energy systems (RES) within the building over its lifetime (Hernandez and Kenny, 2010).

Net Zero Energy Cost Optimal Buildings

The EPBD recast requires Member States to ensure that minimum energy performance requirements of buildings are set with a view to achieving cost optimal levels using the comparative methodology framework established by the European Commission (EPBD, 2010; EU, 2012). The cost optimal performance level should get updated and calculated every 5 years. This mechanism will benefit from the technological evolution, energy performance improvements, and cost reductions (Kurnitski, 2013). All member states must link cost optimality and the nZEB principle in a consistent way and facilitate its convergence until 2021.

3 PRINCIPLES OF NZEB DESIGN

The zero energy buildings and zero carbon buildings goals seek maximum efficiency derived from the notion of neutralizing the resource consumption and define this as zero energy consumption. Worldwide, there are several concepts and solutions for high-performance buildings. Global initiatives such as the 2030 Challenge in the United States or the nearly Zero Energy Target mandated through the European Energy Performance Building Directive (EPBD) are aimed at reducing harmful emissions from buildings. There are also voluntary green ratings and certification systems such as Leadership in Energy and Environmental Design (LEED), BREEAM, or Minergie that aim to upscale an industrial creation of NZEB. In the previous chapter, several definitions of NZEB were explored with variations in requirements depending on the building typology, climate, and geographic context. Therefore, in this section we will present the basic principles of NZEB design.

3.1 Trias Energetica

Although there is little documentation on principles of NZEB design, there is a generic set of rules that would apply to all high-performance buildings: *Trias Energetica*. The *trias energetica* set of rules is a three-step approach to design high-performance buildings. This three-step approach was developed in 1979 in the Netherlands under Kees Duijvstein and is considered as the basic rule of thumb for sustainable building design (Korbee et al., 1979).

The three steps are (see Fig. 2.4):

1. Reduce the demand for energy by avoiding waste and implementing energy conservation measures (ECMs).
2. Use sustainable sources of energy instead of finite fossil fuels.
3. Use the cleanest possible fossil energy as efficiently as possible.

Only when a building has been designed to minimize the heat loss in heating dominated buildings, or minimize the heat gain in cooling dominated buildings, should the design team focus on renewable energy solutions on-site, including solar systems or heat exchange and recovery systems. This rule of thumb emphasizes the importance of architecture, bioclimatic design, and energy efficiency (EE).

3.2 NZEB Design Principles

There are four principles to design NZEB. These principles can be implemented on different levels and can include detailed metrics such as embodied energy and environmental impact. Table 2.1 lists a series of measures that can be applied for NZEB based on the four NZEB

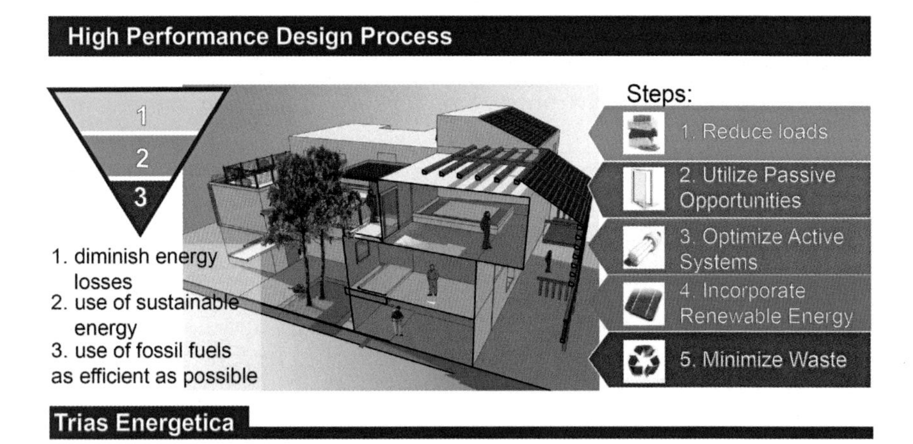

FIGURE 2.4 The fundamentals of NZEB design.

TABLE 2.1 The Four Principles of NZEB Design

First NZEB principle	Second NZEB principle	Third NZEB principle	Fourth NZEB principle
Reduce energy demand	Improve indoor environmental quality	Provide renewable energy share	Reduce primary energy and carbon emissions
Reduce internal loads Reduce building envelope loads Reduce HVAC equipment energy consumption	Set up minimum fresh air per person Enable Natural Lighting Set up a maximum occupant density	Produce energy from renewable sources on-site Introduce renewable energy delivered from nearby or offsite Avoid double counting	Reduce the primary energy demand Reduce the carbon emissions related to delivered energy

design principles. Depending on the building typology and climatic context, design teams need to apply these four principles to identify the best fit to purpose measures that respect these steps or principles, which are listed below:

- First, reducing the energy demand for all newly constructed buildings. The energy demand value is for the sum of the demands of the building, space heating, space cooling, DHW, auxiliary energy, ventilation, lighting, and appliances.
- Secondly, improving indoor environmental quality (IEQ) allowing maximum thermal comfort and avoiding overheating. This includes air quality control through mechanical ventilation.
- Thirdly, fixing a percentage of renewable energy demand to be covered by renewable energy annual balance. It is also important to amend additional measures to address energy matching and storage issues.
- Fourthly, reducing the overarching value for primary energy consumption and carbon emissions per year. It is also important to amend additional measures to address mobility and materials' embodied energy issues.

In this context, bioclimatic or climate responsive design together with building EE are an integral part of the UN's sustainable development Goal 7, which aims to "ensure access to affordable, reliable, sustainable, and modern energy for all." Finally, the *trias energetica* steps need to be associated with performance thresholds described in the following section. The importance of performance thresholds is to set up minimum acceptable and fit-to-purpose variances for specific building performance metrics. The *trias energetica* steps and ECM indicated in Fig. 2.5 need to be

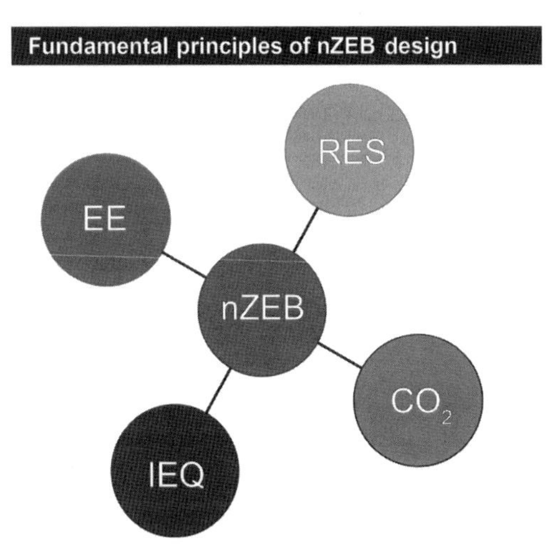

FIGURE 2.5 Principles of nZEB design.

coupled to measurable thresholds in order to be able to validate and evaluate the building performance. During the design phase, building performance simulation and energy modeling tools can verify those performance thresholds to assess the building design. After the building construction, building performance monitoring should be used to trace the preset performance thresholds to close the building energy and carbon gap.

To sum up, the *trias energetica* principle together with the 4 NZEB design principles are founding principles for ultra-EE buildings' design, IEQ, and renewable energy generation. Those principles are recognized worldwide and lead to the selection of design solutions sets, systems, and products that are biased toward energy conservation, meeting the total primary energy uses and assuring IEQ. The optimal balance between demand and supply, or between energy saving and energy generation will be always challenging and depends on the local context and cost effectiveness of the different applications (Attia and De Herde, 2010a). Chapter 3, NZEB Performance Thresholds, elaborates further on NZEB design principles and other indicators and thresholds that are not included in this section. The next section will demonstrate different successful concepts and standards that embrace those principles and formulated design packages of NZEB for different building types in different climates.

4 CONCEPTS, TYPOLOGIES, AND STANDARDS FOR NZEB

Broadly applied concepts for NZEB are foundational to efforts by governments and private entities to make NZEB mainstream in the

Architectural, Engineering, and Construction (AEC) market (Kurnitski, 2013). Today there are several achieved paths for NZEB of all types, sizes, and climates. NZEB are no more design frameworks or solely demonstration projects. NZEB include a wide range of mainstream building typologies based on formal standards that reflect a universal trend of market uptake. This section presents the major concepts, types, and standards on various NZEB.

4.1 NZEB Concepts and Standards

Passive House

As a continuation of the history of NZEB, discussed earlier in Section 2.1, the German Passive House (PH) concept is regarded as the most famous concept to achieve NZEB in heating dominated climates. The PH was first conceived in a townhouse development in Kranichstein in Darmstadt Germany in 1994. Based on a funded research project, the concept aimed to reduce the building energy demand by a factor of 10. Five basic principles form the cornerstone of any PH building. This includes a peak load limit of 15 kWh/m^2 per annum (4.75 kBtu/ft^2 per annum) for annual heating and cooling demand each, an envelope airtightness of 0.6 volume ACH (0.033 CFM/ft^2) at 50 Pa and compliance with the requirement for the nonrenewable primary energy demand of $\leq$120 kWh/m^2 per annum (38 Btu/ft^2 per annum). A PH building is highly insulated with high-performance windows and thermal bridge free design (see Fig. 2.6). By addressing remaining needs with on-site renewable energy generation, the concept paved the way for NZEB. Today the PH concept has been formulated into a standard under the PHI in Germany and is voluntarily adopted across Europe. Since 2010, the PHI is active in North America resulting

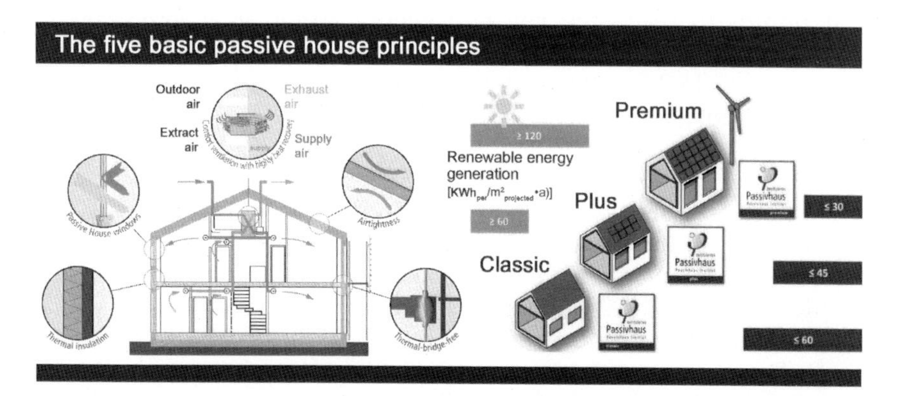

FIGURE 2.6 Basic Passive House principles and the three classes of standard certification (Passipedia, 2017).

in the creation of the PHIUS. With the ambitious European 2020 Climate-Energy Framework, many heating dominated countries including Belgium, Denmark, and Sweden have adopted the standard as obligatory in their own climate. Case studies and examples can be found in Chapter 12, NZEB Roadmap and Tools.

Since 2016, the PHI defined three new classes for PHs. With the use of the Passive House Planning Package (PHPP), which is a tool for calculating and validating the new PH designs, a new evaluation procedure, focusing on renewable primary energy (RPE) serves as a basis of the new classes. As shown in Fig. 2.6, heating demand and cooling demand should not exceed 15 kWh/m^2 per annum, while the overall demand for RPE will be used instead of the primary energy demand, which was previously used. In the case of the PH Classic category, the maximum threshold should be 60 kWh/m^2 per annum. The category of PH Plus is more stringent as it should not exceed 45 kWh/m^2 per annum for RPE. Also the PH Plus must generate at least 60 kWh/m^2 per annum of energy in relation to the area covered by the building. For PH Premium, the energy demand is limited to just 30 kWh/m^2 per annum, with at least 120 kWh/m^2 per annum of energy being generated on-site by the building (Passipedia, 2017).

Active House

With the increased popularity of PH, Velux the Danish Window manufacturer created a competing standard called Active House. The Active House building standard incorporates similar PH design strategies, but focuses on the occupants' well-being encouraging natural daylighting and natural ventilation. The concepts feature automated controls to bring in fresh air when needed and automated blinds and exterior awnings to control shading. The PH and Active House require RES, such as solar water heaters, PV panels, or geothermal heat pumps to reach the net zero energy target. It is important to mention that both concepts emerged mainly for residential buildings and are now looking forward to dominate the commercial building typologies too.

Energy Performance of Buildings Directive

The Directive on the energy performance of buildings (EPBD) is the Directive of the European Parliament and Council on EE of buildings. The Directive came into force in 2003 and had to be implemented by the EU Member States by 2006. It was inspired by the Kyoto Protocol which commits the EU to reduce CO_2 by 8% by 2010, to 5.2% below 1990 levels. The directive came into force in 2006 and requires member states to comply with Article 7 (Energy Performance Certificates), Article 8

(Inspection of boilers), and Article 9 (Inspection of air conditioning systems) within three years of the inception date, the deadline being early 2009 (EU, 2002). According to the EPBD, the energy performance of buildings should be calculated on the basis of a methodology, which may be differentiated at national and regional level. That includes, in addition to thermal characteristics, other factors that play an increasingly important role such as heating and air-conditioning installations, application of energy from renewable sources, passive heating and cooling elements, shading, indoor air quality, adequate natural light and design of the building. The methodology for calculating energy performance should be based not only on the season in which heating is required, but should cover the annual energy performance of a building. That methodology should take into account existing European standards.

The EPBD requires all new buildings to be nearly zero-energy by the end of 2020. All new public buildings must be nearly zero-energy by 2018. The EPBD targeted to support the European Commission in its activities to guide the member states on how to interpret the requirements for nZEB, develop a common reporting format on nZEB to be used by member states, evaluate the adequacy of measures, and activities reported by member stated in their national plans. A comprehensive database on nZEB projects of all typologies and climates was developed by Wuppertal University and allows the cross analysis of the experiences and data from realized NZEB (Schimschar et al., 2013). Several studies conducted under the IEE program revealed the importance to assess boundary conditions of different dimensions when designing and defining low and zero energy buildings.

Energy Performance Certificates

Energy performance certification (EPC) is one of the most effective mechanisms developed by the EU under the EPBD framework to promote EE in the market. They provide information for consumers on buildings they plan to purchase or rent (see Fig. 2.7). They include an energy performance rating and recommendations for cost-effective improvements. Certificates must be included in all advertisements in commercial media when a building is put up for sale or rent. They must also be shown to prospective tenants or buyers when a building is being constructed, sold, or rented. After a deal has been concluded, they are handed over to the buyer or new tenant. EU countries must also put in place schemes for the inspection of heating and air-conditioning systems, or take measures that have an equivalent impact on energy savings. Under the EPBSD, all EU countries have established independent control systems for energy performance certificates and inspection reports for heating and cooling systems. For example, France introduced

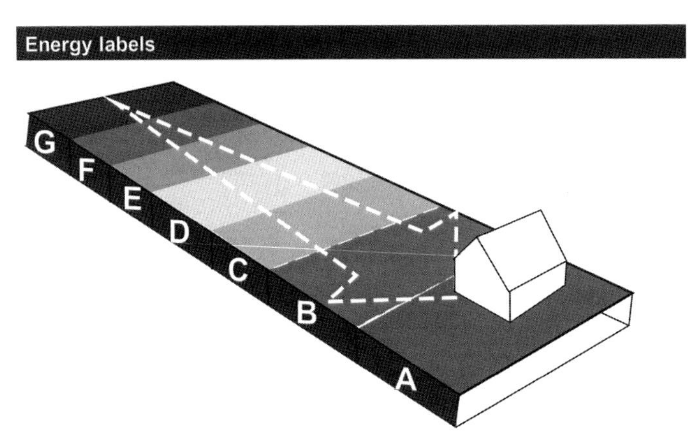

FIGURE 2.7 An EPC shows the energy current and potential energy rating of a property based on charts divided into seven bands ranging from A to G. Each range has a set amount of energy rating out of a maximum of 100 points (being maximum efficiency).

its EPC in 2006 for marketed properties for sale and all newly constructed buildings. Ten years later, the French EPC database comprises more than two million buildings. An EPC is valid for 10 years and an accredited energy assessor must base the assessment on an on-site audit.

2000 Watt Society

The idea of the 2000 Watt Society is a Swiss concept to divide global energy consumption by three and GHG emissions by eight by 2100. An average European citizen consumes an average of 6500 W per person per year every day. The idea behind this concept is to shift to renewable energies while sharing energy resources to reach an average of 2000 W. The Swiss Society of Engineers and Architects promoted this concept in partnership with the Swiss Federal Energy Office to develop the concept and transform it into a label that can be applied to building and district scale. From a theoretical point of view, the concepts are very advanced and aim to neutralize energy consumption by using renewable energy for both electricity and heating. One of the unique aspects of this concept is the inclusion of mobility's impact within the calculation balance. The concepts allow only the use of public transport, electric vehicles, and bicycles. There are already realized in a project in Zurich and two new certified neighborhoods are under construction in Basel.

Factor 4 Buildings

The Factor 4 concept follows the Sustainable Development World Strategy, the Kyoto protocol, and the European energy policy which is to reduce by Factor 4 energy consumption in European countries before

2050. In 1972, the Club of Rome—a global think tank concerned with the future of humanity and the planet—published a book titled *The Limits to Growth* (Meadows et al., 1972). Through computer modeling, *The Limits to Growth* demonstrated that unchecked economic and population growth in a global system of limited resources is unsustainable. As a possible solution to the challenges identified in their book, *The Limits to Growth,* Ernst Von Weizsäcker et al. (1998) of the Wuppertal Institute (Germany) and Amory and Hunter Lovins of the Rocky Mountain Institute (US) published a report—Factor 4: Doubling Wealth, Halving Resource Use—to the Club of Rome in 1995. Factor 4 develops the concept of resource productivity—the amount of "wealth" we can extract from the resources we use. The rationale is to halve global resource consumption while doubling global welfare, thereby achieving a fourfold improvement in resource productivity. Factor 4 means that resource productivity should grow fourfold. The amount of wealth extracted from one unit of natural resources can quadruple. Factor 4 demonstrates how improvements in resource productivity are essential in building a just and sustainable global society. Factor 4 aims at a fourfold improvement in resource efficiency—the amount of wealth (services) generated per unit of resource used—within the Factor 4 area and beyond the zero-energy building target to include resources consumption over the full life cycle.

Zero Energy Ready Home

In 2008, the US DOE launched a program to recognize hundreds of leading builders for their achievements in EE. The DOE ZERH represents a new program to award and train builders for homes on the path toward NZEB. A US DOE ZERH is a high-performance home which is so energy efficient that a renewable energy system can offset all (or most) of its annual energy consumption. The home must comply with the Energy Star for homes certificate and comply with inspection checklists. This includes the high-performance envelope that meets 2012 International Energy Conservation Code levels for insulations and provides comprehensive indoor air quality through full certification in US EPA's Indoor airPlus program (DOE, 2017a; EPA, 2017). The home should also accomplish savings on the cost of future solar PV installations.

Green Certification Programs

Another important example that supports the progressive high-performance building target is the green certification program and rating system. Certification programs such as ENERGY STAR, Green Star, Living Building Challenge, CASBEE, HQE, GRIHA, and DGNB proliferated at the beginning of the new millennium as a means to address sustainability in the built environment. The most two famous rating systems, which have been associated with NZEB design, are BREAAM and LEED. BREEAM (Building Research Establishment Environmental Assessment Method) was conceived by the British Research Establishment and was first used in 1990. The LEED, developed by the US Green Building Council (USGBC) in 1998, provides a suite of standards for environmentally sustainable construction. Where there are prescriptive credits in LEED, these are generally less onerous than in BREEAM. While green certification programs are criticized for being costly, following a check box mentality or flawed in setting priorities, there is no doubt that they changed the AEC industry (Attia, 2014). We should note that rating systems are under continuous development to achieve exceptional performance and every new version is stronger than the previous one. There is a shift away from using calculation only methods to depend more and more on monitoring which is a rigorous approach to improve performance and help building stakeholders to commit to performance-based accreditation. The success of those with the most famous green certification programs and rating systems is related to their ability to develop adaptable and context-sensitive building assessment methods and mechanisms. Emerging rating systems such as the WELL Building Standard are shifting the focus toward occupants' well-being and health, while performance verification through monitoring and postoccupancy evaluation is becoming more central. We expect a proliferation of new rating systems, including SITES and ARC, etc., in the coming years and a diversification of the performance-based green certification programs. Therefore, we recommend to combine green building certification with NZEB targets to provide a third party assessment and validation that helps design teams to achieve the performance targets through a systematic framework during building design, construction, and operation.

4.2 NZEB Types and EUI

There is a wide range of mainstream NZEB types and ownership that reflects the universal trend of NZEB adoption. NZEB types vary according to their layout, size, location, operation process, and function resulting in different EUIs for different building types. Meanwhile, there are two main metrics to assess the primary energy intensity (see further discussion on building occupant density in Chapter 3: NZEB Performance Thresholds and Chapter 10: Occupant Behavior and Performance Assurance). The first metric emerges from the notion to power and condition buildings using primary energy intensity per square meter (foot). The second metric emerges from the notion to serve people using the primary energy intensity per person. Using one of both metrics can change our assessment of high-performance building drastically. By using the EUI per area, we will stay further away from achieving our environmental GHG emission targets. However, today the EUI indicator per area represented in the kWh/m^2 per annum is the commonly used universal metric. The future question is how to reduce the EUI per person without losing comfort, privacy, and quality. For the national definitions of NZEB in EU member states, the performance levels should be specified for the following building types:

1. single-family houses of different types
2. apartment blocks
3. offices
4. educational buildings
5. hospitals
6. hotels and restaurants
7. sports facilities
8. wholesale and retail trade services buildings
9. other types of energy-consuming buildings

The discussion of EUI is important because different NZEB types are getting more and more represented. In the past, most NZEB were small, demonstration-scale projects, but today NZEB have been built across a range of sizes and types in many countries. The vast majority of NZEB are residential or office buildings smaller than 2000 m^2. They represent mainly newly constructed buildings. However, an emerging category is renovated NZEB. The different databases for NZEB suggest more diversity and distribution of different NZEB types including schools, pharmacies, retail stores, and even the high-intensity building types such as hospitals and restaurants (DOE, 2017b; EnOB, 2017; NBI, 2016). There are even plans to construct NZE towers worldwide in 2018. Tracking existing NZEB indicates that they are being built across all climates. For sure, the most concentration of NZEB can be found in the Northern Hemisphere in industrial countries, but there are representative projects in almost all climates worldwide (DOE, 2017b). NZEB can be grouped and classified under three main categories namely: Heating dominated, mixed mode, and cooling dominated buildings. In the near future, it is expected that with the advancement of EE policies and building codes worldwide, NZEB will be more common and get classified according to their loads, type, and EUI which will lead to benchmarking which will inform future designs.

5 APPROACHES FOR NZEB

Several countries strengthened their national policies on building EE in the past few decades. These national plans can include differentiated targets according to the category of building. For example, the EU launched the nZEB energy objective in 2010 and ASHRAE explored the development of maximum technically achievable ultra-low energy use building set presented in ASHRAE 1651-Research Project in 2016. The German PH standard is the oldest concept that rigorously applies the *trias energetica* rule of thumb to achieve ultra-low energy buildings that require little energy for space heating or cooling (15 kWh/m^2 per annum).

5.1 Overview of NZEB Regulations and Policies

In Europe and across the world, there have been two main ways to bring NZEB into the built environment. The first way is a conservative and gradual policy (step-by-step) driven approach seeking to increase the performance thresholds to reach high-performance buildings over time through mandatory building codes. The second way is more progressive (fast and ambitious) based on encouraging high end performance following voluntary green program certifications. The following section explains both ways and provides a brief overview on the current state of NZEB market status.

Gradual Progress

In trying to achieve NZEB, the EPBD introduced EPCs and since 2010 mandated that all new buildings constructed within the EU after 2020 should reach nearly zero energy levels. This European ambitious policy is fully based on a prescriptive-based design approach that started in 2006 and gradually pushed the building EE requirements and renewable on-site production in a progressive way. The EU sustainability strategy is based on large scale deployment of zero energy buildings (nZEB) implying high resource efficiency and renewable energy dependence. The Climate-Energy Framework 2020 sets three key targets to cut 20% in GHG emissions (compared to 1990 levels), increase the EU renewables share by 20%, and improve EE by 20%. The main instrument to achieve those targets in the building sector is the Energy Performance of Building Directive (EPBD) recast that sets the standards for new and renovated buildings across Europe. The Directive 2010/31/EU (EPBD) indicates that EU Member States must ensure that by 2021 all new buildings, and already by 2019 all new public buildings, are nZEB. Accordingly, most member states recently revised their existing rules, regulations, and guidelines as well as starting to set up the means for increasing the penetration of those high-performance buildings by setting up nZEB definitions on the national level.

In 2013, ASHRAE introduced its Standard 189, Standard for the Design of High-Performance, new Green Standard to help architects, engineers, and facility managers to utilize a standard approach to designing, building, and operating Green Buildings. The design requirements were created to result in a 30% reduction in energy consumption compared with standard 90.1. Later, in 2016, the ASHRAE research project 1651P developed the maximum technically achievable energy targets for commercial buildings and coupled it to LEED Rating system. Since 2010, the US General Services Administration's upgraded requirements for LEED Gold certification, as a minimum in all new federal building construction and renovation projects. By using the LEED, the US General Services Administrations can evaluate and measure achievements in sustainable design. With a slightly different indicator, in 2006 the United Kingdom set a zero-carbon level for all new houses built after 2017. The zero carbon homes policy aimed to dramatically reduce CO_2 emissions from housing, which currently makes up nearly a third of the UK's GHG emissions, a figure which could rise to 55% by 2050. However, the policy was scraped by the UK government in 2015.

According to the previous review, the step-by-step progressive approach has been used slowly since the oil crisis in 1970s and introduced stringent performance measures over long periods following a mandatory approach across Europe and other industrial countries. Figure 2.8 compares the gradual evolution of building energy performance requirements in the United States and Europe in the past 50 years. It seems likely that the EU member states are close to reach an energy-neutral building stock by 2050.

Ambitious Leaps

The second way to achieve NZEB is more ambitious, based on embracing ultra-efficient green certification programs, concepts, and standards. There are examples for such ambitious standards including the PH from Germany, Minergie from Switzerland, the Active House from Demark, and the certification programs as described in Section 4.1. This approach is mainly voluntary, but it has been a market driver worldwide. Nongovernmental organizations (NGOs) worldwide adapted those standards and concepts to drive the sustainability and green construction market. With the global concern about health, ecological problems, increased evidence of global warming, and rising prices of crude oil citizens seek ambitious leaps to end the fossil dependence and go beyond the gradual formal regulations for the EE of buildings. The Solar Settlement in Freiburg-Vauban, Germany, completed in 2005, is an example of one of the earliest urban Plus Energy Building initiatives based on a grass root cooperative planning process. The citizen's association "Forum Vauban e.V.," an NGO, applied to coordinate the participation process and was recognized as its legal body by the City of Freiburg in 1995. This is an example which shows how citizens,

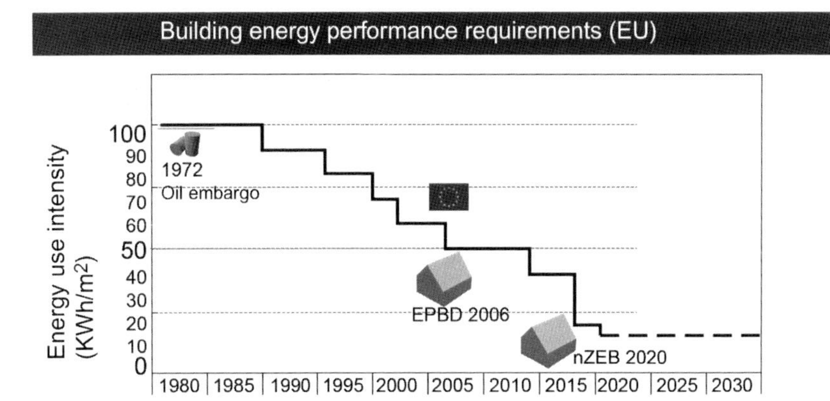

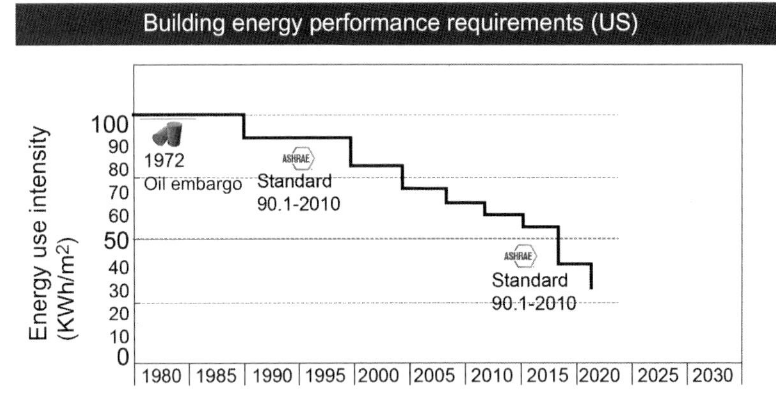

FIGURE 2.8 Evolution of building energy performance requirements in the EU and United States (Attia, 2016).

owners, and developers can exceed current regulations and invest and innovate, almost 25 years before the European commitment and progress toward NZEB. For sure, the gradual progress and slow performance progress of codes is necessary to provide policy consistency for the building stock. However, ambitious leaps can always lead technology and practice in the long-term future. Fig. 2.9 compares both approaches and describes their main advantages.

5.2 High-Tech Versus Low-Tech

In the race against climate change and world population increase, we should seriously consider the impact of urbanization and materials scarcity in the coming years. In this context, the economic and ecological crisis will push the idea of a circular economy and drive our choices

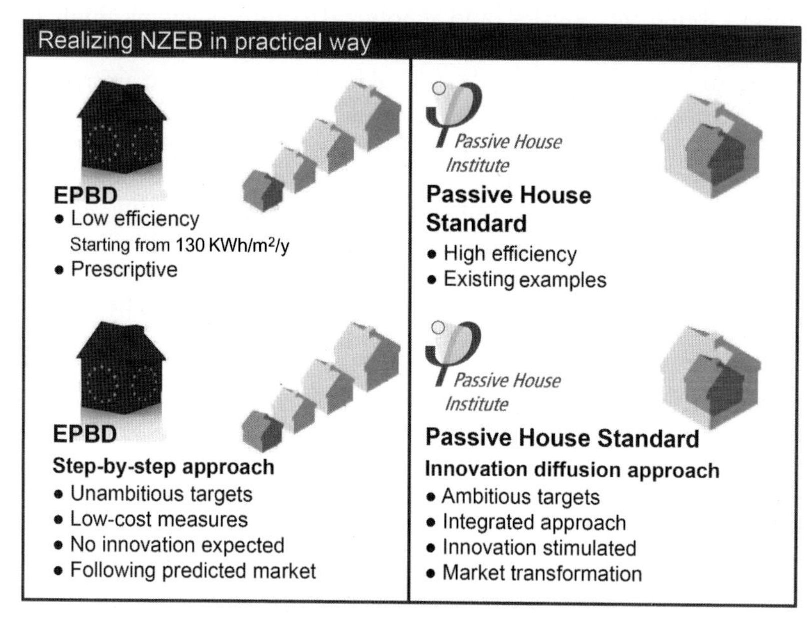

FIGURE 2.9 Comparison of gradual performance improvement approach versus the fast performance leaps approach.

regarding the technological solutions and regenerative design approach to reach NZEB (Attia, 2016). NZEB are systems that can become very complex to meet multiple performance criteria including owner requirements. The simpler the design of NZEB, the better they perform and last with minimal maintenance. There are two major approaches to reach NZEB—high-tech or low-tech approaches. The low-tech design approach is focused on simplicity, passive measures, lean construction materials while using basic building solutions and components (Bihouix, 2017). The low-tech design approach can sometimes result in lower performance when compared to high-tech approaches. Low-tech NZEB depend on less technology, robust, and resilient solutions. Building 2226 in Lustenau Austria is an example for a comfortable high-performance building ensuring a comfortable internal temperature of 22–26°C without powered ventilation, heating or cooling. On the other side, high-tech design approach is more technology-driven with energy-intensive systems and multiple integrated solutions and products to guarantee robust performance and human comfort (Slessor and Linden, 1997; Daniels, 1998). As a consequence, the high-tech design approach leads to lower energy intensity but requires constant follow up and maintenance. The predictive capabilities of building science played a significant role in increasing the role of technology in buildings. With the technological advancement of building materials science, façade engineers, building physics, and emotional control systems, high tech solutions became a principal component of NZEB.

Based on experience in industrial and nonindustrial countries and facing the global urbanization challenges, we confirm that both approaches are vital and require continuous development. The low-tech approach for NZEB will benefit from maximum passive possibilities, including natural lighting, natural ventilation, passive heating, and passive cooling. This approach should depend on simple technologies and solutions including solar chimneys, celling fans, evaporative cooling systems that can be simply controlled and maintained. The low-tech design approach can help increase NZEB uptake due to its simplicity and cost effectiveness that can adapt to the socio-economic conditions in many emerging countries. On the other hand, the high-tech design approach is equally important to achieve further technological breakthroughs and meet occupants' needs through cutting edge technologies. We are on the verge of a new industrial revolution with a focus on smart buildings, smart cities, and automation. NZEB are in continuous development to become the center of this revolution and transformation. Therefore, high-tech design approaches will be the basis for smart living and a highly interactive built environment. Finally, we need to be aware about the advantages and limitations of both approaches and make sure to master the development of low-tech as well high-tech NZEB depending on the context and future plans.

6 DISCUSSION

Setting a framework and definition of NZEB is essential to achieve a mature comprehension of their performance and functional use. In this chapter, we discussed the evolution of NZEB definitions and the associated principles and concepts of its design. Setting a consistent definition is very helpful to guide the design process and allow the design team members to focus on the essential design parameters. On the operational level of building performance, we should always remind ourselves that NZEB are energy-intensive buildings, with occupant-based functional services that have a constantly varying performance. The complexity of a NZEB is caused by its expectation to meet the occupant needs, interact with the grid, generate energy, and perform under ultra-EE restrictions. This makes NZEB complex systems with high intensity energy used and constant varying performance requirements. On the primary energy level, we should take into account the complexity of energy conversion and energy transmission in relation to carbon emissions and national energy policies and energy mix. The constant variation of demand and supply elucidates that NZEB should not be conceived as independent entities. Today, NZEB form less one percent of the European building stock, but tomorrow they will become representative. This means that we need to design them as an integral component of the built environment within a larger system of energy grids and energy stations. On the embodied energy level, and with growing demand to neutralize the carbon emissions of buildings over its whole life cycle, we should take into account the importance embodied energy. By 2030, European member states will be very close to neutralizing the building sector's energy use. Building materials' environmental impact will be the directly following challenge to address for NZEB. All these aspects have been elaborated on and discussed in this chapter and can be grouped under three major categories, namely (1) design, (2) scale, and (3) adaptability. The following section will discuss these three criteria for setting a consistent NZEB definition.

The first criterion for formulating a consistent NZEB definition is to empower and enable design. The design of NZEB is the way the concepts, boundary conditions, and EUI all combine or connect to balance the energy consumption and energy generation while providing comfort and meeting the environmental targets of GHG emissions reduction. Many performance requirements—including EE, IEQ, RES, and carbon emissions—have to connect to achieve robust performance. The value and importance of design boundaries lies in ease of use and informing designers to formulate concepts leading to a NZEB. In this context, the EUI is a key design indicator to assess the performance of NZEB. Worldwide, the most common metric for EUI is energy use per square meter (or feet), except for the 2000 Watt Society concept discussed in Section 4.1. The 2000 Watt Society concept is the only concept that sets a target metric and performance threshold per building occupant and not conditioned space. In the long term, we should question the current EUI metric (in relation to building area) and assess different scenarios and metrics to meet the environmental global targets in relation to the planet population increase. This includes the concepts that set specific target metrics for NZEB. In this chapter, we presented the PH concept that sets the EUI associated with a target metric of 15 kWh/m^2 per annum for heating and cooling demand. Moreover, questioning the EUI metrics should not only focus on energy per space area or occupant, but also for materials. Since the introduction of environmental product declaration in 2015 by the EU, which requires performing LCA for all building materials used in the European market, and with initiatives to introduce a materials passport for newly constructed buildings, the definitions of NZEB have been challenged. Including the embodied energy together with the operational energy for NZEB seems to be in store very soon and will be reflected into national codes. So, for any consistent and current definition, the balance between EE and RES on the one hand and the operational energy and embodied energy on the other hand will need to be taken into account and formulated into a definition framework (Attia and De Herde, 2010b, 2011).

The second fundamental criterion to formulate a NZEB definition is investigating and foreseeing the impact of scaling up. Based on the ideas and experiences discussed in this chapter, any NZEB definition should incorporate a planning policy for grids management, mobility, and spatial planning to investigate the environmental and economic implications of NZEB on the on urban and city scale. Planning policies through successful zoning can reduce the need to travel, and neighborhood layouts can reduce the need for heating or cooling. Large scale spatial planning can help to improve transit options, increase local employment and residential densities, and increase the amount of green spaces in relation to NZE Districts design. Energy services can also be more efficiently delivered at a district scale than at a building scale with distributed renewable or low-carbon energy supplied by local energy service companies. Moreover, there are plans and goals to reduce the GHG emissions on a national level for every country. For example, the EU member states are looking to reduce the overall carbon emissions by 20% and increase the renewable shares in the national energy mix by 20% compared with the 1990 levels. This means that the national mixes will change in the near future and the primary energy factors are expected to be lowered due to larger share of renewable energy. As a consequence, the industry working on green solutions, systems, and technologies is formulating its strategies for products and services development in line with those changes. This will lead to a significant market transformation and any definition should take into account those implications to become in harmony with the energy mix and associated industrial infrastructure and transformations. For example, in Northern European countries there is a tendency to neutralize the energy mix by investing in hydro and wind power while coupling buildings to a flexible energy grid promoting heat pumps for space heating and cooling. Therefore, a NZEB definition should take those considerations into account and become aligned with the building sectors' local context. This refection should include the interaction between different buildings on the district or city level with the grid. The need to stabilize the electric grid and provide storage capacity makes it essential to think beyond the single building scale and site boundaries. The uptake of electric vehicles in the near future should be coupled to a building's energy demand and supply bringing smart buildings and mobility together on the agenda.

The third fundamental criterion required to formulate a definition for NZEB is adaptability. The adaptability of NZEB is a fundamental feature that should be reflected in any future definition. Adaptation and flexibility should be also embedded for long-time building type transformations for multi-functional use. In the same time, NZEB are complex as they often involve novel systems, materials, and products that interact dynamically with the outside environment, internal environment, and occupants. Balancing the occupants' needs, EE and energy generation requires flexibility and constant variation in the NZEB management and use. Operation and maintenance of high-performance buildings are a basic component of NZEB. A robust performance of NZEB without facility management is impossible. Since NZEB are characterized with high EUI they are also highly sensitive facing constant variations including environmental (e.g., climate), human (e.g., occupant behavior), or technical (e.g., HVAC systems). Therefore, they need to be adaptable and flexible during operation and managed by savvy professionals. Meeting occupants' needs while interacting with the grid capabilities is an essential operational task that should be part of the NZEB definition. Energy and IEQ monitoring through measurement and verification is part of a definition as explained in Chapter 10, Occupant Behavior and Performance Assurance. The key feature of NZEB is to provide continuous monitoring on its performance operation and communication with occupants instantly. In the past few years, the response of facility managers and occupants became a new practice in NZEB operation. This means that to maintain a robust performance of a NZEB, it requires responsive and interactive operation and occupant behavior to stabilize the constant variations it faces. Therefore, flexibility and adaptability are a part of any NZEB definition.

Finally, there are significant differences in the progress and implementation of NZEB across the world. Many Western and Northern European countries managed to develop concepts, definitions, and construction technologies of NZEB that are effective and correspond to their climates. However, we are far from having national NZEB definitions worldwide. We believe that NZEB definitions will continue to evolve. The NZEB definitions need to look at a range of criteria, including materials, rather than energy only. Also, the NZEB definition and corresponding policy measures should evolve to address smart buildings and smart cities. We hope that professional and public authorities understand that a definition of development is a process that should involve building stakeholders (policy, industry, investors, building professionals, building users, banks, etc.) and must be consensus based on the national level. We should reflect on the new possibilities that a transforming energy market could bring with electric vehicles and smart grid management. Also, definitions should include protocol quality assurance, including measurement and verification so those definitions translate into consistent building codes while being coupled to training for designers, builders, and workers.

7 LESSONS LEARNED # 2

In conclusion, NZEB definitions require adaptable and context-sensitive formulation. There are several approaches and concepts to reach NZEB. We believe that the performance-based approach is key to achieve technical feasibility and financial cost effectiveness of NZEB. Performance-based and evidence-based design can help to develop local NZEB definitions, concepts, and standards. Definitions provide quality assurance and provide great attention to detail during the design and construction process to achieve certification. Once they are tested and validated through demonstration projects, they can be turned into prescriptive standards and legislations. This is the optimal way to evolve national definitions taking into account the effects of climate change, existing building stock renovation and to bring NZEB concepts into the high-volume market worldwide.

NZEB are complex and require a definition to simplify their design, operation, and monitoring. The founding principles for any NZEB definition are based on setting a proper balance between EE, renewable energy generation, human comfort, and carbon emissions. In this chapter, we proved that any definition of NZEB needs to push the boundaries of design and become contextualized following a consensus-based development process to reflect the national interests, capabilities, and future plans of a country. This process should involve different stakeholders in relation to the energy market and policies, local climate, socio-cultural, and economic status in every country. At the same time, a NZEB definition should include clear target metrics and EUI indicators to facilitate benchmarking, design, and performance assessment of different building types. The essence of a NZEB definition is to simplify the design and make it easier for building professionals to intentionally apply the four design principles (described in Section 3.2) and connect different systems, products, technologies, and solutions to solve problems. Therefore, a definition of NZEB should address three criteria to remain consistent and achieve a robust performance. First of all, a definition should address design to set up the boundary conditions, performance indicators, target metrics, and performance thresholds (see Chapter 3: NZEB Performance Thresholds) for new and existing buildings taking into account multi-objective performance criteria. Second, a definition should deal with scaling-up NZEB and its impact, taking into consideration the urban and grid level, to get the most value of high-performance buildings on the national and regional level. This means being ready to formulae a definition that pivots NZEB into the energy system and involving NZEB into energy planning and policy. Thirdly, a definition for NZEB should address adaptability and flexibility, allowing the building functions and uses to change over time, while maintaining flexible operation and interaction with the grid. A definition should foresee the role of adaptability—this includes meeting occupant needs and assessing the building performance. So, if design is about balance or energy neutrality, and scale is about planning then adaptability is about interaction to guarantee the performance effectiveness of NZEB. Any NZEB definition should allow flexibility in the operation of NZEB by communicating its vital performance signs and enabling its facility managers and users to constantly adapt to meet the designed performance requirements. By setting a consistent definition, we can simplify and inform the design of NZEB, allow them to attain their performance targets through assessment with the flexibility they need to adapt to occupants and local conditions around the world.

References

Attia, S., De Herde, A., 2010a. Strategic decision making for zero energy buildings in hot climates. In: Proceedings of EuroSun 2010.

Attia, S., De Herde, A., 2010b. Towards a Definition for Zero Impact Buildings, Sustainable Buildings CIB 2010, Maastricht, Netherlands.

Attia, S., De Herde, A., 2011. Defining Zero Energy Buildings from a Cradle to Cradle Approach, Passive and Low Energy Architecture, July, Louvain La Neuve, Belgium.

Attia, S., 2014. The usability of green building rating systems in hot arid climates. In: ASHRAE Proceeding of Conference on Energy and Indoor Environment for Hot Climates, February 24–26, Doha, Qatar.

Attia, S., 2016. Towards regenerative and positive impact architecture: a comparison of two net zero energy buildings. Sustain. Cities Soc. 26, 393–406.

Attia, S., Eleftheriou, P., Xeni, F., Morlot, R., Ménézo, C., Kostopoulos, V., et al., 2016. Overview of Challenges of Residential Nearly Zero Energy Buildings (nZEB) in Southern Europe, Sustainable Buildings Design Lab, Technical Report, Liege, Belgium, 9782930909059.

Attia, S., Eleftheriou, P., Xeni, F., Morlot, R., Ménézo, C., Kostopoulos, V., et al., 2017. Overview and future challenges of nearly Zero Energy Buildings (nZEB) design in Southern Europe. Energy Build. 155, 439–458. ISSN: 0378-7788. Available from: https://doi.org/10.1016/j.enbuild.2017.09.043.

Balcomb, J.D., 1992. Passive Solar Buildings, vol. 7. MIT Press, Cambridge.

Besant, R., Dumont, R., Schoenau, G., 1979. The Saskatchewan conservation house: some preliminary performance results. Energy Build. 2 (2), 163–174.

Bihouix, P., 2017. L'âge des Low Tech Vers une civilisation techniquement soutenable. Le Seuil.

Bruno, R., Steinmüller, B., 1977. The Energy Requirements of Buildings. Laboratory Report. Philips, Aachen.

Bruno, R., Hörster, H., Söllner, G., Steinmüller, B., 1979. Optical and thermal properties of windows for passive solar systems. In: ISES Conference Atlanta, May 1979.

Butti, K., Perlin, J., 1980. A Golden Thread, 2500 Years of Solar Architecture and Technology. Van Nostrand Reinhold Company, New York.

Daniels, K., 1998. Low Tech, Light Tech, High Tech. Birkhauser, Boston.

Deviren, A.S., Tabb, P.J., 2014. The Greening of Architecture: A Critical History and Survey of Contemporary Sustainable Architecture and Urban Design. Ashgate Publishing, Ltd., Burlington.

DOE, 2017a. Guidelines for Participating in the DOE Zero Energy Ready Home. Retrieved from <https://www.epa.gov/indoorairplus> (accessed 28.02.17.).
DOE, 2017b. Building Performance Database. Retrieved from <https://energy.gov/eere/buildings/building-performance-database> (accessed 01.08.17.).
EnOB, 2017. Net Zero-Energy Buildings—Map of International Projects. Retrieved from <http://www.enob.info/en/net-zero-energy-buildings/nullenergie-projekte-weltweit/> (accessed 01.08.17.).
EPA, 2017. Indoor airPLUS. Retrieved from <https://www.epa.gov/indoorairplus> (accessed 28.02.17.).
EPBD, 2010. Directive 2010/31/EC, of European Parliament and of the Council of 19 May 2010 on the Energy Performance of Buildings (Recast).
Esbensen, T., Korsgaard, V., 1977. Dimensioning of the Solar Heating System in the Zero Energy House in Denmark, United States, pp. 195–199.
EU, 2002. Directive 2002/91/EC of the European Parliament and of the Council, 16 December 2002, Energy Performance of Buildings. Retrieved from <http://eur-lex.europa.eu/legal-content/EN/TXT/?uri = CELEX%3A32002L0091> (accessed 10.02.17.).
EU, 2012. Cost Optimal: Commission Delegated Regulation (EU) No. 244/2012 of 16 January 2012 Supplementing Directive 2010/31/EU of the European Parliament and of the Council on the Energy Performance of Buildings by Establishing a Comparative Methodology Framework for Calculating Cost-Optimal Levels of Minimum Energy Performance Requirements for Buildings and Building Elements. <http://ec.europa.eu/energy/efficiency/buildings/buildings_en.htm>
Hernandez, P., Kenny, P., 2010. From net energy to zero energy buildings: Defining life cycle zero energy buildings (LC-ZEB). Energy Build. 42 (6), 815–821.
IEA, 2017. The IEA SHC Task 40/ECBCS Annex 52 'Towards Net Zero Energy Solar Buildings (NZEB)'. Retrieved from <http://task40.iea-shc.org/> (accessed 19.08.17.).
Klingenberg, K., Kernagis, M., Knezovich, M., 2016. Zero energy and carbon buildings based on climate-specific passive building standards for North America. J. Build. Phys. 39 (6), 503–521.
Korbee, H., Smolders, B., Stofberg, F., 1979. Millieu voorop bij uitwerking van een global bestemmings plan, TH Delft, afd. Bouwkunde, BOUW, no. 22, 27 Oktober 1979.
Korsgaard, V., 1976. Zero-Energy-House. NP-22388. National Technical Information Service NTIS, Springfield, Virginia.
Kurnitski, J. (Ed.), 2013. Cost Optimal and Nearly Zero-Energy Buildings (nZEB): Definitions, Calculation Principles and Case Studies. Springer Science & Business Media, London.
Passipedia, 2017. The New Passive House Classes. Retrieved from <https://passipedia.org/certification/passive_house_categories> (accessed 01.02.17.).
Marszal, A.J., Heiselberg, P., Bourrelle, J.S., Musall, E., Voss, K., Sartori, I., et al., 2011. Zero energy building—a review of definitions and calculation methodologies. Energy Build. 43 (4), 971–979.
Mazria, E., 1979. Passive Solar Energy Book, Rodale Press, Emmaus, PA
Meadows, D.H., Meadows, D.H., Randers, J., Behrens III, W.W., 1972. The Limits to Growth: A Report to the Club of Rome (1972). Universe Books, New York.
NBI, 2016. List of Zero Net Energy Buildings, NewBuildings Institute. Retrieved from <http://newbuildings.org/wp-content/uploads/2016/10/GTZ_2016_List.pdf> (accessed 01.08.17.).
Nisson, J.N., Dutt, G., 1985. The Superinsulated Home Book. John Wiley & Sons, Hoboken.
Scognamiglio, A., Røstvik, H.N., 2013. Photovoltaics and zero energy buildings: a new opportunity and challenge for design. Prog. Photovolt.: Res. Appl. 21 (6), 1319–1336.

Sartori, I., Napolitano, A., Voss, K., 2012. Net zero energy buildings: a consistent definition framework. Energy Build. 48, 220–232.
Schimschar, S., Hermelink, A., Boermans, T., Pagliano, L., Zangheri, P., Voss, K., Musall, E., 2013. Towards Nearly Zero-Energy Buildings—Definition of Common Principles Under the EPBD. Ecofys, Politecnico di Milano, University of Wuppertal (Unpublished) for European Commission.
Shurcliff, W.A., 1988. Super Insulated Houses; And, Air-to-Air Heat Exchangers. Brick House Publishing Company, Andover.
Slessor, C., Linden, J., 1997. Eco-Tech: Sustainable Architecture and High Technology. Thames and Hudson, New York.
Steinmüller, B., 1979. Thermal Performance of Buildings and Window Systems in the US and Europe. Laborbericht 541. Philips Research Lab, Aachen.
Tabb, P., 1984. Solar Energy Planning: A Guide to Residential Settlement. McGraw-Hill Book Company, New York.
Von Weizsäcker, E.U., Weizsäcker, E.U., Lovins, A.B., Lovins, L.H., 1998. Factor Four: Doubling Wealth-Halving Resource Use: The New Report to the Club of Rome. Earthscan, London.

CHAPTER

3

Net Zero Energy Buildings Performance Indicators and Thresholds

ABBREVIATIONS

AEC	Architectural, Engineering and Construction
ASHRAE	American society of heating, refrigerating and air-conditioning engineers
CEN	European Committee for Standardization
DOE	Department of Energy (United States)
EE	energy efficiency
EIA	environmental impact assessment
EPBD	energy performance building directive
EPC	energy performance certificate
EPD	environmental product declaration
EUI	energy use intensity
EU	European Union
FIT	feed-in tariff
GHG	green house gases
HVAC	heating, ventilation and air conditioning
IAQ	indoor air quality
IEQ	indoor environmental quality
LCA	life cycle assessments
MVHR	mechanical ventilation with heat recovery
NIBS	National Institute of Building Sciences
nZEB	nearly Zero Energy Buildings
NZE	Net Zero Energy
NZEB	Net Zero Energy Buildings
PE	primary energy
PEB	positive energy buildings
PH	passive house
PHIUS	Passive House Institute US
PHPP	passive house planning package
PM	particulate matter

Net Zero Energy Buildings (NZEB)
DOI: https://doi.org/10.1016/B978-0-12-812461-1.00003-4

PMV predicted mean vote
PPD predicted percentage dissatisfied
PV photovoltaic
RES renewable energy systems
SBS sick building syndrome
VOC volatile organic compounds
UNEP United Nations Environment Programme
ZERH zero energy ready home

1 INTRODUCTION

After the discussion of NZEB terminologies, definitions, design principles, concepts, and boundary conditions on a theoretical level in Chapter 2, this chapter will address the key performance indicators and thresholds to design robust NZEB. The overall aim of this chapter is to understand regulations that are moving toward NZEB construction, analyze building performance indicators, and critical target metrics.

We will explore the key performance thresholds and metrics that need to be incorporated earlier on during the design process in strong relation to the major challenges that face NZEB design, construction, and operation. We highlight the importance of performative design as an approach that can enable and empower design teams facing the complexity of NZEB design. The chapter presents guidance and recommendations on the minimum necessary performance thresholds for NZEB and make these explicit. The result presents an overview of the challenges and provides recommendations based on available empirical evidence to further lower these barriers worldwide, but with a focus on the European construction sector. By defining the performance thresholds for NZEB (including the metrics) we offer design teams the possibility of verification and design assessment during early design decision-making phases to instill practical, evidence-based, and quantifiable guidance for high performance buildings in particular NZEB. The chapter provides suggestions for minimum energy efficiency, indoor environmental quality, and renewable energy thresholds to shift the identified gaps into opportunities for NZEB performance-based design.

2 CHALLENGES OF NZEB

The design of NZEB is a challenging problem of increasing importance. The robust design depends largely on addressing three major challenges which are illustrated in Fig. 3.1:

- Energy performance gap
- Sick building syndrome (SBS)
- Design decision stress

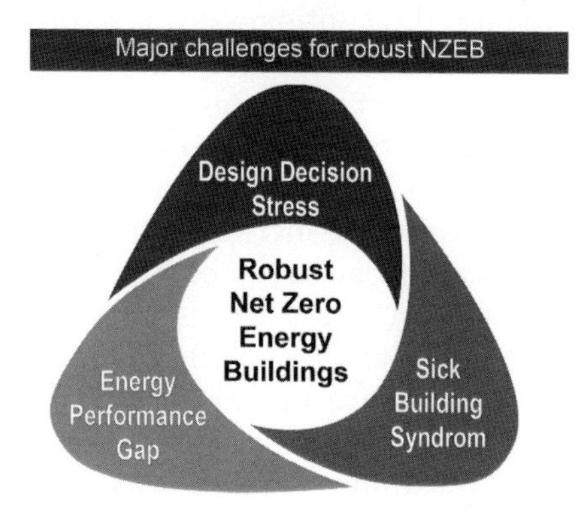

FIGURE 3.1 Major challenges to design robust net zero energy buildings.

Despite the overall increase of NZEB worldwide, constructing NZEB face three fundamental challenges. The first challenge is mainly the performance gap which represents a significant performance difference between the design assumptions and the real building energy consumption (Attia et al., 2013a; Ilmonen, 2015). The gap between theoretically calculated annual energy consumption and the actual measured annual energy consumption in NZEB is a common problem in the building energy efficiency field. The energy performance gap has become the most significant problem associated with high-performance buildings. With the advancement of measuring and monitoring techniques, the building community was enabled to follow-up and track the real performance of buildings in a simplified way. The performance gap became a serious challenge for NZEB design because it reflects the complexity of NZEB and the consequences of poor design, construction, or operation. On the level of design, the energy performance gap is directly related to the lack of setting clear and measurable performance indicators and thresholds together with the use of theoretical simulation or calculations. On the level of construction, the performance gap is related mainly to poor realization and lack of quality assurance. Finally, on the level of occupancy, the gap is related to occupant behaviour, systems controls, and climatic variations. We can add to these main factors with other parameters (i.e., climate change, building services, facility management, plug loads, and occupancy schedules) that create a multitude of variables which influence NZEB performance (de Wilde, 2014). This brings us to the earlier discussion on the importance of design with a number of uncertainties. Therefore, the way we design robust NZEB from setting performance thresholds and targets, to anticipating climate scenarios and understanding occupant behaviour, are very critical.

The second largest challenge associated with NZEB is the SBS. There are many more important facets of NZEB than energy performance, such as indoor air quality, daylighting, occupant productivity, satisfaction, and thermal comfort that could be overlooked when designing NZEB. Time constraints and cost pressure can divert the focus of design teams and as a consequence NZEB can suffer from the SBS. The SBS comprises various symptoms associated with occupants' dissatisfaction with a building. Health problems, increased absenteeism, and comfort-related problems (overheating) are very common in high performance buildings (Attia and Carlucci, 2015). Issues related to occupancy density, indoor air quality, and circadian rhythms regulation are all examples that confirm the importance of empowering building occupants and interact with them before and after design using pre- and postoccupancy evaluations to get maximum possible feedback to create a balance between energy performance targets and thresholds and designing healthy buildings.

The third and final challenge of NZEB design is making informed design decisions. Designers have been faced with a pool of various choices to arrive at the NZEB performance objective. Combining passive and active systems early on is a challenge, as is, more importantly, guiding designers towards the NZEB objective that requires outstanding energy efficiency and indoor comfort performance criteria (Attia et al., 2013b). The complexity of NZEB and the major challenges regarding the performance robustness of NZEB form a choice stress for many design teams. By choice stress we mean the stress associated with making informed decisions. The design process of NZEB requires design teams to spend more effort upfront in the design process. The difference between the NZEB design process compared with conventional building design is the serious decision-making and choice stress early on during the design and construction process (Attia and De Herde, 2011; Attia, 2012). Using performance-based design approaches and setting clear and measurable performance goals and thresholds is the best methodology to design robust and performative NZEB. By prioritizing the performance indicators and thresholds, decision makers can focus on closing the energy performance, achieving healthy NZEB, and assess design scenarios to create more performative designs for NZEB.

3 PERFORMATIVE DESIGN FOR NZEB

Performative Design can create robustness, adaptation, and scale for NZEB. When used strategically, design can help the architectural, engineering, and construction (AEC) industry to create high-performance buildings. Most building professionals don't think about design this way, but NZEB that are well designed are actually well connected and part of an adaptive/flexible and robust system. It is very important to ask the design team the following questions:

1. How can NZEB be designed to stay robust?
2. How do you create the most value from design?
3. How do you use design to innovate?

Although performative design is important, it must be flexible during operation and connect to scaling-up to help achieve NZEB' impact on an urban scale. When performative design, adaptation, and scale connect, building design professionals can create a robust low-impact built environment. A NZEB is a complex system that requires a clear framework of interrelationships based on design principles and performance thresholds. Therefore, design for high-performance buildings refers to design that is strategic, innovative with clear connection to performance-based and evidence-based approaches, and creates a balance between energy savings, energy generation, and indoor environmental quality. For NZEB, designs must create quality across construction and operation. Design is estimated to be 1% of the lifecycle cost of buildings, but it can reduce over 90% of lifecycle energy costs (Hawken et al., 1999). Performative design should ultimately be design that leads the AEC culture and unleashes the real potential of the construction sector. Such performative design is powerful. Once we understand the NZEB design principles and decide to setup the performance thresholds we can unlock the power of design to drive flexibility, robustness, and scale.

3.1 Performance-Based and Prescriptive Design

The performance-based design approach and the prescriptive design approach are the two main ways to achieve NZEB status. There are available concepts such as the Passive House (PH) or the Active House developed for Northern and Central Europe which try to achieve NZEB status by following a prescriptive approach or a mix of both approaches (see Chapter 2 for further explanations of both concepts). The United Nations Environment Programme (UNEP) backs the PH standard and advocates for its adoption (Thorpe, 2016). These concepts have been developed over long research and development periods, sometimes exceeding 20 years. So, all high-performance building designs started with a performance approach to test and validate the design's robustness and flexibility. Building designers tend to rely on the performance-based design to first come up with earlier solutions and then request them to be applied in new designs after being coined as prescriptive approaches.

In the case of the PH, once it was developed, tested, and approved in Germany's temperate continental climate, it almost became a prescriptive design approach that provided a holistic design solution or concept. Although this concept has been slow to take off in North America, it is popular in Europe, with over 18,000 buildings registered to the PH standard (Passive House Database, 2017). However, this does not mean that PH as concept can be applied everywhere or for every building typology. The Passive House and Active House standards have their niches and may be appropriate for particular markets. However, it is very important that every country conducts its own research and development to find the design concepts that reduces the amount of energy used in new buildings as much as possible, while producing as much energy as possible onsite. In this context, the performance-based design approach is the first approach to follow before translating the developed design strategies and measure these into concepts that are based on prescriptive approaches to simplify the design and construction process.

Performative design is a performance-based building design process where the building team is involved in the various phases of the building delivery process to ensure long-term performance-in-use of buildings as an explicit target (see Chapter 4 for further explanation). NZEB are complex systems where the outcome of design solutions is not always easy to assess. NZEB design demands innovation and design optimization to reach optimal or nearly-optimal solutions and measures. In this context, the concept of NZEB as complex and sensitive, requiring attention during design, construction, and operation emerges. NZEB require valid design and powerful simulation tools as well as assessment tools that can be different for the design team, contractor, or facility managers. They require well-defined performance indicators and thresholds and perfect collaboration between the building design and construction team. In order to achieve NZEB status, the conventional building delivery process should be changed and the entire building design team should embrace this change which is associated with risks. Therefore, there is a need to follow the performance-based approach to develop locally adapted NZEB.

Between 2010 and 2018, the AEC industry succeeded to adapt the performance-based design approach through many demonstrations and pilot projects with different building typologies in different heating-dominated countries. This includes residential buildings, office buildings, schools, and retail buildings. However, there are still serious challenges to achieve NZEB status for high rise buildings, towers, hospitals, airports, and other similar high-intensity buildings. More importantly, there is a serious challenge to achieve NZEB status in heating- and cooling-dominated climates. In this context, the performance-based approach is the best approach to support the project delivery process to achieve ultra-efficient buildings for new and complex building typologies in different climates.

Once the NZEB design becomes mature enough to be ready for mainstream, moving from demonstration into high-volume markets, the prescriptive approach emerges as a sound and tangible approach that is fast, less costly, and more systematic during implantation. Prescriptive design approaches for NZEB should consist of evidence-based rules that are simple to apply, verify, and repeat. They should allow replicating regional and local successful concepts, strategies, and solutions to allow the industry to standardize its products and mass produce for market penetrations. Finally, it is important to mention that the prescriptive design approach and the performance-based building design approach are complementary. We start with the performance-based approach to develop and validate robust NZEB concepts, then we coin them into simple prescriptive design approaches.

3.2 Designing for Robust NZEB

Creating a robust NZEB is an ongoing challenge. There are a number of uncertainties associated with NZEB design. Connecting different elements and systems of a building to accommodate the occupants' needs and achieve a neutral energy balance is a complex design problem. A robust design connects building elements, components, and systems together in a robust way so that climate change, other unintended loads, or occupant actions do not result in great variations causing a performance gap test or the SBS. A robust NZEB is designed for optimal performance with minimum performance variation and occupants' complaints over time (Kotireddy et al., 2015).

Therefore, it is essential to increase designers' knowledge about NZEB design problems and options to reduce decisional uncertainty and increase design robustness (Attia and De Herde, 2011; Attia, 2012). The performative design approach is a smart way to create nearly optimal NZEB designs that are soundly robust. The goal we have to reach for NZEB status is not only minimizing energy demand as much as possible, but also developing an optimal design in order to find balance between energy performances, energy generation, and variation of occupant needs and indoor environmental quality (see Fig. 3.2). For NZEB, designers tend to minimize energy demand without taking variations into account, including appropriate and fit-to-purpose HVAC systems' design and sizing in order to meet the heating or cooling energy demand. This is like putting the building on a strict diet that deprives a body in order to appear healthy. The same analogy can be made with buildings when designers neglect small uncertainties (in terms of user

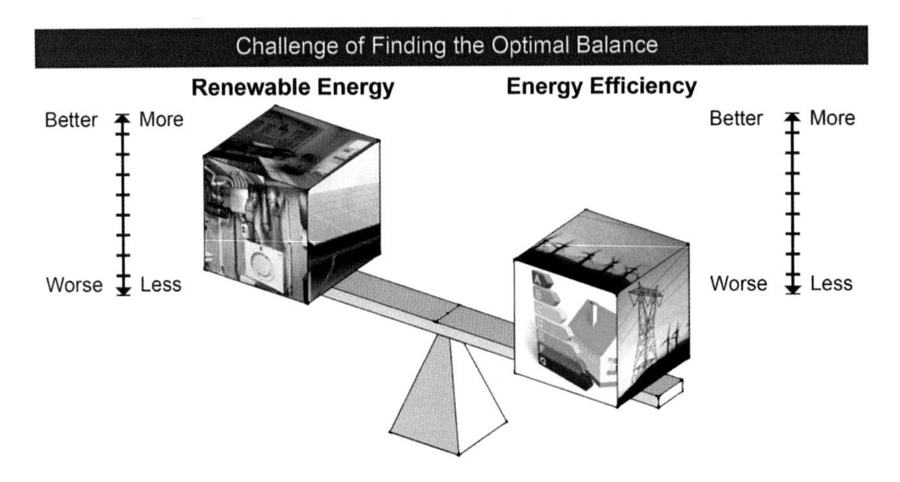

FIGURE 3.2 The challenge of finding the optimal balance between energy efficiency, renewable energy generation and IEQ.

behaviour, systems control, or climatic change) that can strongly influence the energy demand and HVAC systems performance that will not be able to adapt. Therefore, robustness embeds adaptability as a starting point of sustainability and is based on performance designs that are based on a holistic design approach that identifies and validates clear performance indicators and thresholds.

4 NZEB PERFORMANCE INDICATORS AND THRESHOLDS

In this section, we discuss the importance of setting performance indicators and thresholds for NZEB to guide the performative design. The performance thresholds presented in this section have been adopted from various case studies based on deep analyses looking at the social/political/economic and technical backgrounds behind the NZEB development worldwide. The analyses provide guidance on the challenges and constraints regarding the implementation of NZEB and provide an overview that is translated into recommendations and conclusions in Chapter 13 for NZEB design, construction, and operation. The following questions are essential and must be answered before starting the design of a NZEB. We did not rank them in any specific order (see Fig. 3.3):

1. What is the minimum carbon emission threshold for the NZEB?
2. What is the minimum energy efficiency threshold for the NZEB?

FIGURE 3.3 Seven performance thresholds for NZEB.

3. What is the heating/cooling energy needs balance for the NZEB?
4. What are the indoor environmental quality limits for the NZEB?
5. What is the minimum renewables threshold for the NZEB?
6. What is the occupancy density for the NZEB?
7. What is the cost threshold per square meter of squared built up area for the NZEB?

4.1 Carbon Emissions Threshold

Global greenhouse gas (GHG) emissions continue to grow causing the carbon emission gap associated with global warming. The building sector should adopt and adapt solutions that can deliver low impact emission reductions at a large scale and accelerate energy efficiency. NZEB can significantly contribute to carbon emission reduction. Hence, the carbon emissions threshold should use a carbon measuring indicator that can properly reflect building carbon emissions associated with operational energy consumption on the short term. CO_2 emissions are a side effect from fossil (mix) energy burning to produce delivery and consume energy by buildings. When calculating the carbon emissions for NZEB we do not focus on the carbon emissions associated with the end-use energy. End-use energy, which is the energy directly consumed by the building users, includes electricity, gasoline, and natural gases. We should focus, though, on the emissions associated with the harvesting or generation of primary fuels from fossil resources. In other words, we should focus on emissions associated with primary energy. The primary energy demand from nonrenewable energy sources that is supplied to the building for all energy uses arising in the building. Electricity is an example of how carbon associated end-use energy and primary energy differs. Primary energy is the energy that goes into electricity production in plants and utilities, while end-use energy is the amount of electricity that we use in buildings. These are two very different numbers. A typical power plant runs at about 33% efficiency. This means that power plants consume three times as much energy in fuel, as the amount of energy produced as electricity. As a consequence, the carbon emissions should be calculated and associated with primary energy production. Also, we should add transportation that makes up a small portion of primary energy to the emissions associated with primary energy.

For building-associated carbon emissions, the primary energy use for NZEB reflects the depletion of fossil fuels in proportion to carbon emissions (Attia, 2018). The primary energy is proportional to CO_2 emission for operational energy. This proportion is distorted if energy comes from nuclear plants. However, CO_2 emissions are a reliable indicator to ensure the energy performance of buildings. For example, the European Energy Performance Building Directive (EPBD) promotes CO_2 emission reduction in primary energy to achieve the 2050 goals of reduction and decarbonization. To meet the high-performance target of NZEB, the EU suggests that an overarching primary demand and CO_2 emission threshold associated with energy demand should be below 3 $kgCO_2/m^2$ yearly.

In the long term, we need to curb and limit carbon emissions associated with the embodied energy of building materials. This step should be associated with encouraging the selection of sustainable and long-lasting building materials based on an environmental impact assessment (EIA) and lifecycle assessments (LCA). Most of the embodied energy for NZEB is used for the production of the building materials (Attia, 2016). Lasting and continuing usability are the main factors for the energy efficiency of NZEB performance. So far, there are no specified limits of CO_2 emissions embodied in building materials. Rating systems such as Minergie-ECO, BREEAM, LEED, and DGNB are rewarding design teams to take carbon emissions associated with embodied energy into account in green buildings. Since 2016, the EU mandated providing Environmental Product Declarations (EPD) based on LCA for all construction materials. However, the EU is preparing legislation that aims to curb the CO_2 emissions and energy embodied in building materials as set to a limit of kilogram CO_2 per kilogram of materials. Chapter 6 elaborates on this topic and presents insight on the environmental impact of NZEB in relation to operation and embodied energy. Finally, setting a minimum carbon emissions threshold for primary energy and embodied energy of NZEB remains essential. It helps to directly identify the sources of energy and the associated carbon emissions from a global perspective.

4.2 Minimum Energy Efficiency Threshold

To achieve high efficiency in buildings, an ambitious energy and carbon emissions reduction must be required for NZEB using universal indicators (Atanasiu et al., 2011). On the one hand, this would not limit the possibility to adapt the targets/thresholds level of those indicators to local conditions. On the other hand, it would provide a common language/standard across the world, which is essential for the construction industry to develop solutions in a stable and coherent framework for NZEB. Clear definitions of the energy levels and their calculation/ measurement steps are presented in the European Standards. Evaluation of energy efficiency of new buildings and retrofits require the calculation of energy needs for heating, cooling, and hot water and energy use for lighting and ventilation. For example, the same calculation procedure, starting from energy needs and uses and ending with primary energy, is detailed step-by-step in the EU official "guidelines establishing a methodology framework for calculating cost-optimal levels of minimum energy performance requirements. For calculation of primary energy with primary energy factors, EN15603 presents explicit formulas, with degrees of flexibility for Member States. However, due to the lack of policy definition for ultra-low energy buildings, initially different definitions were introduced by business networks and mixed business/policy networks in the recent years" (Annunziata et al., 2013).

There are significant differences in the definition of minimum building energy efficiency threshold performance worldwide. The disparity is mainly due to the climatic, social, technological, and economic variation between countries, and this is partly justified. However, more importantly, the terminology and the definitions are not the same so comparisons are either difficult or impossible. For example, several European countries opt to comply with the PH Standard to guarantee that energy needs for heating and cooling are both below 15kWh/(m^2/ year). However, the PH Standard is sometimes perceived as high-tech building design in construction approaches and hence not feasible across Europe. Therefore, the challenge to implement and comply with NZEB performance requirements is high (Kotireddy et al., 2015). The challenge is not only for new construction, but also for renovation. A more precise definition of the indicators such as the one proposed in Hermelink et al. (2013) and eceee (2015) would help in the future to move towards a stronger framework for the actors of the construction industry, without restricting the flexibility needed in every country or state.

> NZEB should consume ultra-low energy and emit less CO_2 by design, rather than meet higher energy demand with onsite renewable generation.

4.3 Heating-Cooling Balance

The characterization of the balance of heating and cooling energy needs is important for high-performance buildings to limit unnecessary space conditioning systems and distribution. For example, in heating-dominated climates, designers seek to eliminate active cooling by using passive cooling design measures. This can lead to significant cost cuts due to the use of a single active mechanical system while providing simple control and maintenance. In heating-dominated countries, it is possible to achieve relatively easily summer comfort conditions and hence concentrate the largest part of the design effort to reducing the energy need for heating and to dealing with a single active conditioning system to optimize size and costs. However, in mixed mode and cooling-dominated countries, higher summer temperatures and solar radiation result for most building typologies and designs in an equilibrium of heating and cooling energy needs (see Fig. 3.4). The necessity to solve potential conflicts between winter and summer comfort objectives indicates a higher probability of having to install both heating and cooling systems (active, passive, or hybrid) and to bear the associated costs. For example, the study by Badescu et al. (2015) in Romania suggests that an active cooling and heating system should be used when PH buildings are implemented in mixed mode and hot climates.

The implication of a symmetric or quasi symmetric balance of heating and cooling energy needs lead often to the choice of dual active systems with thermal and electric energy demand which can have a large impact on initial cost, operational cost peak loads, and energy supply networks. Passive cooling systems such as earth buried pipes for cooling, ventilation air, evaporative cooling, and night sky radiation are also available,

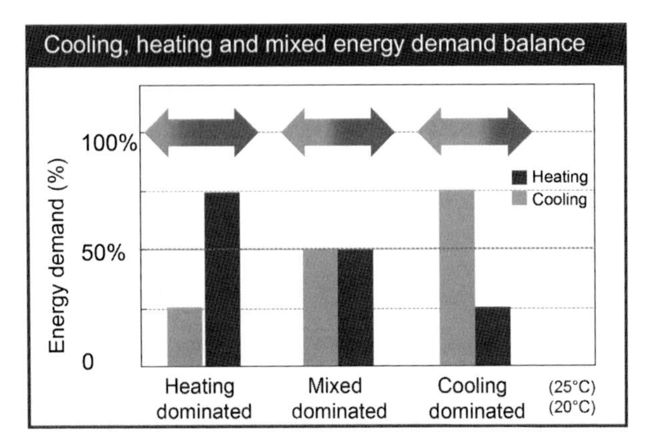

FIGURE 3.4 Cooling, heating, and mixed energy demand balance as an important performance indicator for NZEB design.

but need careful design and adaptation to climate, air, and outdoor conditions (pollution, noise, and mosquitos etc.), which requires highly skilled and savvy architects, engineers, and builders. In warm climates, low energy need thresholds for heating, e.g., 15–30 kWh/(m^2 year), can be met more easily for heating than in cold ones (Schnieders, 2009; Schnieders et al., 2015). This is due to milder weather and shorter extreme climatic cold waves. It is then possible to reduce heating needs even though various design parameters are not optimal (shapes, orientation, insulation, window sizes, and performance of components, to name a few). By reducing the envelop conductivity and infiltration and selecting optimal glazing and window openings, the heating energy demand can be reduced significantly. In this context, aiming at "nearly zero energy heating" targets to achieve the optimum savings is technically feasible. The use of heat recovery ventilation (HRV), also known as mechanical ventilation heat recovery (MVHR), can provide adequate space conditioning with minimum additional energy input and allows heat distribution directly through the air supply. In the case of Mixed-Mode climates, this could then make it possible to reach low heating energy demand values around 5 kWh/(m^2 year). However, in cooling-dominated countries, limiting the energy need for cooling below 15 kWh/(m^2 year) is not always possible due to high solar radiation, high outdoor ambient temperature, and the heat island effect in cities. Therefore, any definition for NZEB should take into account the heating-cooling balance for every climatic zone in every region or country and require energy efficiency thresholds and recommend passive or efficient active systems solutions accordingly. In our review of challenges of NZEB, we will focus on how the heating-cooling balance principle is addressed in current definitions.

4.4 Indoor Environmental Quality Limits (Thermal Comfort)

Fig. 3.5 summarizes the evolution of comfort models over the past 50 years. The available models worldwide are mainly focused on office buildings, partly because of the limited number of surveys in the area of residential buildings. However, the scope of these standards is can be extended to "other buildings of similar type used mainly for human occupancy with mainly sedentary activities and dwelling" (EN 15251, 2007; EN 16798, 2016).

For example, the European Committee for Standardization (CEN) introduced the European standards EN 16798 in 2016 (formerly named EN 15251-2007), which suggests the adoption of the Fanger's PMV/PPD model for mechanically heated and/or cooled buildings and Humphreys and Nicol's adaptive model for buildings without mechanical cooling

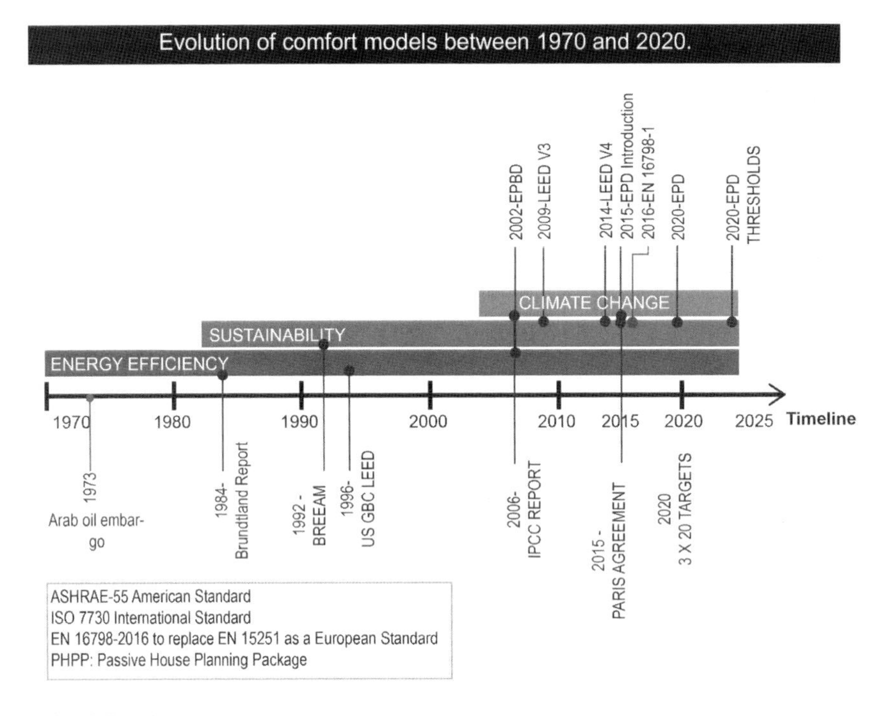

FIGURE 3.5 The evolution of comfort models between 1970 and 2020.

systems (2002). For NZEB, short and long-term comfort indices should be calculated according to EN 16798, in addition to energy performance indexes. The connection between thermal performance and comfort is explicitly discussed in EPBD (eceee, 2015). Also, various organizations have made proposals for comfort targets—e.g., French regulation requires that in air conditioned buildings the set point temperature in summer should not be set below 26°C; CIBSE Guide A defines "overheating" as occurring when the operative temperature exceeds 28°C for more than 1% of the annual occupied hours in the living areas of (free running) dwellings or when the bedroom operative temperature exceeds 26°C for more than 1% of the annual occupied hours (unless ceiling fans are available). The PH standard requires as a summer comfort criterion that "the number of hours in excess of 25°C may not exceed 5% of the time working." This criterion is verified by using a dynamic simulation. However, comfort as defined by PH may be challenging to achieve via the passive techniques traditionally adopted in good quality construction in southern hemisphere countries and may not correspond to the expectations of building occupants based on prevailing climate, clothing habits, and culture (Hermelink et al., 2013; Givoni, 1998; Carlucci et al., 2013; Attia and Carlucci, 2015; Pagliano and Zangheri, 2010). A discussion on the issue of defining comfort objectives took place, e.g., in the European project Passive-on, which involved experts from both northern and southern hemisphere countries with the objective of exploring the adaptation of the PH concept suitable for Mediterranean climates and led to a recommendation to refer to EN 16798, including the option to use adaptive comfort where suitable (Pagliano et al., 2007; Pietrobon and Pagliano, 2014; Causone et al., 2014).

A discussion about sick buildings and risks of overheating has emerged. The number of studies addressing summer comfort in NZEB in warm and hot climates based on measured data is now limited. Some extensive simulation studies find overheating risks in conventional buildings and significant improvements when going to well-designed advanced buildings (Barbosa et al., 2015; Ridley et al., 2013; Mlecnik et al., 2010; Peacock et al., 2010; Unit, 2013). In PHes "summer comfort can be achieved only resourcing to passive improvements, without any active cooling system," while "with common building envelop solutions and construction materials typically used in Portugal, simulations showed long periods of thermal discomfort for the heating season, as well as long periods of overheating during the summer" (Figueiredo et al., 2016). The studies conducted in United Kingdom, Belgium, and the Netherlands (Barbosa et al., 2015; Ridley et al., 2013; Mlecnik et al., 2010; Peacock et al., 2010; Unit, 2013) for different PH projects reported overheating periods during summer. The over focus on energy performance in NZEB can lead to health and comfort problems. Badescu et al. (2015) reported excessive overheating hours in a Romanian case study and recommended the inclusion of active cooling systems for such high-performance buildings. However, the existence of various definitions of overheating and explicit indexes, including the long-term comfort indexes proposed in EN 16798, are rarely used for designing buildings or assessing their actual comfort performance after occupation (Carlucci et al., 2014).

At present, the overheating phenomena in NZEB is often attributed to a combination of air tightness, insulation, and thermal mass, sometimes without other fundamental information offering in the analysis, such as: The presence or lack of solar protection, presence, and quality of the connection to available passive cooling sources like outdoor air in summer nights, soil, or sky vault at night, and presence and quality of means for air velocity control in occupied spaces.

4.5 Renewable Energy Generation Threshold

For NZEB, it is necessary to first reduce energy needs for heating and cooling, and, in the second step, to cover a significant portion of those needs by energy from renewable sources onsite or nearby. For example, in Switzerland the authorization to install summer air conditioning is subordinate to showing that the envelope is well designed to minimize energy needs for cooling (presence and effectiveness of solar shading with g-values optimized based on facade orientation, adequate insulation, and thermal mass, as specified in SIA382), and detailed verification procedures are in place. Thus, energy efficiency is an effective policy tool and, together with cost-effective energy savings, can play a primary role in meeting energy, climate, and economic goals. However, many new constructions worldwide fail to take up ultra EE and renewable energy measures that are cost-effective. Investment in building renewable energy technologies sometimes seems easier to implement and communicate to occupants, investors, and the media. There is evidence that some building owners invest and lean towards RES due to the legal and construction barriers in investing in energy efficiency (Unit, 2013).

On the other hand, the European recommendations for NZEB advise to include a share of renewable energy production onsite including the renewable share of heat pumps (Atanasiu et al., 2011). However, in dense urban areas, renewable energy sources (solar, imported biomass, and so forth) have limitations regarding solar access and pollution associated with burning biomass. For example, 70% of particulate emissions in Brussels, Belgium are due to biomass burning (Bruyninckx, 2016). Similar problems with air pollution from biomass burning is reported in Pianura Padana, Italy and more generally described by the European Environmental Agency. Thus the optimal balance between the minimum threshold performance for EE and the renewables onsite production share for NZEB remains a challenge. The impact of these parameters varies strongly depending on energy cost, legal, environmental, and construction barriers as well as requiring long-term vision that would help overcome these challenges.

> Renewable energy generation is location-dependent and site-dependent.

4.6 Occupancy Density

Evidence presented in Chapter 10 and gleaned from the literature suggests that occupancy density is another indicator that influences the EUI in NZEB. In office buildings, we find flexible work styles that vary between workplaces and meeting spaces. An NZE office today will have a variety of workstation settings and closed work places. Conferences rooms and auditoria by nature will have high occupant density. However, the trend in office spaces is the growth in meeting spaces, collaborative spaces, and other spatial organizations to cater for flexible working styles (Bedford et al., 2013). The effect of these varying functional uses is significant on the overall occupant density. Across Western Europe, the main density can be 10 m^2 per person whereas in North America this value goes up to 12 m^2 per person, while in Hong Kong or Dubai this value can go down to 4 m^2 per person. The critical factor for NZEB design in this context is to foresee the unintended consequences of those varying occupancy densities on the overall EUI. Especially that over time, building occupants' needs change and building owners adapt the occupant density of buildings. Lessons learned from post occupancy studies presented in Chapters 11 and 12 confirm the strong influence of changing the occupancy density. Over the past few years occupant density has risen due to economic pressure according to a study conducted by the British Council for Offices (Bedford et al., 2013). NZEB are mechanically ventilated buildings with specific ventilation requirements. The most significant changes of ventilation requirements are due to occupant density. With higher occupant densities, the delivered air increases and the cooling loads increase in association with latent and sensible loads increase, resulting in overall EUI increase.

The other aspect associated with occupancy density is the overall energy demand and occupancy density of a whole building in relation to its footprint. In dense urban locations, high-rise apartment buildings or office blocks have considerably limited land or roof-top areas. Clarifying this indicator early on during the design process, can examine the ability to match the energy demand with onsite RES or not for NZEB. Such a parameter is very important to relocate the project location or seek an optimal selection and sizing of RES.

Therefore, design teams should take into account the importance of occupancy density and need to design for diversity, allowing densities to be raised locally in buildings. The building capacity is a crucial design performance threshold because people create the demand for indoor environmental quality and building services. Postoccupancy evaluation studies (see Chapters 10 and 11) should be encouraged to close the carbon and energy performance gap and achieve higher productivity and satisfaction of users in NZEB.

> Increasing the energy efficiency for NZEB depends crucially on setting occupancy densities during the design phase in consistency with occupancy densities during building operation.

4.7 Cost Threshold

In order to design and construct NZEB, cost effectiveness is essential. The cost of NZEB per square meter of built up areas is a crucial performance threshold for developers, owners, and building teams. Using cost analysis tools and cost modeling, design teams should focus on cost modeling to come up with a rough estimation of cost per square meter for NZEB' different typologies early on in the design stage (Hamdy et al., 2017). Design trams should exercise cost control strategies for NZEB procurement, design, construction, and operation (Leach et al., 2014). NZEB have proven to be technically feasible, but in many countries relatively cost-neutral (see Fig. 9.3). The additional investment cost in Belgium, Germany, France, or Italy can vary significantly from −15% to +30% (Sartori et al., 2015; Moreno-Vacca and Willem, 2017; Heymer et al., 2016; Lalit and Petersen, 2017). The majority of NZEB realized worldwide consist of residential small projects, while the minority are commercial large-scale projects. In both cases, buildings require high insulation levels, technical installation of HVAC and heat recovery systems, as well as more capital intensive RESs such as PV. It is expected in the coming years that the PV installations' cost will continue to drop beside the impact of the mass generation of HVAC systems (heat pumps, heat exchangers) adapted to NZEB. Thus, the investment cost will be reduced with large-scale and massive NZEB constructions (see Chapters 9 and 12).

5 NZEB STATUS WORLDWIDE

According to the seven performance indicators and thresholds presented at the beginning of Section 4, our study goes further in analyzing the situation worldwide regarding NZEB status. Europe's extraordinarily ambitious targets are assuring the growth of NZEB construction year after year (European Commission, 2010). The Zero Energy Ready Home project in the US resulted in more than 15,000 ultra-efficient homes representing a new level of home performance with rigorous requirements that ensure outstanding levels of energy savings and comfort. In this section, we present the results after interviewing national experts in the investigated countries (namely Denmark, France, Germany, Italy, United Kingdom and the United States) for which we had access to representative information and insight. For this chapter, we present representative responses to the seven questions that reflect the most significant findings. The complete results are partially based on the work of Attia et al. (2017).

5.1 Denmark

The Danish government aims to reduce the total GHG emissions in Denmark by 40% by 2020. The Danish Building code introduced ambitious targets to achieve NZEB status through progressive performance classes. The performance "Class 2015" and "Class 2020" set up an energy consumption framework for minimum energy efficiency thresholds. "Class 2015" required that the overall requirements for building energy demand should not exceed 20kWh/m^2 per annum for heating, ventilation, and hot water (Thomsen, 2014). This is a 50% reduction of energy consumption compared with 2006. "Class 2020" requires that the overall requirements for building energy demand should not exceed 25 kWh/m^2 per annum for heating, ventilation, cooling, hot water, and lighting (Danish Energy Agency, 2014). A maximum demand is defined for total heating, ventilation, cooling, and hot water and lighting for nonresidential buildings. Regarding the primary energy conversion factors, electricity will count for 1.8 biofuels, fossil fuels by 1, and district heating by 0.6 by 2020. Compared to 2015, the primary energy factors have been lowered because it expected that renewable energy will make up a larger proportion of the Danish energy mix. The minimum renewable energy generation threshold for NZEB in Denmark is expected to be 50% by 2020 (Thomsen, 2014).

Finally, the indoor environmental quality limits must be documented through calculation for residential and commercial buildings, where the operative temperature should not exceed 26°C for more than 100 hours annually and a temperature of 27°C must not be exceeded for than 25 hours annually. For NZE commercial buildings there are prescriptive stringent requirements to comply with the Danish Working Environment Authority.

5.2 France

- France is facing a paradigm shift; moving from a pure energy approach with the RT2012 to an environmental assessment approach with Life Cycle Assessment (LCA) for the new building thermal regulation entitled RE2018 and RE2020.
- RE2018 promotes Positive Energy Buildings (BEPOS *Bâtiment à énergie positive*), targeting all building consumptions (energy use including electrical household appliances and computers among others, and not just "conventional" consumption), and will strive to balance building energy consumption from nonrenewable resources with renewable energy resources—future energy mix, including renewable electricity (FR, 2015).

- The new regulations set a standard for Low Carbon Buildings, taking into account the environmental impact of building material selection, construction, and operation, including energy systems, GHG emissions, and the transport means used to access the building. Both these approaches enable defining a label E + C- (Energy positive and Low Carbon) with several levels of graduation that is being tested.
- Designers should be committed to passive cooling, optimizing the bioclimatic design indicator *Bbio*, which addresses compactness, window surfaces, orientation, thermal inertia, and airtightness to name a few. Single-family and multiple-family dwellings should be designed without using "active" cooling systems (FR, 2015).
- The French building thermal regulation RE2018 will be based on standard EN 16798 concerning the adaptive comfort and ISO 7730 standard with a focus on summer comfort.
- With the BEPOS label, French regulations seek to maximize the rate of onsite photovoltaic production.

5.3 Germany

The Federal German Government developed an increasingly stringent regulation framework to reach nZEB by 2020 and NZEB by 2030. The German regulations are developed by the Energy Conservation Regulation (EnEV) and are focused on fossil fuels conversion losses and sets up maximum permissible primary energy demand for heating, hot water, ventilation, and air conditioning. The definition of nZEB standard focuses on "KfW efficiency houses" under the label of KfW Efficiency House 40, 55, and 70. The number indicates the value of primary energy consumption (QP) in relation (%) to a comparable reference building according to EnEV requirements. An Efficiency House 40, e.g., does not use more than 40% of the annual primary energy consumption (QP) of the corresponding reference building. In 2016, Germany changed its energy calculation methodology to become more in line with European Standards.

Renewable energy generation is mandatory for new buildings according to the Renewable Energy Heat Act. The minimum amount of renewable energy generation is regulated by this act requiring the use of solar heating. The minimum share can be from solar thermal heating (a minimum share of heating energy need of 15%), biomass (solid and liquid at least 50%, gaseous at least 30%), geothermal energy, and environmental heat (at least 50%), waste heat, combined heat, and power generation. Combinations of renewable energies with substitute measures are permitted.

- The heat requirement in the building stock as well as the entire primary energy requirement are to be reduced by 80%, in contrast to previous formulations and the requirement by the EPBD 2010.
- The zero energy or emission targets can no longer be confined only to new buildings and single housing buildings. The targets are extended to cover the renovation of existing building stock (BMWI; BMU 2010; BMWI 2011).
- In December 2014, the Federal Cabinet adopted the Action Program for Climate Protection 2020, with which the Federal Government wants to ensure that Germany has reduced its GHG emissions by 40% by 2020. The building sector, and especially the renovation of existing building stock, will play a considerable to role to achieve this goal Musall (2016).
- In addition to the EPBD, the EN 15217 is adopted for NZEB and building certification.
- DIN EN 15603 (Energy efficiency of buildings—Total energy demand and consumption) is considered as the basis for building assessment together with DIN V 18599 which provide the general framework for the calculation of energy efficiency.
- DIN EN 15603 2008 includes the reference to primary energy and CO_2 emissions to be used to determine the effects of the transformation chains of energy carriers. According to this standard, an annual net end primary energy balance by the difference of weighted final energy import and export of all energy carriers is represented by a numerical indicator [DIN prEN 15603 2013].

5.4 Italy

- In Italy, a NZEB is defined as a building which has a better performance (in terms of energy needs for heating and cooling, total PE demand, RES, etc.) than a baseline (virtual) building, which has the same shape, function, window/wall ratio, and specified baseline properties (e.g., U value, g-value, etc.).
- Some of the set properties of the baseline building (e.g., U-value) depend on surface-to-volume, so there is no explicit fixed value in kWh/(m^2 a) for being classified as a NZEB. The baseline U-values are higher than those recommended for PH certification (0.15 W/m^2K) and there is no explicit requirement on air-tightness. In general, the baseline building results have energy needs for heating higher than a PH (Pagliano and Zangheri, 2010).
- The requirements for a retrofitted building to be labeled as a NZEB are essentially the same as for a new building according to the

Decreto 26 giugno 2015. Night ventilation during summer is generally an effective cooling strategy in the Italian climate, while external temperature is relatively low. Residential buildings in rural and quiet urban areas can rely on natural ventilation, which could be ameliorated by openings and window design. However, in parts of urban areas, due to noise and pollution, nocturnal natural ventilation in residential buildings might need to be substituted by mechanical ventilation, thus requiring a sizing of the ventilation system to achieve higher ACH that would be required for winter alone. Tertiary buildings, not occupied at night, might profit in summer and mid-seasons of natural night ventilation to evacuate solar and internal heat gains irrespective of their location, with the addition of some simple antiintrusion protection at windows.

- The risk of overheating might be relevant throughout the country for newly built NZEB in case their energy concept is borrowed from northern design experiences and no night ventilation (or other connection to cool sources) is applied. Nevertheless, large improvements in the ability of the building fabric to decouple interior from exterior conditions is highly needed. The health consequences of the combination of high temperature and low quality building envelope are quite heavy (Robine et al., 2008; Santamouris and Kolokotsa, 2015; Sakka et al., 2012).
- The energy needs for cooling are expected to grow in coming decades due to climate change, unless significant improvements of the envelope of existing buildings and passive cooling measures for buildings and cities are applied (Ascione et al., 2016; Guarino et al., 2016; Santamouris, 2016)—this increase in energy needs for cooling is also expected in high-performance buildings (Pagliano et al., 2016).

5.5 UK Carbon

Under the Climate Change Act, the UK aims to achieve at least an 80% reduction in carbon emissions from homes by 2050. The UK's target for all new homes to meet the Zero Carbon Standard from 2016 comes in advance of the EPBD target for all new buildings in the EU to be "Nearly Zero-Energy Buildings" from 2020. In 2007, the UK government introduced a policy stating that from 2016 all new homes constructed must meet a Zero Carbon Standard. The UK approach to "Zero Carbon Home"—level 6 is an example of a governmental definition using the zero-energy emission. The Zero Carbon Home approach is based on offsetting carbon emissions in a two-stepped approach. Firstly, the "carbon allowance" approach is a mix of obligatory energy efficiency measures coupled to onsite generation (UKGBC, 2014). Secondly, the "allowable solutions supply" approach, which extends onsite generation by nearby or offsite generation (UKGBC, 2014). The Zero Carbon Home definition excludes plug loads such as appliances, computers, and TVs to name a few. The Carbon Compliance Limit is expressed in $kgCO^2/m^2$ per annum to provide a clear link with the government's carbon reduction strategy. The primary energy demand calculation is incorporated in the UK National Calculation. Proposals for Carbon Compliance levels were published in 2011, and suggest a minimum standard of 10–14 $kgCO^2/m^2$ per annum for dwellings. However, for nondomestic buildings, the United Kingdom has no definition yet. Methodologies, but no indicators for NZEB have been set. There is no specific energy efficiency or renewable energy thresholds defined by the UK government (DCLG, 2015 and DECC, 2015).

The advantage of the UK Zero Carbon standard is that it advocates for high-performance buildings to get assessed beyond the building and site, but tackles GHG emissions to the ultimately relevant global scale, which is beneficial for the planet. The British approach sets a focus on greening the national grid in parallel with greening building properties. The approach is ambitious and remains theoretical because it is difficult to achieve. Increasing the renewable energy share of the UK energy mix and at the same time installing RES for new buildings, is costly and divides effort to place initial investment in the right place. On the opposite side, the EU NZEB target is focused mainly on energy efficiency to reduce buildings' energy demand and consumption as much as possible, regardless of the used energy source. This helps buildings in an effective way to make NZEB mainstream sooner.

5.6 United States

In 2015, the Department of Energy (DOE) in the US developed a common national zero energy definition with supporting metrics and guidelines. With the help of the National Institute of Building Sciences (NIBS) and the Institute's High Performance Building Council DOE, and NIBS selected the term "Zero Energy Building (ZEB)" to represent ultra-efficient buildings balanced by renewable energy production over a year. They identified the need for additional definitions for related groupings of buildings including "Zero Energy Campuses," "Zero Energy Communities," and "Zero Energy Portfolios" to expand the reach of the ZEB concept. Since 2008, DOE Builders Challenge program has recognized leading builders for their achievements in energy efficiency—under the name of Zero Energy Ready Homes representing a new level of home performance, with rigorous requirements. Based on the 13693 Executive Order mandate, all Federal Buildings must become zero energy by 2020. Five years later there is a plan to convert 1% of existing buildings to achieve zero energy. In 2012, DOE recognized the Passive House Institute US's (PHIUS) work and entered a collaborative agreement with PHIUS to copromote Zero Energy Ready Home (ZERH), and the PHIUS + passive building program in the US market. Also, ASHRAE developed The 2020 Vision on NZEB and describes the boundary conditions when buildings will produce as much energy as they use. So far, there are no minimum or maximum energy performance thresholds for NZEB mandated in any state. Federal government agencies and many state governments are beginning to move towards NZEB in a voluntary approach. States such as California set policy goals of all new residential constructions to be NZEB compliant by 2020, and all commercial building to be NZEB by 2030.

6 RECOMMENDATIONS FOR NZEB PERFORMANCE THRESHOLDS

The market of NZEB in Europe, United States, and Asia is in the predevelopment phase. The next step of the market development process would be the take-off phase before the acceleration phase and finally reaching the stabilization phase. In order to balance the presented overview on the state of implementation of NZEB and their technical and societal barriers, we present a group of recommendations. The recommendations provides future perspectives for policymakers, funding agencies (including the national banks and the European Investment Bank), and building stakeholders with regard to the transition phases and development of NZEB. We classify and group a series of recommendations under five major topics.

Technical Development

A prerequisite for any technical development should be based on a common regional framework and terminology. We recommend the systematic use of ASHRAE, EN, or ISO definitions of energy levels in both technical and policy documents in order to facilitate the work and reduce the costs of design and construction firms. Countries and states should make clear and explicit requirements for low energy needs for heating and cooling in their national implementation of the NZEB concept. Reaching NZEB status requires that we change our rules-of-thumb and design assumptions of the real potential of bioclimatic architecture and passive design in mixed-mode and cooling-dominated climates. We need new and different concepts that are geoclimatically developed with respect to climate sensitivity, while avoiding overheating risks. This includes developing the definitions and performance requirements

TABLE 3.1 Suggestion for NZEB Performance Threshold in Selected Countries

Country	Min. energy efficiency		Primary energy $kWh/m^2.a$	RES share (%)	Carbon emissions $kgCO_2/m^2.a$
	Energy need for cooling	Energy need for heating			
	$kWh/m^2.a$	$kWh/m^2.a$			
Denmark	15	15	120	100	25
France*	5–20	5–20	50	50	3–10
Germany	5–15	5–15	120	100	–
Italy	15	15	120	50	–
United Kingdom	15	15	120	50	10–14
United States	30	20	120	50	–

*In France, 26°C is for air-conditioned buildings. For non-air-conditioned buildings, the inner temperature should be lower than the reference indoor temperature (TIC REF) during 5 days. TIC REF is determined by standard computation and depends on the climatic zone.

for NZEB in southern hemisphere countries. As part of our research, we present in Table 3.1 some performance threshold suggestions for nZEB in southern Europe's member states. We suggest the minimum EE and RET production across the investigated countries. Table 3.1 lists the suggested performance thresholds based on the input provided by national experts. To complement Table 3.1, we recommend using the adaptive comfort model EN 16798-1 in Europe and ASHRAE 55-2016 adaptive comfort model in the United States. We need to develop NZEB that are energy efficient, but also healthy and comfortable, while meeting relevant IEQ requirements. We advise countries and states to continue developing national adaptive comfort requirements backed by field measurements and surveys in relation to fuel poverty. It is crucial to reach a consensus on the definition of NZEB for new buildings and to retrofit existing buildings (Attia et al., 2017).

Organizational—Harmonizing and Sharing

Countries need to harmonize their NZEB standards and share their experiences with other countries. The creation of regional or national NZEB observatories will help to create a database of monitored NZEB. Energy Performance Certificates (EPS) and monitored data on real case studies need to be collected, quantified, and shared. This can help in generating and consolidating regional knowledge and expertise. Also, we should take into account other building-related standards and norms in order to address indoor environmental quality and the environmental impact assessment of materials. Private rating systems and standards, such as the Passive House, Active House, LEED, DGNB, and BREEAM or others, can also encourage the holistic design/build/operate approach through integrated project delivery processes, which is very important to foresee that future regulations are related to well-being and environmental product declaration. We should consider the knowledge transfer between countries as a good starting point to increase the knowledge uptake and accelerate the implementation of NZEB.

Organizational—Infrastructure

On the organization level, we recommend to take strong action for developing the necessary human and industrial infrastructure for NZEB implementation. The empowerment of scientists, professionals, industrial stakeholders, policy makers, and local authorities makes them able to embrace the transition. For example, we recommend empowering building researchers and allowing them to develop monitoring based concepts and definitions of NZEB and long-term case study analyses and reasoning (Garthley et al., 2017). This includes the generation of new weather files (using up-to-date hourly climate series and simulations of future climates with regional models), climate comparisons, guidelines for passive cooling and efficient active cooling systems, grid interaction models, and preparation for the following deep renovation challenge. This step involves creating an industrial strategy to empower the local industry to produce and supply buildings with ultra-efficient products and materials. We need new and different building design concepts that are geoclimatically developed with respect to climate sensitivity and the technological state of progress. Locally or regionally manufactured building components, products, and materials should lead market transformation.

Legislation and Enforcement

Legislation should be based on evidence-based policy strategy. Legislation must require permits and certification for renovation as well as new constructions which are able to trace renovation status, which is presently subject to extreme uncertainty, and building stock energy performance. This includes ensuring quality of construction through quality checks, compliance procedures, and proper commissioning. A project like QUALICHECK (2015) is a good start to achieve the reliability of EPC declarations and the quality of the works. The project ensures better enforcement and refurbishment in the frame of the revised EPBD, to create regulatory conditions to ensure better IEQ. This can also be achieved by ensuring regular inspections and continuous commissioning of passive systems and technical building systems of NZEB to maintain the envisaged IEQ parameters, EE, and RES production

Educational—Awareness

More attempts are needed to raise awareness of energy neutral buildings and to discuss the strategic approach of SMEs to develop a

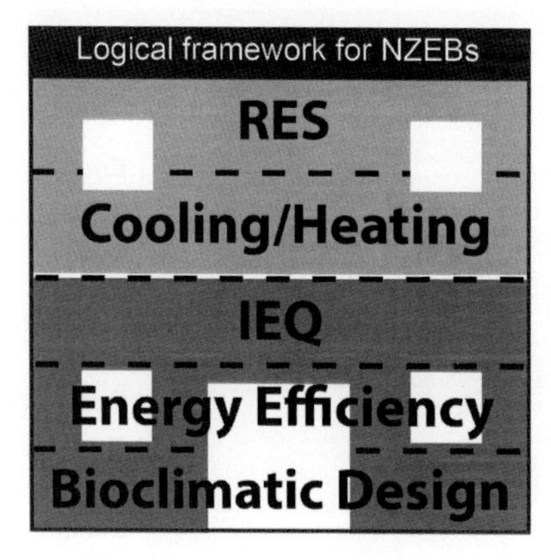

FIGURE 3.6 Logical framework for NZEB design.

suitable conceptual model for NZEB. Professional education can bring advanced concepts and technologies to SMEs. We recommend better preparation for building professionals and providing vocational training while simplifying the design and construction process of NZEB. This includes educating professionals and one-shop service providers and builders. Networking and raising awareness can bring various forms of strategic alliances in addition to a strategic framework for improving NZEB quality and profitability for SMEs (Fig. 3.6).

7 DISCUSSION

There are serious technical challenges that hinder the design of robust NZEB. In this chapter, we discussed the energy and emissions gaps, SBS, and decision-making stress associated with the design of these high-performance buildings. NZEB are complex and have many different connected elements and systems, and therefore require a performative design. We introduced performative design as a powerful approach to integrate components, systems, and solutions of NZEB and to get the details right. An approach that can help building designers to intentionally use design to solve problems, connect to flexibility during operation. Performative design must be associated with the establishment of NZEB performance indicators and thresholds during early design phases, which necessarily lead to the development of robust NZEB performance. By selecting a definition for NZEB in Chapter 2 and answering the seven questions on performance thresholds in Section 4 of this chapter, we set the foundation for early design decisions. NZEB can be designed using the seven major performance indicators indicated in Fig. 3.7 through quantifiable measures and metrics.

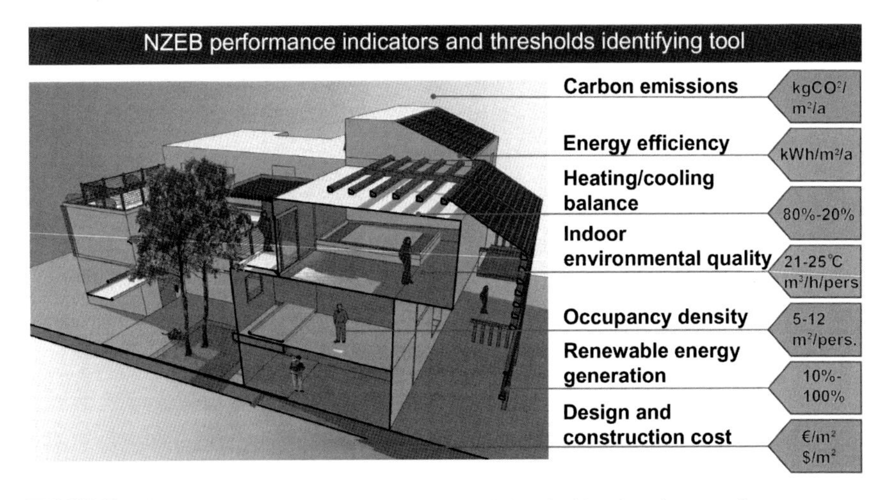

FIGURE 3.7 NZEB performance indicators and thresholds identifying tool.

Fig. 3.7 should be used as a tool to guide the design process early on by answering every question to the fullest possible extent.

Determining the feasibility of a NZEB design concept is based on the building minimum energy efficiency and site renewable energy potential. During the NZEB design, designers should not only focus on operational energy and the associated emissions (PE), but also the embodied energy of the building materials. During the project conceptualization, the building EUI should be used to set up the energy needs and available renewable energy generation in strong relation to IEQ, occupied density, and the project's financial budget. In this chapter, based on Table 3.1, we provide suggestions for the optimal balance between energy demand and energy generation in relation to IEQ to achieve a robust NZEB. The design scenarios assessment and comparison should be closely done through a performance-based design approach using powerful simulation tools in relation to real monitored demonstration projects. It is important to underline that the seven performance indicators and thresholds need to be part of an overall holistic design approach. We should not forget that the presented thresholds in Table 3.1 are all minimum thresholds for national regulations. This should not block ambitious initiatives and pilot projects to go beyond these threshold values and even achieve Positive Energy Buildings (PEBs). PEBs are the apex of regenerative building design and sustainable built environment (Attia, 2018). PEBs are expanding the scope regarding high-performance buildings with positive impact.

As we consider energy efficiency and carbon emissions in this chapter, we did not discuss the synergy of NZEB with the grid and the influence of transportation. We should not neglect the interaction between these NZEB and PEBs with the grid and the kind of stress they might cause in their functioning requiring the upgrade of current grids to become smart to deal with peak demand, peak generation, storage, and exchange. Chapter 8 will discuss these topics in depth and provides insight on the influence of embracing the suggested indicators, thresholds, and metrics in this chapter.

In the EU, NZEB have received increasing attention and have become part of the energy policy in 28 countries. Despite the specified performance by the EPBD, every country has the freedom to set its own performance indicators and targets, which creates confusion and decision-making stress to national and local authorities, builders, and NZEB designers. In this context, NZEB should be designed to the extent that is under the informed and controlled decision-making of designers. The seven performance indicators shown in Fig. 3.7, were defined based on lessons learned from several NZEB projects, including one presented in Chapter 11. They provide designers with consistent guidance on the minimum performance thresholds for a NZEB design for which they understand the key influential parameters. In this chapter, we identified the key performance indicators and thresholds that are useful design enablers to inform design teams to set an optimal balance between energy demands, IEQ, and onsite energy generation for robust NZEB and PEBs considering future climate scenarios.

Finally, the seven performance indicators and thresholds will enable the verification of claimed NZEB. The effectiveness and robustness of the applied design solutions should achieve comfort and high IEQ and, at the same time, achieve environmental policy targets. Performance monitoring and postoccupancy evaluations should take place using measurement and verification techniques to feed back the design loop to close the emission and energy performance gap (between as designed and as built), and inform the design decision-making.

8 LESSON LEARNED # 3

The conceptual design of NZEB should be developed by answering the seven major design questions related to performance indicators and carry out trade-offs between them. A proper balance between these seven performance indicators and their thresholds should be made to meet the performance requirements of NZEB over the buildings' life span. The fit-to-purpose defined performance thresholds, by design teams and decision makers, should be based on the feasibility to achieve specific performance values relative to the context and optimization of various design scenarios. Taking these performance thresholds and integrating them with the context of the project will be next crucial step.

Context is very important to influence decision-making because it includes climate, local industry, occupant behaviour, economic status, legislations, and local knowledge. In this sense, context is everything for a NZEB. Context brings complexity and makes it harder for designers to select the thresholds for the key performance indicators of NZEB. What is expected from building designers and experts is to use design to make complicated NZEB simpler. Using the performative design approach, designers can identify the key performance indicators and thresholds that we discussed in Sections 4 and 5 simplifying, standardizing and integrating those performance requirements in their design concept to make them easier to understand, easier to construct, and easier to operate.

References

Annunziata, E., Frey, M., Rizzi, F., 2013. Towards nearly zero-energy buildings: the state-of-art of national regulations in Europe. Energy. Available from: http://dx.doi.org/10.1016/j.energy.2012.11.049.

Ascione, F., De Masi, R.F., de Rossi, F., Ruggiero, S., Vanoli, G.P., 2016. Optimization of building envelope design for NZEB in Mediterranean climate: performance analysis of residential case study. Appl. Energy 183, 938–957.

Atanasiu, B., Boermans, T., Thomsen, K.E., Rose, J., Aggerholm, S., Hermelink, A., et al., 2011. Principles for nearly zero energy buildings. Retrieved from http://www.forskningsdatabasen.dk/en/catalog/2186082897 (accesses 15.07.17.).

Attia, S., 2012. A Tool for Design Decision Making-Zero Energy Residential Buildings in Hot Humid Climates, PhD Thesis. UCL, Diffusion universitaire CIACO, Louvain La Neuve978-2-87558-059-7.

Attia, S., 2016. Towards regenerative and positive impact architecture: a comparison of two net zero energy buildings. Sustain. Cities Soc. 26, 393–406. ISSN 2210-6707, https://doi.org/10.1016/j.scs.2016.04.017.

Attia, S., 2018. Regenerative and Positive Impact Architecture: Learning from Case Studies. Springer International Publishing, London978-3-319-66717-1.

Attia, S., De Herde, A., 2011. Early Design Simulation Tools for Net Zero Energy Buildings: A Comparison of Ten Tools. International Building Performance Simulation Association, Sydney, Australia, November 2011.

Attia, S., Carlucci, S., 2015. Impact of different thermal comfort models on zero energy residential buildings in hot climate. Energy Build. Volume 102, 1 September 2015, Pages 117-128, ISSN 0378-7788, https://doi.org/10.1016/j.enbuild.2015.05.017.

Attia, S., Gratia, E., De Herde, A., 2013a. Achieving informed decision-making for net zero energy buildings design using building performance simulation tools. Int. J. Build. Simul. Tsinghua-Springer Press, Vol 6-1, P3-21, 10.1007/s12273-013-0105-z.

Attia, S., Hamdy, et al., 2013b. Computational Optimisation for Zero Energy Buildings Design Interviews Results With Twenty Eight International Expert. International Building Performance Simulation Association, August, Chambery, France.

Attia, S., Eleftheriou, P., Xeni, F., Morlot, R., Ménézo, C., Kostopoulos, V., et al., 2017. Overview of challenges of residential nearly Zero Energy Buildings (nZEB) in Southern Europe. Energy Build. 155, 439–458.

Badescu, V., Rotar, N., Udrea, I., 2015. Considerations concerning the feasibility of the German Passivhaus concept in Southern Hemisphere. Energy Efficiency 8 (5), 919–949.

Barbosa, R., Barták, M., Hensen, J.L., Loomans, M.G., 2015. Ventilative cooling control strategies applied to passive house in order to avoid indoor overheating.

Bedford, M., Harris, R., King, A., Hawkeswood, A., 2013. Occupier Density Study. British Council for Offices (BCO), London.

BMWI; BMU, 2010. Bundesministerium für Wirtschaft und Technologie (BMWi); Bundesministerium für Umwelt, Naturschutz und Reaktorsicherheit (BMU) (26.09.2010): Energiekonzept für eine umweltschonende, zuverlässige und bezahlbare Energieversorgung. Berlin.

BMWI, 2011. Bundesministerium für Wirtschaft und Technologie (BMWi) (03.08.2011): Forschung für eine umweltschonende, zuverlässige und bezahlbare Energieversorgung. Das 6. Energieforschungsprogramm der Bundesregierung. Berlin.

Bruyninckx, H., 2016. Belgian air quality one of the worst in Europe. The Brussels Times 01.02.2016.

Causone, F., Carlucci, S., Pagliano, L., Pietrobon, M., 2014. A zero energy concept building for the mediterranean climate. Energy Procedia 62, 280–288. Available from: https://doi.org/10.1016/j.egypro.2014.12.389.

Carlucci, S., Pagliano, L., Zangheri, P., 2013. Optimization by discomfort minimization for designing a comfortable net zero energy building in the Mediterranean climate, doi:10.4028/www.scientific.net/AMR.689.44.

Carlucci, S., Pagliano, L., Sangalli, A., 2014. Statistical analysis of the ranking capability of long-term thermal discomfort indices and their adoption in optimization processes to support building design. Build. Environ. 75, 114–131.

Danish Energy Agency, 2014. Denmark's National Energy Efficiency Action Plan (NEEAP) Energy Efficiency Policies in Europe, Case Study Danish Building Code – Denmark http://www.energy-efficiency-watch.org/fileadmin/eew_documents/EEW3/Case_Studies_EEW3/Case_Study_Danish_Building_Code_final.pdf.

DCLG, 2015. 2010 to 2015 government policy: energy efficiency in buildings, Department for Communities and Local Government UK.

DECC, 2015. Energy Efficiency Strategy: The Energy Efficiency Opportunity in the UK. Department of Energy & Climate Change. UK.

de Wilde, P., 2014. The gap between predicted and measured energy performance of buildings: a framework for investigation. Automat. Construct. 41, 40–49.

eceee, European Council for an Energy Efficient Economy, 2015. EPBD consultation, review, possible revision and nZEB: A methodological discussion and proposals by eceee, Available from http://www.eceee.org/policy-areas/Buildings/resolveuid/ed0a87b89f9343aa83738561f214d3b8, Accesses 15.11.2017.

European Commission, 2010. 2020 Climate & Energy Package. European Commission, Retrieved from http://ec.europa.eu/clima/policies/strategies/2020/index_en.htm, Accesses 15.11.2017.

EN 15251, 2007. Indoor Environmental Input Parameters for Design and Assessment of Energy Performance of Buildings Addressing Indoor Air Quality, Thermal Environment, Lighting and Acoustics, in, European Committee for Standardization, Brussels, Belgium.

EN 16798-1, 2016. Indoor Environmental Input Parameters for Design and Assessment of Energy Performance of Buildings Addressing Indoor Air Quality, Thermal Environment, Lighting and Acoustics, in, European Committee for Standardization, Brussels, Belgium.

Figueiredo, A., Kämpf, J., Vicente, R., 2016. Passive house optimization for Portugal: overheating evaluation and energy performance. Energy Build. 118, 181–196.

FR, 2015. Règles techniques applicables aux bâtiments faisant l'objet d'une demande de Label Bepos-effinergie 2013. Version 3, 6 pages. Septembre 8th 2015.

Garthley, M., Mischler, F., Geyer, F., Nejc, J., Vlainić, M., Csaczar, C., et al., 2017. EmBuild - Empower Public Authorities to Establish a Long-Term Strategy for Mobilizing

Investment in the Energy Efficient Renovation of the Building Stock. European Union Research and Innovation Programme under Grant Agreement No 695169.

Givoni, B., 1998. Climate Considerations in Building and Urban Design. John Wiley & Sons, New York.

Guarino, F., Longo, S., Tumminia, G., Cellura, M., Ferraro, M., 2016. Ventilative cooling application in Mediterranean buildings: impacts on grid interaction and load match. Int. J. Ventilat. 1–13.

Hamdy, M., Siren, K., Attia, S., 2017. Impact of financial assumptions on the cost optimality towards nearly zero energy buildings - a case study. Energy Build. Volume 153, 24 August 2017, Pages 421-438, ISSN 0378-7788, https://doi.org/10.1016/j.enbuild.2017.08.018.

Hawken, P., Lovins, A.B., Lovins, L.H., 1999. A road map for natural capitalism. Retrieved from https://books.google.be/books?hl = nl&lr = &id = nO9MMxTtaeMC&oi = fnd&pg = PA250&dq = hawken + 1999 + building&ots = KxAPp6S_jv&sig = MCuDiXK06LH-E_VJKrHGtBxiHBW4 (accessed 01.08.17.).

Hermelink, A., Schimschar, S., Boermans, T., Pagliano, L., Zangheri, P., Armani, R., et al., 2013. Towards Nearly Zero-Energy Buildings: Definition of Common Principles Under the EPBD Ecofys. University of Wuppertal, Politecnico di Milano–eERG.

Heymer, B., Pless, Sh, Hackel, S., 2016. Zero Net Energy Building Cost and Feasibility. Webinar.

Ilmonen, L., 2015. Development of a quality assurance tool to minimize performance gap in NZEB.

Kotireddy, R., Hoes, P., Hensen, J.L.M., 2015. Optimal balance between energy demand and onsite energy generation for robust net zero energy buildings considering future scenarios. In Proceedings of IBPSA Building Simulation 2015. Hyderabad, India.

Lalit, R., Petersen, A., 2017. R-PACE: A game-changer for net-zero energy homes. Rocky Mountain Institute (RMI), Boulder, CO.

Leach, M., Pless, S., Torcellini, P., 2014, March. Cost Control Best Practices for Net Zero Energy Building Projects. In Preprint. Presented at the iiSBE Net Zero Built Environment 2014 Symposium.

Mlecnik, E., Visscher, H., Van Hal, A., 2010. Barriers and opportunities for labels for highly energy-efficient houses. Energy Policy 38 (8), 4592–4603.

Moreno-Vacca, S., Willem, J., 2017. Hyper-Efficient Building Workshop: Strategies for Complying with Local Law 31 Day 2, 30/08/2017, AIA NY, NYC, USA.

Musall, E., 2016. Klimaneutrale Gebäude – Internationale Konzepte, Umsetzungsstrategien und Bewertungsverfahren für Null- und Plusenergiegebäude. Bergische Universität Wuppertal, Wuppertal, Germany.

Nicol, J., Humphreys, M., 2002. Adaptive thermal comfort and sustainable thermal standards for buildings. Energy Build. 34 (6), 563–572.

Pagliano, L., Zangheri, P., 2010. Comfort models and cooling of buildings in the Mediterranean zone. Adv. Build. Energy Res. 4 (1), 167–200.

Pagliano, L., Zangheri, P., Pindar, A., Schnieders, J. 2007. The Passivhaus standard in Southern Europe. Paper presented at the 2nd PALENC Conference and 28th AIVC Conference in the 21st Century, Crete Island, Greece.

Pagliano, L., Carlucci, S., Causone, F., Moazami, A., Cattarin, G., 2016. Energy retrofit for a climate resilient child care center. Energy Build. 127, 1117–1132. Available from: https://doi.org/10.1016/j.enbuild.2016.05.092.

Passive House, 2017. Passive HouseDatabase. http://www.passivhausprojekte.de/ (accessed 01.08.17.).

Peacock, A.D., Jenkins, D.P., Kane, D., 2010. Investigating the potential of overheating in UK dwellings as a consequence of extant climate change. Energy Policy 38 (7), 3277–3288.

Pietrobon, M., Pagliano, L., 2014. Mediterranean passive house solutions towards nearly zero energy buildings in Italian regions. Paper presented at the 18th Passivhaus Conference, Aachen.

QUALICHeCK, 2015. Overview of existing surveys on energy performance related quality and compliance, http://qualicheck-platform.eu/2015/06/report-status-on-the-ground/ (accessed November 2017).

Ridley, I., Clarke, A., Bere, J., Altamirano, H., Lewis, S., Durdev, M., et al., 2013. The monitored performance of the first new London dwelling certified to the Passive House standard. Energy Build. 63, 67–78.

Robine, J.M., Cheung, S.L.K., Le Roy, S., Van Oyen, H., Griffiths, C., Michel, J.P., et al., 2008. Death toll exceeded 70,000 in Europe during the summer of 2003. Comptes Rendus Biologies 331 (2), 171–178. Available from: https://doi.org/10.1016/j.crvi.2007.12.001. ISSN 1631-0691. PMID 18241810.

Sakka, A., Santamouris, M., Livada, I., Nicol, F., Wilson, M., 2012. On the thermal performance of low income housing during heat waves. Energy Build. 49, 69–77. Available from: http://doi.org/10.1016/j.enbuild.2012.01.023.

Santamouris, M., 2016. Cooling the buildings – past, present and future. Energy Build. 128, 617–638. Available from: https://doi.org/10.1016/j.enbuild.2016.07.034.

Santamouris, M., Kolokotsa, D., 2015. On the impact of urban overheating and extreme climatic conditions on housing, energy, comfort and environmental quality of vulnerable population in Europe. Energy Build. 98, 125–133. Available from: https://doi.org/10.1016/j.enbuild.2014.08.050.

Sartori, I., Nori, F., Herkel, S., 2015. Cost analysis of nZEB/Plus energy buildings. REHVA J. May 2015.

Schnieders, J., 2009. Passiv Houses in South West Europe: A Quantitative Investigation of Some Passive and Active Space Conditioning Techniques for Highly Energy Efficient Dwellings in the South West European Region. Passivhaus Institute.

Schnieders, J., Feist, W., Rongen, L., 2015. Passive Houses for different climate zones. Energy Build. 105, 71–87.

Thomsen, K.E., 2014. Danish plans towards Nearly Zero Energy Buildings. Rehva J.

Thorpe, D., 2016. UNEP backs passivhaus to help meet climate targets. The Energy Collective Retrieved from: http://www.theenergycollective.com/david-k-thorpe/2393110/unep-backs-passivhaus-to-help-meet-climate-targets, accessed 01 August 2017.

UK GBC, 2014. A Housing Stock Fit for the Future: Making Home Energy Efficiency a National Infrastructure Priority. Green Building Council, UK.

Unit, E. E. I, 2013. Investing in Energy Efficiency in Europe's Buildings: A View From Construction and Real Estate Sectors, Committed by GPBN. BPIE, WBCSD.

Further Reading

ANSI/ASHRAE Standard 55—ThermalEnvironmental Conditions for Human Occupancy; American Society of Heating, Refrigerating, and Air-Conditioning Engineers (ASHRAE): Atlanta, GA, USA, 2010.

Baccini, M., Biggeri, A., Accetta, G., Kosatsky, T., Katsouyanni, K., Analitis, A., et al., 2008. Heat effects on mortality in 15 European cities. Epidemiology 19 (5), 711–719.

CHAPTER

4

Integrative Project Delivery and Team Roles

ABBREVIATIONS

AEC Architectural, Engineering, and Construction
BIM building information modeling
BPS building performance simulation
HVAC Heating, Ventilation, and Air Conditioning
IPD integrative process delivery
PH Passive House
MEP Mechanical, Electrical, and Plumbing
NREL National Renewable Energy Lab
NZEB Net Zero Energy Buildings
OPR Owner Project Requirements
RFP request for purpose
RFQ request of qualification
ZERH Zero Energy Ready Home

1 INTRODUCTION

Lessons learned from case studies often tell us about the decision-making stress and risk associated with NZEB design, construction, and operation (see Chapter 12: Roadmap for NZEB Implementation). Stories architects and building engineers rarely share are about the shock of the unexpected or sometimes about the regret over missed design opportunities. We talk often about our great designs, but rarely about the process we followed to achieve this design. We should always remind ourselves that NZEB are highly energy-intensive systems with a functional performance, strongly dependent on occupants' needs and behavior, facing constant variations and changes due local or external conditions. They are not the conventional buildings we know all about.

DOI: https://doi.org/10.1016/B978-0-12-812461-1.00004-6

NZEB design, construction, and operation is beyond the capacity of most building professionals. Even enhanced by technology, our experience with NZEB design is limited and we have just two reasons that we may nonetheless fail. The first is ignorance—where we may err because science has given us only partial understanding of NZEB and how they work. Discussions, presented in the previous chapter (Chapter 3: NZEB Performance Thresholds), show how we cannot eliminate the energy performance gap and sick-building syndrome. We haven't learned how to do that. The second type of failure associated with the design process is ineptitude. In this case, knowledge exists, but we fail to apply it correctly. This is related to the decision-making stress associated with the design of high-performance buildings. Consider, as an example, the undersized ventilation system of 11 newly built Passive House (PH) Certified Schools in Flanders, Belgium that forced students to follow their classes outdoors for a week during a heat wave in September 2016.

The situation of the Architectural, Engineering and Construction (AEC) industry at the start of the 21st century indicates that we accumulated know-how of NZEB without making it accessible and manageable. Avoidable failures are common and persistent, across many fields, from aviation, medicine, automobile, to finance and governance. However, the volume of complexity of what we know about NZEB exceeded our individual ability to deliver its benefits correctly and reliably (performance robustness). That means we need a different strategy and project delivery process for overcoming the performance gap, sick-building syndrome and decision-making stress. We need to create multidisciplinary design teams, equipped with dynamic building performance simulation (BPS) tools to support decision making. We need to use an integrative design process that orchestrates design decisions and makes them in the right sequence, with nothing dropped. We need a process that reduces risks and reminds us of the minimum necessary steps and make them explicit. We need a process that allows us to offer the possibility of verification and quality assurance for the design, construction, and operation of NZEB. Therefore, this chapter will articulate key differences between net zero and conventional projects from a work flow and cost delta perspective. We will identify the design process and summarize roles in the integrated, high-performance design process needed to achieve net zero energy performance. Based on lessons learned from the design-build team and the design-build process for new NZEB, we will provide a better understanding and guiding overview for future projects delivery and implementations.

2 INTEGRATIVE PROCESS DESIGN (IPD)

One of the hardest things in managing a NZEB is to see all the issues related to the design, construction, and operation process. The more complex a NZEB, the more difficult it is to plan and manage the process. Therefore, the only way to connect all the systems and building components, manage the design team, and respect the milestones deliverable is follow proofed process. Processes that will encourage maximum communication, empower the wisdom of the group, track the project progress, and guide stakeholders to deliver a healthy and robust NZEB. In the following paragraphs, we will present the complex nature of the NZEB design, construction, and operation process as well as the main constituents of IPD.

2.1 Complexity of NZEB Design Process

Thinking about failures and lessons learned presented in Chapter 11: NZEB Case Studies and Learned Lessons, as a small sample of the difficulties we face in building engineering, makes me think that the problem we face for NZEB design, construction, and operation is ineptitude—we need to make sure that we apply the knowledge we have consistently, fit-to-purpose, and correctly. Just making the right selection of a Heating, Ventilation, and Air Conditioning (HVAC) system choice among the many options for space conditioning in a high-performance building can be difficult even for expert mechanical engineers. Furthermore, whatever the chosen systems, each involves abundant complexities and pitfalls. Careful studies have shown that, for example, the relation between the building geometry or morphology and the potential energy generation, by integrating photovoltaic on building envelope, should have it designed and verified during the earliest design phases (Attia, 2016a,b). After that, the possibility to reach the net zero energy targets falls off sharply. In practical terms, this means that, during the earliest concept development phase and during the first design charette, design teams must decide on the building energy balance for the whole year and the used space conditioning systems in relation to the nature of the renewable systems and peak load. A charette is a kind of intensive workshop with key project stakeholders who sit together early on in a steam cooker like environment to discuss the influence of different design decisions associated with cost and feasibility studies in detail. This involves translating the program into a mass, estimating the photovoltaic roof surface area, estimating the building demand and supply, propose an HVAC system, discuss the decision with the building team, obtain agreement from the owner to proceed, a value engineer to confirm there are no conflicts with grid policies or permission problems that have to be accounted for, and then get the design process started.

However, the likelihood that this intensive steam cooker approach will occur within the first 90 minutes of the first design charette meeting is probably low. This is not an unusual example. Based on lessons learned from case studies presented in Chapter 11: NZEB Case Studies and Learned Lessons, the authors have been trying for some time to understand the source of our greatest difficulties and stresses to design high-performance buildings or performance-based green buildings (Attia, 2012). The reason for those difficulties is not cost or regulations. It is the complexity of bringing indoor environmental quality (IEQ), energy efficiency, cost, construction quality, performance validation, and assessment all together through building design, construction, and operation to achieve robust performance. This is a challenging task that is associated with decision-making stress and decision-making risk.

Stress and decision making are intricately connected (Starcke and Brand, 2012). The effect of stress on design decision-making is relevant to building engineering. Design decision-making stress is defined as the stress caused by overwhelming information that hampers a designer from making the fit-to-purpose design decision. The design decisions should be achieved in a short time period embracing multi-objective building performance criteria including occupants' well-being, energy neutrality, environmental control systems, construction materials, and performance assessment. Know-how sophistication of building technologies has increased remarkably across the AEC industry, and as a result so has our struggle to deliver building projects. As indicated in Chapter 9, Construction Quality and Cost, regarding the construction quality and Chapter 11: NZEB Case Studies and Learned Lessons, regarding the lessons learned from case studies, the AEC is conservative and shows a serious, slow-innovation uptake. The industry is crippled by fragmentation and systematic delivery and technical errors, requiring mastery complexity and of large amounts of modernization through knowledge management. This means that the existing knowledge needs to be applied correctly to avoid errors during the whole project delivery process. We need to encourage more experience and training to fight ignorance. At the same time, we need to reduce decision-making stress by properly planning the design process and setting up collaborative and multidisciplinary teams to prevent ineptitude. This involves BPS techniques and tools that will be presented in Section 5 of this chapter. Learning what we should not do is equally as important as what we should do.

Designing a high-performance building is risky and it should not be like that. With the increasing interest to consider the health impacts and avoid sick-building syndrome, many building owners and occupants are looking for healthy NZEB (Dodge Data & Analytics Media, 2016). Including health impacts and providing IEQ for NZEB while already embracing a performance-based approach increases the risk of errors and failures. As the profession is approaching a performance-based era, architects, engineers, and contractors are legitimately worried about the increasing responsibility and associated risk with NZEB. With the increasingly popular process of subcontracting the design and construction liability, performance robustness is highly risky. Based on the experience from the US NREL Research Support Facility (see Chapter 11: NZEB Case Studies and Learned Lessons) the client asked the design team to stay longer and more closely involved with their NZEB project. Builders as a consequence, had to create contingency plans to their bid to adequately cover the increased risk of such a performance-based project. Architects had to get closer to occupants and facility managers were found responsible for occupants' actions (Attia, 2016a,b).

Finally, we can state that building professionals are facing a complex built environment and challenges by higher decision stress and decision risk. Therefore, they must follow systematic and stepped approaches that are evidence based to provide protection against potential failures and decision risks. Following an integrative process delivery process will assure the design team members have explicit minimum necessary steps and milestones. An integrative process delivery offers the possibility of verification and instills a discipline of higher project delivery performance. The next paragraph will explain further and identify the characteristics of integrative process delivery process for NZEB.

2.2 Conventional Versus Integrative Process Project Delivery

In order to understand and learn about the integrative process delivery process of NZEB we will need to describe first the conventional project delivery process. This is an important intellectual exercise to identify clear the difference between both approaches.

Conventional design processes are sequential; the work is done step by step (Attia et al., 2013a,b). At first, architects design the building form, facade articulation and orientation, general aesthetic features, window area and placement. Only then, do engineers design the HVAC system in the context of the previously design envelope. Once the design is over, a builder is chosen by bid. The construction will then be executed in respect of the drawings, blue prints, and models created by the design team. This method has significant advantages concerning the human resources management by minimizing dispute risks in an efficient way. By nature, this process is rigid and linear. It decreases interaction and communication between the team members. Thus, collaboration is basic in such a process which only requires performance verification for a single point in time during commissioning. The late involvement of the general contractor (already after the design was set) potentially has serious effects on the building performance robustness. In a conventional process the contractor is mainly focused on controlling the costs, deliver on project delivery time with a large focus on stability and building safety. However, if the building performance is not within the target to costs estimations and constructability issues are not verified, it can result in costs growth during construction or occupancy. Occupants are almost always excluded from the process without any postoccupancy evaluation or soft landing (see Chapter 10: Occupant Behavior and Performance Assurance). Fig. 4.1, shows the conventional design process illustrating the relation between effort and time. As we can see the most effort is spent after the concept development.

On the opposite side of the conventional design process, the Integrative Process Delivery (IPD) process is iterative, synergetic to achieve results that are based on whole building systems across their entire life cycle (ANSI, 2012). Stakeholders in an integrative process will get involved from vision, through completion of construction and throughout building operation. The IPD requires the complete involvement of all at the early stage of design and during all the steps of the project through a multidisciplinary and collaborative environment. It is only possible if everyone knows when they must intervene and what are they asked to do. An integrative process is a comprehensive approach to building systems and equipment. Project team members look for synergies among systems and components, mutual advantages that can help achieve high levels of building performance, human

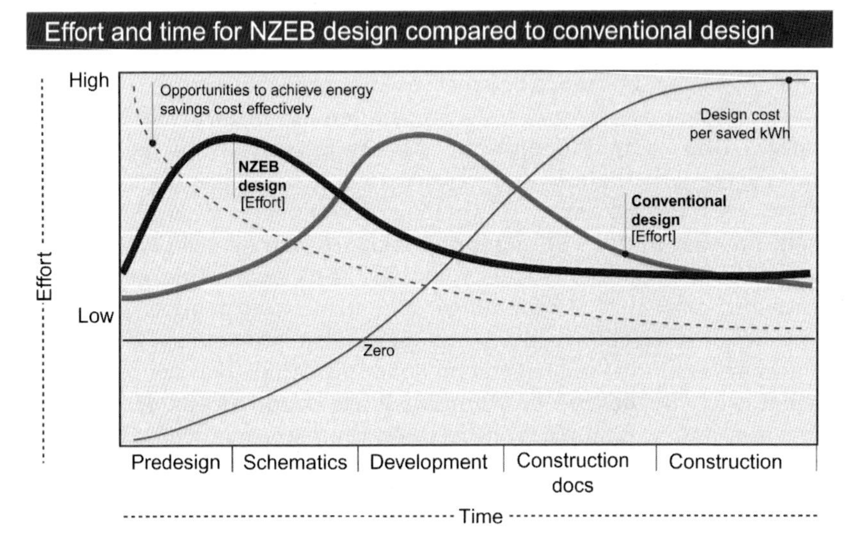

FIGURE 4.1 The conventional design process illustrating the relation between effort and time.

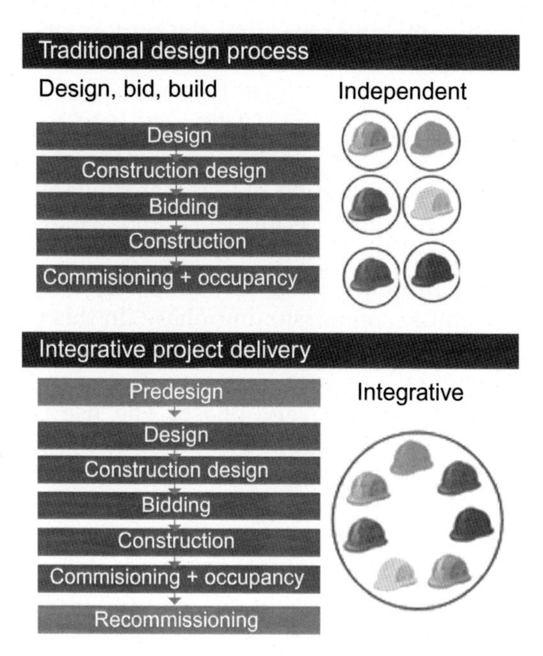

FIGURE 4.2 The additional steps of project delivery process for NZEB.

comfort, and environmental benefits. The major project delivery phases of an integrative process should follow the steps below and in Fig. 4.2:

1. Predesign
2. Conceptual Design
3. Schematic Design
4. Design Development
5. Construction Documents
6. Design Assist (pre-construction testing)
7. Commissioning
8. Occupant operation (monitoring, and postoccupancy evaluation)
9. Recommissioning

An integrative process comprises three phases. The first phase of IPD is the discovery phase. This phase can be seen as the enlargement of the predesign phase. During the discovery phases, the owner's aspirations get clarified and the performance goals are identified and translated into cost-effective environmental goals. In this phase, the design team can check the certification or rating system available points if possible. During a minimum of at least two design charettes, the stakeholders should develop a project vision, define the green goals of the building, set priorities, research green building strategies and technologies, and review laws and standards. This phase requires rigorous and intensive planning and decision-making regarding the performance targets, standards and certification compliances, milestones, contract nature and team responsibilities (Attia and Bashandy, 2016). Also, the design team has to check for the rating system available points, reducing the load of siting and orientation, translating the program into a mass in relation to natural, ventilation, natural lighting, passive heating, and passive cooling, investigating the envelope performance and window to wall ratio and initially selecting an HVAC system (Fig. 4.3).

The second phase, design, and construction begins with a schematic design and ends with the commissioning phase. In this phase, unlike its conventional counterpart, the design development incorporates the optimization and integration of users, systems, business structures, and practices identified in the discovery phase. The use of performances simulation for analysis is essential in this phase to introduce major and minor design improvements and optimize the building's outcome. Section 5 will elaborate on the types of modeling practices that are necessary during this phase.

The third phase is the most undermined phase in a conventional design process. The occupancy, operations, and performance assessment phase is focused on feedback. The integrative process highlights and embraces this phase to measure and verify the real building

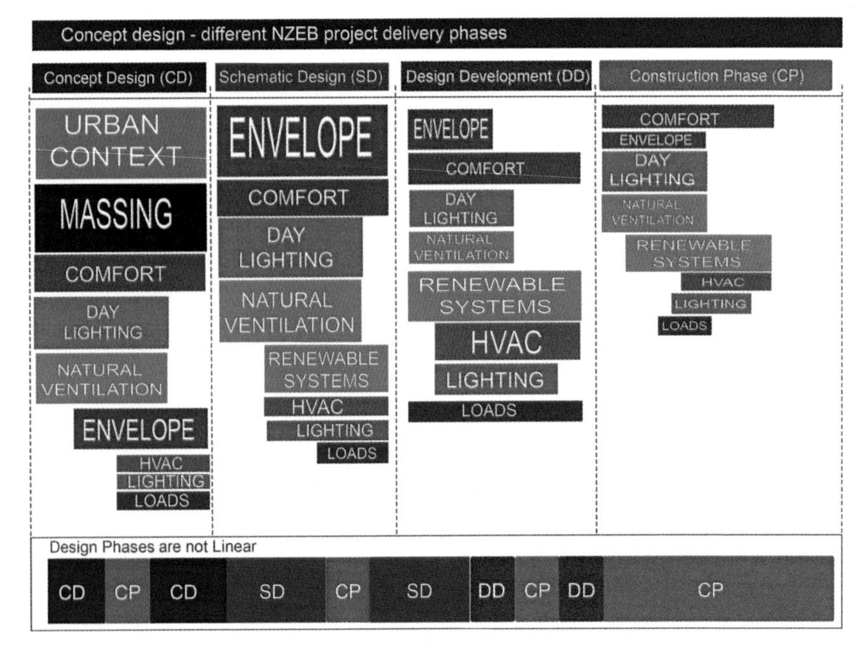

FIGURE 4.3 The different NZEB project delivery phases in relation to the key design measures. Source: *Adapted from Integral Engineering.*

performance and sets up corrective measures to ensure the performance targets are achieved. Measurement and verification of building systems and occupants' feedback are essential for building performance robustness, human comfort, and environmental benefits. There must be formal quality assurance specifications and responsible members with assigned roles. Planning performance goals would be useful in calibrating expectations among the design team and guide the measurement and operation of the energy use or NZEB performance (Scheib et al., 2014). Chapter 10, Occupant Behavior and Performance Assurance, provides sufficient insights based on experience and best practice to plan and realize this important phase.

3 INTEGRATED TEAM AND COLLABORATION

The process of creating a design team that collaborates in harmony must be started as early as possible for a NZEB design. The diversity of expertise promotes synergies and in turn can lead to adaptive and innovative solutions. This requires setting a clear project delivery process with key milestones, deliverables, and responsibilities. Assigning responsibilities to design team members and unified by the team behind the performance-based NZEB target is the first entry point to achieve success. In the following paragraph, we will provide an explanation on the significance of the way to establish a collaborative multidisciplinary team. The second idea that we will present will focus on the importance of maturity and experience for design teams and its effect on NZEB project delivery success.

3.1 Collaborative Multidisciplinary Team

The concept of IPD has been developed from the need to create alternatives to conventional design process. It consists of regrouping all the stakeholders (architects, engineers, contractors, suppliers, etc.) into a single team which will remain united from the outset to the end of the project. A key success factor to design, build, and operate a NZEB is to have a multidisciplinary work team. When representatives from all relevant areas of expertise, mainly from the client, builder, and design team sides as shown in Fig. 4.4, are brought together, the decisions made by the team encompass the full range of perspectives. There are many important stakeholders that need to collaborate early on as a team including the client, project manager, architect, mechanical engineer, contractor, cost estimator, and facility manager. Those stakeholders need to lead the whole building process during its life cycle and to see

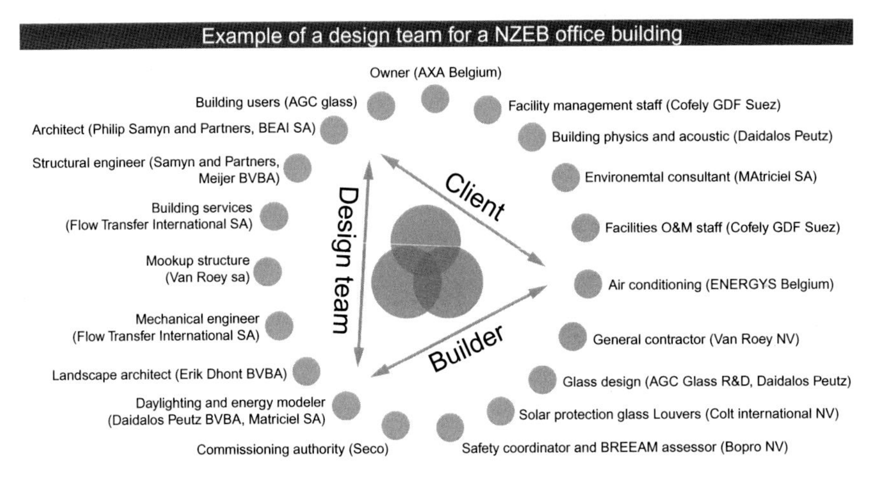

FIGURE 4.4 An example for a NZEB office building with the three major stakeholders (Attia, 2016b).

the pitfalls that commonly happen during the timeline transition moments moving from the design phase to the construction phase, and moving from the construction phase to the operation phase.

Multidisciplinary teams are responsible for the success of collective design, construction, and operation of NZEB. A multidisciplinary team is therefore an effective organizing approach to allow individuals to process different information, generate knowledge, and reflect on their expertise collectively to make informed decisions that bear on complex design problems. The diversity of the team and its representation of key stockholders and significant project phases can stimulate a harmonious pattern of design decisions. The multidisciplinary team diversity can handle intervening design parameters and is adapted to the performance-based design approach that looks for synergies to achieve cost-effective environmental goals for NZEB.

Collaboration between design team members can be successful when the process communication is defined. Collaboration is vital and communication can help all team members to understand and overcome the challenges of consorting by hearing directly other people's expertise. There are cultural differences and expectations often encountered by different design team members from different backgrounds. For example, if the builder is focused the cost, time, and quality triangle the architect might be focused on the architectural quality and innovation of design. Therefore, planning a harmonious and fruitful collaboration between team members can face several challenges. NZEB require a one-team approach and this can be achieved through good planning. Open, respectful and communication based on trust are the key to a successful NZEB project. The process of stimulating collaboration must start as early as possible during the predesign phase. Communication promotes emotional intelligence between team members, connects building strategies to create an NZEB that is designed together as a group, with key team members building on one another's thoughts to an ultimately successful project management. With the design and engineering of larger and more complex NZEB types, communication can foster collaboration between multidisciplinary teams.

Table 4.1 identifies the different team member's roles and responsibilities for a NZEB project:

3.2 Team Experience and Building Performance Coordinator

Excellence at creating complex NZEB requires a minimum level of experience and practice. The level of mastery associated with the design of an NZEB can lead cost reduction and performance robustness. The first interesting factor for developing the expertise of a design team is the duration the team already has been together. In comparison with the automobile industry, the AEC industry has much work to do to learn to improve the building performance-based design process, specifically during early design phases. For example, the Toyota design engineers in Japan worked intensively together for several years developing the Prius (Sobek and Liker, 1998; Daft, 2012). The development started from the design concept, to the Mock-up, to the decision hall where the team assesses the model looking at it from the customer angle. For this Prius, a small team consisting of one or two experts from each area of automobile engineering expertise was established (Sobek and Liker, 1998). The advantage of a long-lasting team is the creation of a sense of mutual commitment and frank communication, besides the continuous incorporation of many design changes for the model updates. The Toyota car model design teams work together for many years. The Toyota Prius team worked together for more than 20 years. The longer a design team works together, the better it gets and the more confidence it gains. On the other hand, the long team experience helped to optimize the car design during its whole life cycle guaranteeing maximum client satisfaction through continuous feedback and postoccupancy evaluation. An experienced team will avoid following a trial and error approach and operate with a mixed team, or set up a consortium of partners (Attia, 2012). According to the study of Moreno-Vacca (2017), on the extra cost of PH certified building in Brussels, he proved that an experienced team is a determinant factor to keep the cost of a PH the same or cheaper when compared with the cost of conventional building. The study of NREL for NZEB points toward the statement and advises

TABLE 4.1 Description of Different Team Member's Roles and Responsibilities in a NZEB Project

Owner representatives	Guides the owner or developer to develop the Owner Project Requirements (OPR) or Design Brief, Basis of Design and contracts and review substantial documents. The owner representative is responsible to define the NZEB performance target, minimum code compliance, standard compliance or even rating systems adoption
Owner project manager	Leads the project delivery process type and structure. The project manager is responsible for the overall project planning through initial operations on the daily basis
NZEB coordinator	An owner representative or project manager who participates in project meetings and reviews substation documents. The NZEB coordinator is responsible for the follow up; control, and validation of the integration and implementation of performance targets (see Section 3.2)
Architect	Responsible for the building design and guides the design team including Mechanical, Electrical, and Plumbing (MEP) engineers and other consultants, in the project IPD
Engineers	Responsible for structure and civil engineering and MEP engineers responsible for systems, installations and building services design. They investigate the energy efficiency options with the energy consultant and energy modeler to compare the design alternative and systems choice for the whole building
General contractor	The general contractor or builder is responsible for the execution of the buildings and provides cost estimation based on the value engineers' calculations. Together with cost estimators and value engineers, the general contractor investigates strategies and solutions that reduce costs and improve energy performance early and throughout the process
Energy or sustainability consultant	Takes the first step in the design process to guide and inform the whole design team on the environmental performance targets and performance expectations for a high-performance building. He or she advises the team on the state of the art solutions and technologies for all building components
Energy modeler	Is responsible for translating the building characteristic into a virtual simulation model and explore and investigate the implications of different design alternatives. The energy modeler creates a code compliant energy model and informs the team about the design interventions necessary to achieve the performance goal
Commissioning Agent	Reviews all drawings and project documents for energy related issues including envelope insulation and airtightness and systems installations performance. The commissioning agent checks and validates the proper equipment installations at the end of the construction process
Facility manager	Maintains the building's performance operation and reports on equipment performance and occupant behavior

to procure a project team that demonstrates experience and provides best value (NREL, 2014). Unfortunately, the design and construction process of most conventional project delivery processes are linear and chopped. The Design-Build for the integrative project delivery method is not older than 20 years and is not commonly used in the AEC industry. The following section will elaborate on the advantages of the Design-Build contracts. Experience from several project process analyses shows that design teams in the AEC industry (and in most cases team members) stay in the same office or company for a short time (Attia et al., 2013a,b; Attia and Bashandy, 2016; Attia, 2016b). Therefore, it is vital to assemble a collaborative multidisciplinary team with experience in high-performance buildings to ensure delivery of a robust NZEB.

In parallel with team experience, a NZEB requires a coordinator. Similar to the LEED Administrator, or BREAAM Assessor and PH Coordinator, the role of the NZEB coordinator is to become proactive with the team and apply robust quality control processes to the project regarding the energy performance targets. The NZEB project coordinator should be experienced and knowledgeable about NZEB projects and processes. This person will be responsible to manage and assign team roles, coordinate and document building solutions and performance throughout the design, construction, and operation process.

4 BUILDING PERFORMANCE-BASED CONTRACTS

Project acquisition and delivery methods have evolved in the past 30 years. In this section, we aim to better understand the considerations to deliver cost effective, energy efficient and, above all, simply designed NZEB. Historically, until the end of the 19th century, clients hired master builders to design, engineer, and construct a building. In ancient Egypt, the code of Hammurabi (1800 BC) forced that master builders assumed absolute responsibility for the design and construction of a project. From that time and up until the 20th century, the master builder model reigned the project contracting methods similar to a process that we call now design-build (Design-Build Institute of America, 2009).

With the evolution of project planning and management, including project schedules and budgets, design and construction services became more specialized (Pless et al., 2011, 2013). According to Konchar (1997) the master builder model resulted in the emergence of the design-bid-build project contracting and delivery. However, the aggressive energy efficiency targets and the intention to close the energy performance gap and provide high quality buildings' performance, the contracting methods are required to be changed to allow the design and construction teams to achieve the high-performance targets. The following section will explain the most fit to purpose project contracting methods for NZEB.

4.1 Project Contracting Methods and Liability

One of the most successful contracting methods is the Design-Build process or the performance-based Design-Build process. It provides the best price for a project, enables the design team's creativity in developing the most cost-effective, integrated, and high-performance building. It often limits the classical conflict between the builder, cost estimator, and value engineer on one side, and the design team and sustainability or energy efficiency consultant on the other side. A Design-Build contract will allow the owner to deal with one single legal entity. The Design-Build team will deliver an operational building that satisfies the owner's design criteria or OPR. The contract distributes the risk between the design and construction teams and makes them responsible to manage the risk commonly. The nature of this contract allows setting performance-based targets to NZEB in a flexible and innovative way. The owner does not rely on plans and specifications to describe the project scope, but rather focuses on the performance targets and project budget. This allows the Design-Build to focus on simple, ultra-efficient and cost-effective solution sets with the high freedom to explore creative paths. Control and accountability is allocated to the Design-Build team to design, build, and deliver the project as one entity. The project meets the contractual obligations specified in the OPR and RFP within a firm fixed-price and fixed-time schedule (Design Sense Inc., 2008). Also, the Design-Build team should have incentive to stimulate perfection, save money, save time, and increase the profit margin under the fixed-price contract. The team remains under a continuing responsibility with reduced liability—an equally shared overall risk for the project among the team members. The risk associated with a performance-based Design-Build contract can lead to higher costs for inexperienced teams. Therefore, team experience, discussed in Section 3.2, is foundational to reduce risk and remain competitive regarding cost.

At the same time, the contract language requiring energy target validation should be followed by energy performance assurance. Chapter 10, Occupant Behavior and Performance Assurance, discusses the importance of allowing owners to get feedback on the energy performance throughout commissioning and occupancy, compare the results to model predictions, and provide feedback to the design team regarding the fitness of installation or control strategies through facility management. This includes specific articulation of monitoring and verification protocols and setting end use budgets as a point of reference for comparison with real performance.

Opposite to the Design-Bid-Build contracts the Performance-Based Design-Bid contract avoids the separate design and construction contracts that can jeopardize the building performance. The Design-Build process forces the value engineering process to adhere to the early design decisions that clearly specifies all minimum project requirements, performance targets and measurement and verification protocols. According to Pless et al. (2013) there are six major advantages of performance-based Design-Build contracting and delivery process:

Single responsibility: The conventional Design-Bid-Build process consists of several points of responsibility with a designer developing the plans and specifications and a builder who works on its implementation. This process is linear and separated and does not allow the design team or architect to have contact with the general contractor during the design concept development resulting in errors or shortcomings. The Design-Build contract promotes productive collaboration, refines the decision-making process, and aligns stakeholders to unify around one focus target.

High performance quality: Single responsibility allows streamlining the building quality and committing to the performance targets and specification as early as possible in the process. When design teams are assembled early on in the process, design synergies are stimulated, resulting in the increase of the project's added value. By meeting the owners' expectations and requirements, the design team excels in exceeding performance criteria without additional cost or significant time.

Cost savings: Cost savings and budget control is the most effective when an integrated design team collaborates to design and build a project from beginning to the end of the project delivery process. For NZEB, the IPD process encourages maximum collaboration between architects and mechanical engineers, which improves cost control and budget management. As a result, the IPD enables the design team to integrate passive design measures and ultra-efficient energy conservation measures, which as a consequence, reduce the building loads and equipment capacity for active space conditioning and ventilation (see Chapter 9: Construction Quality and Cost). It also allows the design team and general contractor to select a cost-effective construction system and building materials. Value engineers collaborate more effectively with the design team instead reporting to the owner or general contractors during constructability reviews. These simple collaborations make the design team and builder work more efficiently. The collaboration makes the building project delivery process closer to the best practices of the automation and aviation industry resulting in significant cost reduction (Watanabe, 2007).

Time savings: When the design is finished in a Design-Bid-Build project, the owner launches a bidding process to find a successful bidder.

This process takes time and is a result of the conventional linear project delivery process. On the opposite side, the Design-Bid contract allows an overlap between the design team and contractor in the presence of the owner. The collaborative nature of the concept development phase allows for maximum communication between the team members resulting in a significant time saving and capital costs. The delivery time reduction is mainly due to the alignment of the design process with the budget and schedule.

Risk management: The risk management in a Design-Bid-Build contract is in the hands of the owner or owner management team. The change orders caused by errors and omissions in the systems sizing or construction documents are fundamentally the responsibility of the owner representative. However, in a Design-Build contract, the risk management is in the hands of the Design-Build team. The contract capitalizes responsibility of the design and construction teams and forces them to anticipate and mitigate those risks.

Recognition and Innovation: The performance-based Design-bid contract bridges the gap between designers and builders and empowers the whole team to collaborate and integrate the best fit-to-purpose ideas in a project. Shared responsibility and early collaboration guarantees building robust performance without following a hierarchical or prescriptive working mode. Owners and the Design-Build team succeed collaboratively to develop creative and innovative solutions, which is the basis for robust performance recognition by occupants and building professionals.

4.2 Project Acquisition and Team Assembly

The project acquisition process and team assembly process starts by the setting the performance-based targets and creating the integrated design team. The owners should start by setting a firm price or estimated budget for the project. With the help of the owner representative team or consultant, the owner sets the target performance requirement of a NZEB and establishes the program. The owner expectations should be translated into OPRs and transformed into a RFP that ranks the owner requirements. The second step is to invite Design-Build teams to propose ideas and solutions that can achieve most of the owners' requirements. The Design-Build team should brainstorm ideas and concepts and evaluate the different project goals. All team members should be defined with the corresponding responsibilities. The third step is to select a Design-Build team to complete the NZEB project for a fixed price in respect to the performance targets and minimum proposed requirements.

The Design-Build Institute of America (DIBA) summarizes seven best practices that relate directly to ensuring energy performance-based procurement process steps (Leitner, 2008) for owner's executive management, as shown in Fig. 4.5:

Best value procurement: Provide the project scope through a list ranking the performance requirements and their priority. The owner's needs will be ranked in at least three areas—mission critical (must be provided), highly desirable (should be provided), and if possible (optional) (Pless et al., 2013). This allows the owner to receive the most possible proposed solutions that fit within the budget.

Two-phase solicitation: Use an initial request of qualifications (RFQ) to assess the experience and record of the soliciting teams and then provide the RFP to a short listed set of design teams.

Short-list three qualified teams: This increases the chance of winning Design-Build teams and allows the reviewing committee to choose the best value team and foresee the potential risks while moving forward in the project delivery process.

Interim interviews during competition: Allow the Design-Build teams to pose their questions through public interviews and provide answers to all teams to reduce the RFP gaps and risks before the project proposal review phase.

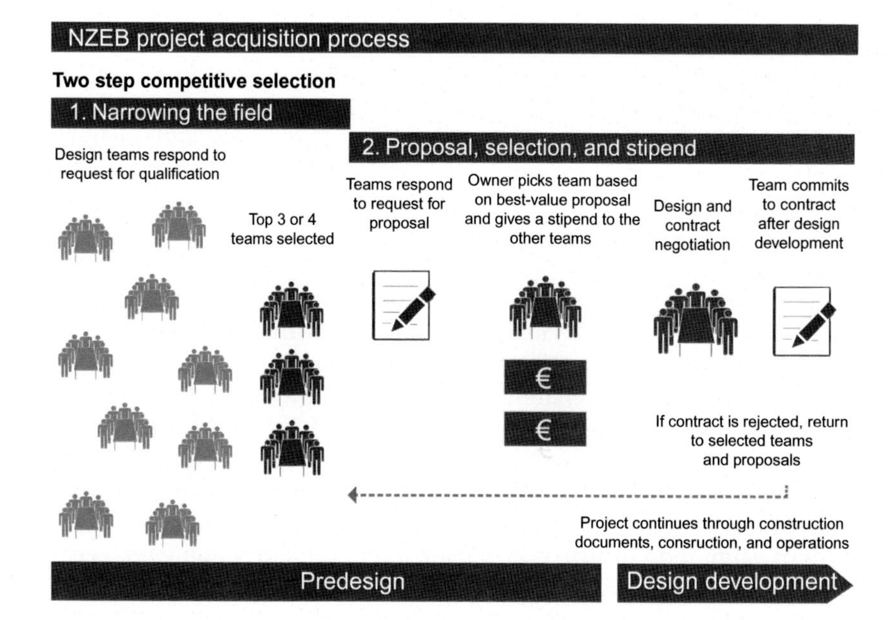

FIGURE 4.5 The project acquisition process as suggested by NREL professionals (Pless et al., 2013).

Stipends to unsuccessful teams: Request conceptual design as part of the proposal and provide stipends to offset their cost and fees. This approach allows the owner to benefit and share ideas from nonselected teams and move them to the winning teams.

Award Incentives: Motivate the subcontractor behavior by offering incentives with typical monetary value of 2%–3% of the total subcontract. This is a best practice that ensures the owner has an influence during design and construction and encourages improvement of performance.

Performance specifications: The performance-based approach was discussed in Chapter 2, Evolution of Definitions and Approaches, and Chapter 3, NZEB Performance Thresholds. The aggressive energy efficiency targets of NZEB require creativity and design decision flexibility. By defining performance specification instead of prescriptive specification, the team can achieve a high-performance building design without significant cost or risk.

During the previously mentioned project acquisition process, the team assembly should take place. An integrated multidisciplinary design team should comprise architect, engineers, builder, cost estimators, and all stakeholders mentioned in Table 4.1. The team assembly process should include key subcontractors who should work together in an iterative and collaborative way. The team assembly can be implemented in two steps.

The first step is to assemble a team, which should happen during the predesign phase. During the predesign phase, an independent client team can help the owner develop the key project requirements and set up the performance targets. This can be the first step to prepare the RFP and review the project proposals and teams. The owner should hire a professional Design-Build service that can help write the program, OPR and RFP. A design charette with different stakeholders can help the owner to define his/her goals more clearly and set up a ranking of the owners' needs.

The second step is to assemble the team, which can happen after the teams respond to the RFP. This step involves recruiting experts and providing training if necessary to the whole Design-Build team. This step involves further design charettes to meet the owners' representatives and collaborate with the project manager. The key service related to implementation and subcontracting should be discussed during this phase and the energy or sustainability consultant together with the energy modeler should be enabled to provide design decision support during design and construction. Cost estimation and value engineering together with third part commission should be clarified at this stage and assigned to team members.

5 BUILDING INFORMATION AND PERFORMANCE MODELING

The building information modeling (BIM) and performance modeling of NZEB is a challenging problem of increasing importance. The NZEB objective has raised the bar of building performance, and will change the way buildings are delivered. The building design community at large is galvanized by mandatory codes and standards that aim to reach neutral or zero-energy built environments (EU, 2009; ASHRAE, 2008). At the same time, lessons from practice show that designing a robust NZEB is a complex, costly, and tedious task. The uncertainty of project delivery for NZEB is high (Attia and De Herde, 2011; Attia et al., 2013a,b). Combining passive and active systems early on is a challenge, as is, more importantly, managing building information and guiding the Design-Build team toward the objective of energy and indoor comfort of the NZEB. The integration of passive and active design aspects during the early design phases is extremely complex, time consuming, and requires a high level of expertise, and software packages that are not available (see Fig. 4.3). At this stage, architects and engineers are in constant search for a design direction to make an informed decision. Decisions taken during this stage can determine the success or failure of the design. In order to design and construct such buildings, it is important to assure BIM and informed decision-making during the early design phases for NZEB. This includes the integration of BPS tools early on in the design process (Hayter et al., 2001, Charron et al., 2006).

BIM and building performance modeling are ideal to lower such barriers. Both techniques can be supportive when integrated early on in the architectural design process. Simulation in theory handles dynamic and iterative design investigations, which makes it effective for enabling new knowledge, analytical processes, materials and component data, standards, design details, etc., to be incorporated and made accessible to practicing professionals. In the past 10 years, the BIM and building performance modeling discipline has reached a high level of maturation, offering a range of tools for building performance evaluation (Hensen and Lamberts, 2011). Most importantly, they open the door to other mainstream specialism, including architects and engineers, during earlier design phases. Chapter 12: Roadmap for NZEB Implementation elaborates on the use of modeling as a tool for NZEB implementation.

The process of NZEB design can be described as a successive layering of constraints on a building. Every new added decision, every defined parameter, is just one more constraint on the designer. At the start of the NZEB design process, the designer has many decisions and a relatively open set of goals. By the end, the building is sharply defined and heavily constrained. For high-performance buildings, high constraints are imposed due to environmental and energy requirements. The constraints provide a useful anchor for ideas. Conceptual early design stages of NZEB can be divided into five substages: (1) Specifying Performance Criteria, (2) Generating Ideas, (3) Zones-Layout Design, (4) Preliminary Conceptual Design, and (5) Detailed Conceptual Design. Substages 2–5 do not always follow a sequential linear order. The design process goes into a cyclic progression between those substages in which each substage elaborates upon previous constraints.

5.1 Barriers to Integrating Building Modeling During Early Design Phases

Experience learned from constructed NZEB, based on postoccupancy evaluation, shows that the design of high-performance buildings is not intuitive, and that building modeling tools are a fundamental part of the design process (Attia, 2016a; Attia and Bashandy, 2016). The nature of the aggressive goals of NZEB requires the early creation of energy models during preconceptual and conceptual design phases. Recent studies on current barriers that face the integration of BPS tools into NZEB design are summarized below (Athienitis et al., 2010). Fig. 4.6 illustrates the barriers of decision making during the early design stages of NZEB design.

Geometry representation in simulation tools: Architects work in different ways through sketches, physical models, 2D and 3D computer generated imagery, and analytically—and thus have different requirements for representing and communicating their design form.

Filling input: The representation of input parameters in the language of architects is a challenge in many tools. There is a clear separation between the architects' design language and the building physics language of most tools. This difference is often addressed by using reduced input parameters or using default values. However, filling in the design parameters is an overlooked issue among BPS tools developers.

Informative support during decision-making: Designers cannot easily predict the impact of decisions on building performance and cost. The building delivery process of NZEB requires instantaneous feedback and support to inform decision-making for passive and active design strategies. The disadvantage of most existing tools is that they operate as postdesign evaluation tools. Therefore, the informative support should be comprehensive enough to include geometry, envelope, and systems.

Evaluative performance comparisons: During the early design stages, benchmarking and the possibility to compare alternatives is more

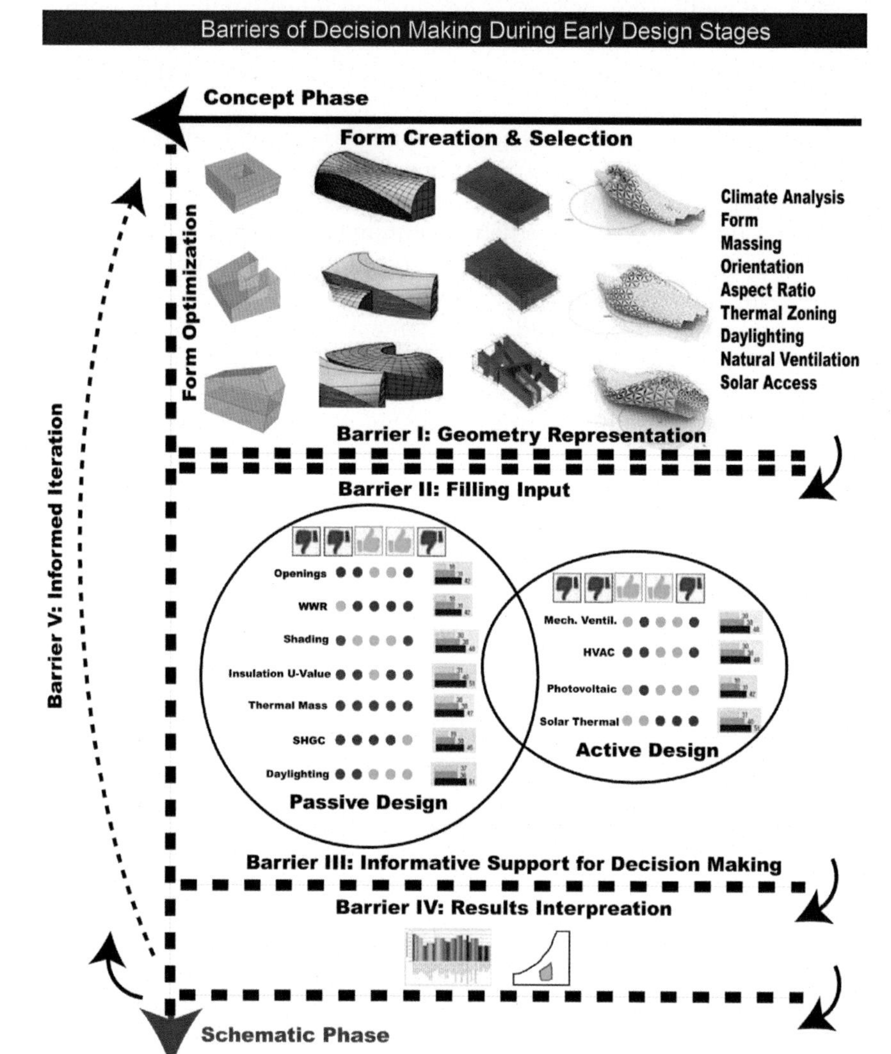

FIGURE 4.6 Barriers of decision making during early design stages.

important than evaluating absolute values. The ideas generation phase is iterative and comparative. Most existing tools do not emulate this process and focus on postdesign evaluation.

Interpretation of results: The representation of simulation output and its interpretation is frequently reported as a barrier among architects and engineers (Donn et al., 2009; Attia et al., 2009). Analytical results presented in tables of numbers or graphs are often too complex and detailed, providing an excessive amount of information. The output representation often lacks variety and visual qualities. Analysis and simulation results should be displayed within the context of the 3D geometric model (Marsh, 2004).

Informed iteration: The most important barrier facing architects and engineers is cycling informed iterations for concept development and optimization. In the past, architects and engineers iterated back on the design for functional and aesthetic optimization purposes. For NZEB, they have to iterate for performance optimization purposes. This requires an understanding of building physics and performance. Architects and engineers need a fundamental understanding of basic building physics that allows them to interpret the simulation feedback and drive them to iterate back to the concept. Therefore, it is highly recommended that architects and engineers work with building performance models and the predesign phase allows them to explore the optimal potential of project outcomes.

5.2 Energy Modeling Tools Review

Very few energy modeling tools can help support the design decision-making for NZEB during early design phases. NZEB design strategy addresses a design duo: First maximum energy efficiency and then the delivery of energy required from renewable systems. Almost no tool listed in Table 4.2 helps to answer this. A critical look at the

TABLE 4.2 Classification of BPS Tools Allowing Design Evaluation and Design Guidance

	Evaluative			Informative	
Support (Technique)	Postdecision Evaluative	Geometry Plug-in	Predecision Evaluative (Para & Opt.)	Predecision Informative (Parametric)	Predecision Informative
Iterations	High	High	Medium	Low	Low
Renewable Systems			SolarShoeBox Energy 10		
Energy Efficiency	EnergyPlus TRNSYS ESP-r IES VE EQUA	OpenStudio IES VE-Ware	Vasari MIT Advisor BDA Desgin Inent HEED Solar House Sunrel	Sefaira DesignBuilder jEPlus iDbuild	BeOPT OptiPlus OptiMaison
Daylighting & Facades		SunTools	COMFEN NewFacades Diva		

existing tools in relation to the NZEB design process shows that several barriers exist in integrating the current building energy modeling at this stage. Therefore, future tools should allow both strategies in order to develop NZEB and supplement the intuitiveness of the design process with analytical techniques and simulation methods.

Over the past few decades, a large number of building energy modeling tools has been developed to help engineers during late design phases. Such tools were developed to produce data concerning buildings' numerical modeling, simulating the performance of real buildings. Those energy modeling tools require a complicated representation of the building alternatives that require specific and numerical attributes of the building and its context. Those tools can be classified under a main group named "evaluation tools" as shown in Table 4.2. The examples in Table 4.2 are meant to be indicative, not exhaustive.

Evaluation tools include energy analysis computer tools. Although by being evaluative they produce results that do not actually provide any direct guidance as to how the NZEB design should be improved or the performance objective achieved. The use of evaluation tools in NZEB design is based on a postdecision trial and error approach, where the simulation results are compared to a desired value. If the results are not satisfactory, the design is modified and the process is repeated. This approach is cumbersome, tedious, costly, and forces architects and engineers to rely on simulation experts during the early design stages. Recently, some plug-ins were developed to facilitate the geometry input and link architectural forms of visualization and 3D representation with the evaluation tools. However, evaluative tools embed most integration barriers.

On the other hand, during the past decade, a range of design tools has been available to help architects and engineers in the design of more energy efficient buildings. Those tools are labeled "guidance tools", which were developed to facilitate decision-making prior to design. They range from quite simple predecision evaluation and analysis tools to parametric and optimization decision tools that aim to inform the design and integrate BPS during the early design process. Table 4.2 shows that most developed guidance tools are predecision evaluative tools. Despite their remarkable capabilities, most of those tools have not been transferred effectively to the AEC community, and in particular to architects during the early design stages. The uptake of most of those tools among architects and engineers is very low, and does not allow continuity with the design process (Attia et al., 2009; Ochoa and Capeluto, 2009; Petersen and Svendsen, 2010). While they are quite useful to lower the "input filling" barrier, they could not lower the "informative support during the decision making" barrier. Currently, few nonpublic tools exist that support design predecisions, including jEPlus, Sefaira, and iDbuild that allow parametric analysis or BEopt that allows optimization analysis (Christensen et al., 2005; Zhang and Korolija, 2010). The potential of parametric tools is very high to bridge the "informative support" barrier because they can provide constructive feedback with very little iterations, and at the same time allow a wide range of solution space. This is in contrast with optimization tools that reduce the solution space to a minimum.

In order to address these shortcomings, we identified the requirements of a tool that can be used for the design of NZEB during early design processes.

In order to support decision-making during the early design phases it is important to include an informative tool for the early design phases that can model the complexity of the design. Informed decision-making, or informed design choice, forms an essential basis for the design of NZEB. This concept is based on providing knowledge prior to the decision-making to influence the decision attitude. To date, however, no such tools and validated measure of informed design choice has been developed for NZEB design.

Decision aids have developed significantly in their sophistication, both in terms of their scope and the technologies used. They are tools developed to help designers make decisions, particularly in areas of performance uncertainty of NZEB design and the range of building energy modeling tools now in existence. Early energy modeling aids were evaluative, based on postdesign evaluations, and catered for HVAC engineers. Later BPS tools were more guiding the design and catering for architects, but still evaluative. More recently, BPS decisions support tools have become more informative aiming to aid before making a decision. This includes parametric analysis automated optimization techniques. However, this mainly caters for engineers during later design stages.

There are extensive bodies of literature that examine the effects of BPS tools as informative decision aids (Attia et al., 2012). By reviewing this work systematically, we found that BPS improved decision-making in a number of ways:

- Increasing designers' knowledge of the design problem and options
- Reducing decisional uncertainty
- Increasing the design robustness

Therefore, we find it essential to include building information and building performance modeling in the early design stages of NZEB design. The available tools and programs are diverse, but NZEB design requires intensive modeling that should focus on key performance indicators (Attia and De Herde, 2011; Attia et al., 2013a,b). The use of modeling should not only be done during the design development phase, but also during the predesign phase to help clarify energy goal definitions and influence the priorities of design decisions.

6 DISCUSSION

Before the 19th century, clients would hire a Master Builder who designed, engineered and oversaw the construction of building projects. However, by the middle of the 20th century the Master Builders were dead and gone. The diversity and sophistication of building technology advancement in every stage of the project delivery phases had overwhelmed the abilities of any individual, including architects to master them. As a first reaction to the increasing complexity, architectural and engineering design split off from construction. Then, with time, each domain became further specialized and split off, until we have architects on one side often with their own areas of subspecialty, and engineers on another. Builders too got fragmented in their multiple divisions, ranging from scaffolding contractors to finish painters. The field looks like a fragmented picture with all its specialists and super specialists. As a consequence, the AEC industry became very slow to modernize and provide high quality building design, construction and operation service when compared with the automobile or aerospatiale industry. The common linear project delivery process is tedious, costly, and disperses team responsibility. Design is separated from budgeting and scheduling, design teams end up in adversarial relationships, with value engineering disputes on overruns cost leading to limiting the performance expectations. The sick-building syndrome and energy gap, discussed earlier in Chapter 3, NZEB Performance Thresholds, are potentially a result of this conventional project delivery process.

Today, high-performance buildings including NZEB are helping the AEC industry to find its way to increase its added value and quality of the built environment. The performance-based Design-Build contracts and the IPD process are changing this situation. Design and performance requirements are becoming aligned with budgets and schedules. Conflict between team members and the flaws caused by the linear project delivery process that offers little coordination have been solved through the IPD process that is followed by multidisciplinary, experienced, and collaborative teams. The new best practice of performance-based Design-Building contracts have been discussed in this chapter. In order to achieve a NZEB, projects owners need to plan an integrative, cost effective, and short project delivery process. The key components for the IPD can be grouped under four major categories, namely (1) performance-based contracting, (2) multidisciplinary and experienced design teams, (3) BIM and performance modeling, and (4) development of process management and performance assurance plan. Next, we will discuss these four criteria for managing a successful NZEB project delivery process.

First of all, the owner or owner representative has to define and articulate the performance targets and minimum requirements for the new building construction or renovation. In Chapter 2, Evolution of Definitions and Approaches, we defined NZEB and in Chapter 3, NZEB Performance Thresholds, we presented different performance criteria and NZEB's minimum or maximum performance thresholds. Both chapters are the foundation for this chapter where we aim to articulate the key components of rigorous project delivery process for NZEB. At this stage, we need to emphasize the importance of formulating those targets by the owner or owner representative into written contract. There are several ways to translate the OPR into a RFP, but the most important is to anchor it in a performance-based contract of Design-Build a multi performance criteria and goals ranking structure RFP. This is a key to guarantee the commitment of all stakeholders involved and to provide a reference for the process and expected performance of NZEB (DBIA, 2013). To make this commitment, owners seek third-party certification through rating systems including LEED or BREAAM or standards such as the PH Standard.

Secondly, collective decision-making of different stakeholders must be enabled as early as possible in the design stage. The owner, builder, and designer have to come together in the form of experienced multidisciplinary design teams to guarantee continuous, fruitful, and direct communication throughout the whole design process. The experience of the team is very important. The success of the car manufacturer Toyota is related to the continuity of the design team who work for several years together and collect feedback and document lessons learned from finished car designs to enhance and optimize new car models and production lines. An experienced team can avoid many mistakes, reduce risk, increase trust, and finally add value in short-time cost effectively. The trial and error project delivery cannot be tolerated for NZEB. An experienced multidisciplinary team can form beneficial synergies and innovative solutions for NZEB.

Third, BIM and performance modeling are the only guides for design optimization. Traditional rules of thumb and guidelines are not very helpful during NZEB design. Performance models of daylight, ventilation, energy, comfort, and cost eliminate the trial and error approach and can provide detailed and accurate performance analysis. With the BIM technology information can be more easily exchanged between the design and construction teams. Therefore, the use of BIM and performance modeling need to be integrated in the design process.

Finally, the owner representative and project manager should set up a framework and process map that describes the expected project development process performance assurance plans. This can be articulated in the RFP, through milestones, and performance expectations during operation. Facility managers must be included in this process to prepare for commissioning, soft landing, performance metering, and monitoring. The design sustentation schedule and performance assurance plans must be included in the RFP so that the Design-Build team understands the time commitment required to implement a NZEB.

7 LESSON LEARNED # 4

In conclusion, when measurable energy efficiency targets are a core requirement of a project and are defined through a performance-based Design-Build contract at the beginning of a project, owners can expect a robust energy performance to meet the NZEB target. Linear tendering with specifications does not encourage innovation. However, performance-based contracts drive behavior. They guarantee the commitment of all stakeholders and quality assurance of the building performance. Also, we have to include the right stakeholders with the right mindset when creating design and construction teams. A decisive leadership and performance-based decision-making is the only way to achieve informed decision-making that make the NZEB a success. The owner and top management must be behind the NZEB target and embrace it as an innovation process. Team collaboration and communication early on can make the building a real business case and overcome the cost challenge.

Lessons learned from practice show that the use of third-party rating systems, and certification or standards can increase trust, accelerate and guarantee the successful implementation of NZEB. The OPR and RFP are very important documents that can guide the whole building delivery process from predesign to operation. For most developers and owners, NZEB are new and intimidating because they are performance-based facilities. The AEC industry is notoriously conservative and designers and builders need time and experience before embracing new concepts and strategies to shift their working load and focus on early design stages. Therefore, experience and technical expertise emerge as crucial elements together with the commitment to written performance criteria, targets, and thresholds to achieve successful and robust NZEB. The good planning and organization of long-lasting experienced and multidisciplinary teams stimulates trust, communication, collaboration, innovation, and commitment.

The quality assurance of performance in operation requires rigorous planning of performance targets, process milestones, timelines, responsibilities, team roles, programs, modeling programs, and budget, measurement, and verification protocols.

References

ANSI, 2012. Integrative Process, Consensus National Standard Guide, for Design and Construction of Sustainable Buildings and Communities. American National Standards Institute, ANSI, Washington, DC.

ASHRAE, 2008. AHSRAE Vision 2020. ASHRAE Vision 2020 Ad Hoc Committee, <http://www.ashrae.org/doclib/20080226_ashraevision2020.pdf> (accessed 10.10.10.).

Athienitis, A., Attia, S., et al., 2010. Strategic design, optimization, and modelling issues of net-zero energy solar buildings. In: EuroSun 2010, Graz.

Attia, S., Beltrán, L., De Herde, A., Hensen, J., 2009. "Architect friendly": a comparison of ten different building performance simulation tools. In: Proceedings of the 11th International IBPSA Conference, Glasgow, Scotland.

Attia, S., De Herde A., 2011. Early design simulation tools for net zero energy buildings: a comparison of ten tools. In: International Building Performance Simulation Association, November 2011, Sydney, Australia.

Attia, S., 2012. A Tool for Design Decision Making-Zero Energy Residential Buildings in Hot Humid Climates (Ph.D. thesis). UCL, Diffusion universitaire CIACO, Louvain La Neuve, ISBN 978-2-87558-059-7.

Attia, S., Gratia, E., De Herde, A., Hensen, J., June 2012. Simulation-based decision support tool for early stages of zero-energy building design. Energy Build. 49, 2–15. ISSN 0378-7788, doi:10.1016/j.enbuild.2012.01.28.

Attia, S., Walter, E., Andersen, M., 2013a. Identifying and modeling the integrated design process of net zero energy buildings. In: High Performance Buildings—Design and Evaluation Methodologies Workshop, BBRI, June, Brussels, Belgium.

Attia, S., Hamdy, M., et al., 2013b. Computational optimisation for zero energy buildings design interviews results with twenty eight international expert. In: 13th Conference of International Building Performance Simulation Association, August, Chambery, France.

Attia, S., 2016a. Towards regenerative and positive impact architecture: a comparison of two net zero energy buildings. Sustain. Cities Soc. 26, 393–406. ISSN 2210-6707, http://dx.doi.org/10.1016/j.scs.2016.04.017.

Attia, S., Bashandy, H., 2016. Evaluation of adaptive facades: the case study of agc headquarter in Belgium. In: Bellis, J.L.I.F., Bos, F.P., Louter, Ch. (Eds.), ChallengingGlass 5—Conference on Architectural and Structural Applications of Glass. Ghent University, Belgium, ISBN 978-90-825-2680-6.

Attia, S. 2016b. Evaluation of adaptive facades: case study, Al Bahar Towers, UAE. In: 2nd Qatar Green Building Council Conference, 13–15 November, Doha, Qatar.

Charron, R., Athienitis, A., Beausoleil-Morrison, I., 2006. A tools for the design of zero energy solar homes. In: ASHRAE. Annual Meeting, Chicago, vol. 112, no. 2, pp. 285–295.

Christensen, C., Horowitz, S., Barker, G., 2005. BEopt: Software for Identifying Optimal Building Designs on the Path to Zero Net Energy. ISES, Orlando, FL.

Daft, R., 2012. Organization Theory and Design. Nelson Education, Toronto.

DBIA, 2013. NREL—Integrating Measureable Energy Efficiency Performance Specifications into Design-Build Acquisition and Delivery. Design-Build Institute of America, Washington, DC (accessed 20.05.14.).

Design Sense Inc., 2008. Appendix A: Conceptual Documents. Research Support Facilities Request for Proposal, Page 3 of 299, Introduction, SolicitationProcess, no. 6. Golden, Colorado.

Design-Build Institute of America, 2009. About DBIA: What is Design-Build? Retrieved 15.02.17 from Design-Build Institute of America <http://www.dbia.org/about/design-build/>.

Dodge Data & Analytics Media, 2016. U.S. Building Owners Will Help Drive the Construction Industry to Create Healthier Buildings. Retrieved from <http://www.construction.com/about-us/press/US-Building-Owners-Will-Help-Drive-the-Construction-Industry-to-Create-Healthier-Buildings.asp> (accessed 15.02.17).

Donn, M., et al., 2009. Simulation in the service of design—asking the right questions. In: Proceedings of IBPSA 2009, Glasgow, pp. 1314–1321.

European Union Parliament, 2009. Report on the proposal for a directive of the European Parliament and of the Council on the energy performance of buildings (recast) (COM (2008)0780-C6-0413/2008-2008/0223(COD)).

Hayter, S., Torcellini, P. et al., 2001. The energy design process for designing and constructing high-performance buildings. In: Clima 2000/Napoli 2001 World Congress.

Hensen, J., Lamberts, R. (Eds.), 2011. Building Performance Simulation for Design and Operation. Spon Press, London.

Leitner, K., 2008. Subcontract # AFJ-8-77550-01. Research Support Facilities: Phase I—Preliminary Design, Phase II—Design Development and Construction. NREL, Golden, CO.

Konchar, M., 1997. A Comparison of United States Project Delivery Systems. Pennsylvania State University, Architectural Engineering, University Park, PA.

Moreno-Vacca, 2017. Beauty at Low Cost and the Passive House Movement, Center for Architecture and AIANY Committee on the Environment (COTE). Video retrieved from <https://www.youtube.com/watch?v = Fto-4tvTPi8> (accessed 15.02.17.).

Marsh, A., 2004. Performance analysis and concept design: the parallel needs of classroom & office, Welsh School of Architecture. In: Between Research and Practice Conference, ARCC and EAAE Transactions on Architectural Education, Dublin.

NREL, 2014. Cost control strategies for zero energy buildings. In: High-Performance Design and Construction on a Budget. U.S. National Renewable Energy Laboratory, Department of Energy. Retrieved from <http://www.nrel.gov/docs/fy14osti/62752.pdf> (accessed 10.02.17).

Ochoa, C., Capeluto, G., 2009. Advice tool for early design stages of intelligent facades based on energy and visual comfort approach. Energy Build. 41, 480–488.

Petersen, S., Svendsen, S., 2010. Method and simulation program informed decisions in the early stages of building design. Energy Build. 42 (7), 1113–1119.

Pless, S., Torcellini, P., Shelton, D., May 2011. Using an energy performance based design-build process to procure a large scale low-energy building. In: ASHRAE Winter Conference <http://tinyurl.com/md6ts2z>.

Pless, S., Torcellini, P., Scheib, J., Hendron, B., Leach, M., 2013. How-To Guide for Energy-Performance-Based Procurement. Department of Energy, Building Technologies Program, Report number: TPP-5500-56705.

Scheib, J., Pless, S., Torcellini, P., 2014. An energy performance based design-build process: strategies for procuring high-performancebuildings on typical construction budgets. In: Proceedings of the ACEEE 2014 Summer Study on Energy Efficiency in Buildings, vol. 4, pp. 306–321 <http://aceee.org/files/proceedings/2014/data/papers/4-643.pdf>.

Sobek II, D.K., Liker, J.K., 1998. Another look at how Toyota integrates product development. Harv. Bus. Rev. 76 (4), 36–47.

Starcke, K., Brand, M., 2012. Decision making under stress: a selective review. Neurosci. Biobehav. Rev. 36 (4), 1228–1248.

Watanabe, K., 2007. Lessons from Toyota's long drive. Harv. Bus. Rev. July/August, 74–83.
Zhang, Y., Korolija, I., 2010. Performing complex parametric simulations with jEPlus. In: 9th SET Conference Proceedings, Shanghai, China.

Further Reading

O'Brien, W., Athienitis, A., Kesik, T., 2009. The development of solar house design tool. In: 11th IBPSA Conference 2009, Glasgow, pp. 1397–1404.

CHAPTER

5

Occupants Well-Being and Indoor Environmental Quality

ABBREVIATIONS

AEC Architectural, Engineering, and Construction
ASE annual sunlight exposure
EPBD Energy Performance of Buildings Directive
BMS Building Managements Systems
BPS building performance simulation
CDD cooling degree day
CIBSE Chartered Institution of Building Services Engineers
CRI color rendering index
GHG greenhouse gas
HDDs heating degree days
HVAC Heating, Ventilation, and Air Conditioning
IAQ indoor air quality
IEQ indoor environmental quality
IESNA Illuminating Engineering Society of North America
PMV Predicted Mean Vote
PPD Percentage People Dissatisfied
NO_2 nitrogen oxide
NZEB Net Zero Energy Buildings
MERV minimum efficiency reporting value
RFP request for proposals
SBS sick building syndrome
sDA spatial daylight autonomy
WHO World Health Organization
UGR Unified Glare Using

Net Zero Energy Buildings (NZEB)
DOI: https://doi.org/10.1016/B978-0-12-812461-1.00005-8

1 INTRODUCTION

Many resources are available to guide decision-making for indoor environmental quality (IEQ). However, those resources are diverse and detailed in a way that makes the issue of IEQ very complicated. There is a market trend for health and well-being within our Architectural, Engineering, and Construction (AEC) industry. There is a significant uptake of rating systems and standards on IEQ in construction projects during the past 5 years. Occupants and building owners sometimes value their own health and well-being above energy savings. At the same time, there is a global trend in increasing air pollution levels and a global trend in the awareness about the effect of particulate matter and exposure to air pollutants. In this context, we position this chapter on occupants' well-being and IEQ. We are on the verge of a new paradigm that will change the AEC industry toward user-centered and quality spaces. We have an oversupply of square meter buildings and an undersupply of quality buildings. Therefore, the market is looking for facilities with high IEQ. Buildings are becoming services rather than products. Buildings are becoming services that support life and the well-being and productivity of people. In this chapter, we explore the main challenges to achieve IEQ, and define the meaning and importance of occupant well-being in relation to physical, financial, and perceptual parameters. For example, literature confirms the indoor air quality (IAQ) in buildings is between three and eight times worse than outdoor air quality (Hermelink, 2016). We are losing between 3 and 4 years of our life span from inhaling fine dust, thus indoor air plays a significant role as a health determinant (Pope et al., 2009). Therefore, we propose design target values for IEQ in NZEB and elaborate on quality assurance through commissioning and monitoring, to make sure those design targets are achieved during design, construction, and operation of future NZEB. Finally, we discuss the changing work modes and the hybrid use of spaces combining professional activities and living activities in cities and how this impacts the perception of IEQ and indoor design. We conclude in Section 7 with the lesson learned from demonstration projects and case studies to achieve excellent IEQ and maximum occupant control in NZEB.

2 CHALLENGES TO ACHIEVE INDOOR ENVIRONMENTAL QUALITY

IEQ of NZEB depends on several external factors related to the projects' outdoor environment and building design and use. By default, NZEB are low energy consumption buildings that are dependent on occupation density and occupants' functional use in the building. This means that the building use and energy use intensity are associated with continuous variations. Those variations can be characterized as building-dependent, including occupant behavior, building envelope, or HVAC systems response. On the other hand, those variations can be characterized as environment-dependent, including outdoor air pollution, road noise levels, heat island effect, or climate change. Both the building-dependent and environment-dependent variations are various and are considered as risk factors to achieve health and well-being in NZEB. Therefore, in this section we will explore those risks to make designers, builders, and owners aware about the influence of the outdoor and indoor risk factors that influence the overall building performance and IEQ of NZEB.

2.1 Risks Associated With Outdoor Environment

Outdoor environment is an important factor that cannot be neglected for IEQ. We are witnessing increasing urbanization worldwide. Cities are getting denser and busier in order to host more and more economic and social activities. The urbanization phenomenon is increasing the risk of air and noise pollution, and microclimate alteration and therefore the negative influences of extreme urban climate and pollution intensification have become increasingly important. We need to understand the risks associated with the outdoor environment of any NZEB. Humans need air exchange for a healthy indoor environment. Next, we will provide an overview on the most influential risks associated with the outdoor environment that need to be addressed when designing and operating NZBEs.

Poor air quality: Outdoor air quality is one of the major environmental health concerns worldwide. The World Health Organization (WHO) that monitors air pollution in all the regions of the world confirms that more than 80% of people living in urban areas are exposed to poor air quality levels (WHO, 2016). As people spend 60%–90% of their time in indoor environments (dwellings, offices, schools, etc.) they are exposed to illegal levels of air pollution that can cause serious long-term problems (Jantunen et al., 2011). There is growing evidence found in medical literature that air pollution is linked to heart disease, lung cancer, worsening asthma, and poor lung development in children and lead to more than 100,000 deaths in Europe in 2016 (WHO, 2014, 2016). Outdoor air pollutants mainly consist of NO_2, SO_2, O_3, CO, HC, and particulate matters of various particle sizes. For example, the performance criteria for offices and housing in European Union countries states that exposure to nitrogen dioxide (NO_2) should not exceed 288 μg/m over 1-hour average, and 40 μg/m over a long-term average. Outdoor ambient NO_2 levels in many European cities are often four to five times above this. NO_2 is directly attributable to annual deaths of thousands in Europe. In urban areas, pollutants are mainly emitted from on-road and off-road vehicles, but there are also contributions from power plants, industrial boilers, incinerators, petrochemical plants, aircrafts, ships, and so on, depending on the location and prevailing winds. This means that "fresh air" supplies in our cities are often far from fresh.

Outdoor air quality is not only affected by airborne chemicals, fumes, or gases. Air pollutants include dust, particulate matter, mosquitos, and flies. Many cities Worldwide are located in sandstorm areas. Another important factor that can affect outdoor air quality is atmospheric pollution. Atmospheric pollution includes dust, particulate matter, pollens, mosquitos, and flies. Pollution caused by these factors pose grave danger to building users and should be controlled in any NZEB. Projects located in regions with occasional violent dust and sandstorms should have high measures of indoor environmental control. An example is the recent outbreak of endemic Zika in some countries, including in the Mediterranean area. Removing standing water and installing window screens for mosquito control is essential in the curtilage of buildings. In this case, NZEB would need insect management plans. Building occupants in a NZEB must be notified about the risks and adapt their behavior in nonair conditions or mixed-mode buildings.

High noise levels: Noise is another major environmental health concern worldwide. People living near noisy roads or close to airports could be exposed to a higher risk of high blood pressure. Noise also provides an unwanted distraction resulting in a major cause of dissatisfaction and productivity decrease among building occupants (Shepherd et al., 2013). Researchers collected information from 41,000 people from Denmark, Germany, Norway, Spain, and Sweden and found that traffic noise is associated with an increase of high blood pressure and hypertension (Tofield, 2017). This means that noise shares many of the negative effects associated with air pollution on human health.

Climate change and extreme weather events: Episodes of extreme heat and cold are becoming more frequent in cities all over the world. Since the year 2000, hot summers with long stretches of high temperatures and humidity sustained for weeks has been experienced in many cities worldwide. A recent report by the WTO-UNEP predicted a decrease for the heating degree days (HDDs) and increase for cooling degree days (CDDs) by 2040 (Scott et al., 2008). Keeping this climate change scenario in mind, the urban microclimate is becoming sensitive to health risk and heat related mortality. Evidence from literature indicates that frequency and severity of extreme weather events has increased markedly in the past 20 years and will continue to increase unless drastic action is taken to reduce greenhouse gas (GHG) emissions. As a consequence, energy regulators and companies have issued warnings that the electricity networks in many countries are not equipped to supply enough power to cover the demand from air conditioners during heat waves. To cope with climate change and extreme weather events, the design of NZEB should prepare buildings for such scenarios and risks that will occur more frequently in the future.

Heat Island effect: The characterization of the microclimate across worldwide cities proves the presence of Urban Heat Island (UHI). The UHI effect is the phenomenon of significant temperature variation between urban city centers and its surrounding rural areas. The temperature difference is larger at night than during the day, and is most apparent when winds are weak. The main cause of UHI effect is from the modification of land surfaces using mainly concrete and asphalt in urban design. Compactly planned and deep street canyons can block and weaken prevailing wind leading to the reduction of the air dispersion capability (Cheng et al., 2009; Li et al., 2009, 2010). UHI has proved to have an important effect on comfort and energy consumption of buildings. Literature provides sufficient evidence that the main factors that contribute to UHI are large surfaces of materials with low albedo and high admittance, tall buildings in narrow streets, reduced vegetation, limited shade, and heat-generating anthropogenic activities (Akbari et al., 2001; Synnefa et al., 2007; Hart and Sailor, 2009; Sailor, 2014; Santamouris, 2013). The presence of UHI mitigation policies, including green infrastructure, street vegetation, or combining vegetation with cool pavements and cool roofs can reduce the average air temperature by 1–3°C during hot summer days. Therefore, it is important to take into account the UHI effects and associated heat stress that can influence the IEQ in NZEB.

Finally, the previously mentioned risks can become magnified depending on the level of interaction between outdoor and indoor air. Many studies indicate that IAQ is affected by outdoor air (Leung, 2015). If the outdoor air pollution and noise levels are high, natural and mixed mode ventilation and cooling strategies are a poor choice for NZEB. Occupants often complain in offices that rely on opening windows. The traffic is noisy, where day-time noise levels of 60 decibels higher or night-time noise levels of 50 decibels or higher can often drive the decision to air condition or mechanically ventilate buildings. Added to that is the risk of extreme heat waves or the presence of mosquitos or flies. Therefore, we recommend to perform a serious annual site analysis and climate characterization to make fit-to-context design decisions.

2.2 Risks Associated With Building Design and Use

Risks that influence the IEQ of NZEB are not only associated with the outdoor environment. There are serious risks related to decisions taken by the design-build-operate team, the design and construction quality of NZEB, and occupant behavior. The nature of NZEB as ultra-efficient buildings makes it difficult to optimize the design decision related to the envelope, systems, or controls while maintaining a good IEQ. Architects and engineers often fail to characterize the real weather and climatic conditions of their project site or commit design errors related to building glazing surface or systems sizing. At the same time, the constant variation of occupant behavior and the intensity of use of building spaces increase the uncertainty of the way the facility will be used. Failing to predict the occupant use and behavior, or failing to engage with and educate occupants to maintain the performance requirements of the building can lead to poor IEQ. Therefore, in this section, we will provide an overview of those risks that are mainly associated with the building design and construction, or users.

Climate characterization: Designers often use climate classification systems, indicators or daily comfort deviation thresholds to human comfort to assess a certain climate. However, with climate change and intensive urbanization, we are faced with microclimate alterations worldwide, which jeopardize the accuracy of those classifications. For example, the Köppen climate classification is one of the most popular ways to understand the climate of a new project site location. However, the Köppen classification is sometimes far from real world measurements with frequent errors to group locations like Cairo into the same climate group as Dubai, or Oslo into the same climate group of Denver. The degree day index is another commonly used index determining the energy required to achieve comfort (Roshan et al., 2016). However, the most commonly used indicator by building engineers are the HDDs and CDDs. These represent the total mean deviation of daily temperature from human comfort temperature. In temperatures higher than threshold temperature, there is a need to cool the environment, and in temperatures lower, there is a need to heat the environment which are called cooling and heating requirements, respectively (WMO, 1991). To date, a variety of HDD and CDD data are available in the literature and standardization for varies cities worldwide. However, most HDD and CDD thresholds classifications are generic and there is no specific data for specific locations (Roshan et al., 2016). For example, North America presents design temperatures that are much more challenging than Europe's (with some exceptions, for example, the Pacific Northwest) (Klingenberg et al., 2016). Many cities are significantly colder than Europe during the winter, while the number of HDDs is almost the same (Attia et al., 2016). Madison, WI, United States has a colder design temperature than Oslo, Norway, while the total HDDs are almost 2000 HDDs lower than in Oslo (Klingenberg et al., 2016). Therefore design temperature is inadequately correlated.

As a consequence, designers fail to properly characterize the local climate and weather. Therefore, there is a serious necessity to conduct fundamental climate characterization before designing NZEB. This should be based on recent and climate sensitive indicators and temperature thresholds representing the real microclimate of the project site. Accurate climate characterization will provide architects and engineers with validated information that can help them to implement robust NZEB in order to achieve maximum IEQ inside the buildings.

Design optimization and overheating risk: Overheating is an example of a risk frequently associated with ultra-energy efficient buildings (Attia et al., 2016). Two factors may cause overheating in NZEB. The first factor is associated with poor envelope and HVAC systems design, while the second is associated with poor prediction of internal heat gains and occupant density. Lowering the demand of appliances and distributing and selecting low-power use equipment should be on the checklist of any NZEB design team in order to reduce overall heat gain. In residential buildings, overheating can affect occupants' well-being and cause increased wakefulness and impaired sleep quality leading to decreased productivity (Yanase, 1998). Numerous studies prove that overheating can be avoided by proper design.

As mentioned in Section 5 of Chapter 4, Integrative Project Delivery and Team Roles, the complexity of NZEB and difficulty to assure its performance robustness and IEQ make it necessary for any design team to refine the design and go through a design optimization process. The use of BPS tools is essential to develop fit-to-purpose design solutions and systems services that can guarantee the IEQ. However, there is a risk that the design team oversees one or more of the influential design parameters. Table 5.1 lists a group of common design errors associated with NZEB designs that can influence the IEQ.

Table 5.1 summarizes the results of analyzing more than 20 NZEB case studies. The lessons learned from these case studies point to the responsibility of the design team and facility manager to avoid these errors. Problems associated with office buildings design liability are related more to envelope design, while problems associated with dwelling design reliability are associated more with systems design. Design optimization of NZEB must take place to guarantee the energy performance, cost and IEQ at the same time.

Occupant density and use: As mentioned in Section 6 of Chapter 4, Integrative Project Delivery and Team Roles, on occupancy, changing the use of occupancy or occupant density can influence the performance of NZEB. In the case of IEQ, these changes can drastically influence indoor environment conditions. Postoccupancy evaluations of NZEB indicate that the nature of building use and occupancy density change, and tenants adapt the spaces in time to the expected functional and

TABLE 5.1 Common Errors Associated NZEB Design, Construction, and Operation and Their Consequences on the IEQ

Design Error	Consequence
ENVELOPE	
Window design (glazed area)	Overheating + thermal stress
Artificial lighting design	Productivity, glare, biological functions (migraines, headaches, eyestrain, etc.)
Shading design	Overheating + glare
HVAC SYSTEM	
Ventilation rate	Overheating + poor air quality
	Draft
Systems and ducts sizing	Overheating + poor air quality
Systems selection	Overheating + dry air
CONSTRUCTION	
Design of construction details	Heat losses, condensation, infiltration, noise
Construction quality	Heat losses, condensation, infiltration
CONTROL	
Design of occupant control	Discomfort, dissatisfaction, complaints, SBS
Building Managements Systems (BMS)	
Selection and distribution of sensors	Overheating, poor air quality

hierarchical work order and methods of their employees (Attia and Bashandy, 2016; Judd et al., 2013; Barthelmes et al., 2016). The risk in this regard remains on the shoulders of the design team that should meet all the functional and comfort requirements of the building, estimate occupant behavior, and take into account the uncertainty in the building energy use and its influence on IEQ. A rigorous, occupant-use analysis should take place to test the different occupant-related scenarios.

Construction and operation: IEQ should be also maintained during construction and operation. IEQ of NZEB does not only depend on setting up performance targets and incorporating these in the design phase. To ensure IEQ, a procedural framework should be planned, designed, and followed during construction and operation. Validation and quality assurance of IEQ should be achieved through energy management systems that monitor the IAQ, thermal comfort, lighting, and acoustic environment. Source control and filtration maintenance protocols should be designed early on in the process and assigned to facility managers. The risk associated with neglecting the quality assurance of the indoor environment can lead to poor construction quality, including air leakage, poor insulation, and thermal bridges (see Table 5.1). As NZEB are airtight and well-insulated, their construction quality must be very high to prevent infiltration, heat transfer, or deterioration of HVAC system.

3 OCCUPANTS' WELL-BEING IN NZEB

In this section, we are mainly concerned with the reasons why it is necessary to achieve well-being in NZEB. According to the WHO: "Health is a state of complete physical, mental, and social well-being and not only merely the absence of disease or infirmity." At the same time, any energy efficiency declaration is meaningless if it is not associated with IEQ performance-based standards. NZEB should help improve health, sleep, comfort, and performance of its occupants. NZEB design should encourage healthy, productive, and active lifestyles while reducing occupant exposure to harmful pollutants. According to Eurostat (2018), 507 million European citizens spend 80%–90% of their time indoors (living and working). At the same time, there is a relation between IEQ in buildings' and occupants' well-being. The cost of buildings with poor IEQ is high. The annual absenteeism rate in the United States is 3% in private sector and 4% in the public sector (US DOL, 2016). Poor mental health costs UK employers £30 billion a year due to lost productivity and recruitment during absenteeism (ACAS, 2014). It is estimated that roughly 30% of European households suffer from humidity problems (or being unable to keep their dwellings comfortably warm) and as a consequence, the occupants suffer from heat stress and health problems. Approximately, 6% of all EU households (13 million) report a lack of daylight. Around 19% of all EU households (41 million) report that their dwelling are not comfortably cool in summer (Hermelink and Ashok, 2016).

We are on the verge of a new paradigm that will change the AEC industry. The function and use of buildings is in a phase of transition. Instead of developing the building as an object or product, with an abrupt relation between the owner and the occupant, we are witnessing a change. The change can be summarized in the tendency to develop the building as a service or quality space. The idea that a building is an independent product/object that gets passed from an owner to a user is coming to an end. The future of the AEC industry is coupled with flexible and high-quality facilities that can host human activities. Tenants buy or rent a square meter of service and no longer a square meter of buildings. The Edge Building in Amsterdam is an example of a facility with flexible workspaces, providing sitting desks, work booths, meeting rooms, or concentration rooms. Using an app, the building knows the

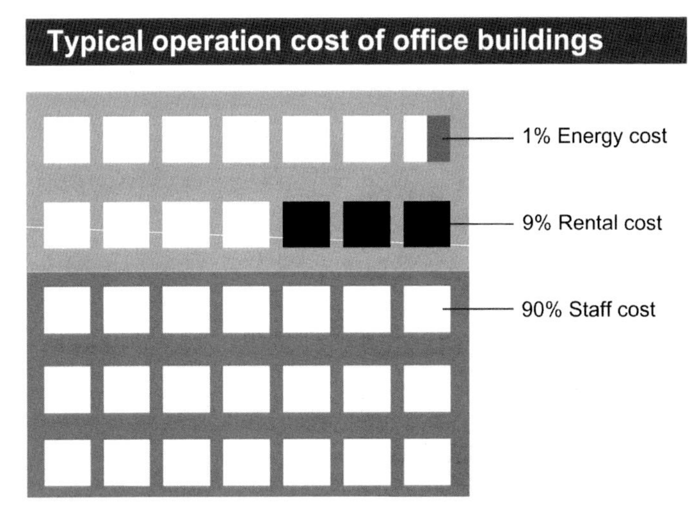

FIGURE 5.1 Typical operation cost of an office buildings. *Adapted from World Green Building Council (WGBC), 2014. Health, Wellbeing and Productivity in Offices: The Next Chapter for Green Building. Retrieved from <http://www.worldgbc.org/news-media/health-wellbeing-and-productivity-offices-next-chapter-green-building> (accessed 21.03.17.).*

occupants' preferences for light and temperature, while it allows the building users to control their environment accordingly. This new concept is called hot desking and is based on the efficient use of space and high IEQ. Therefore, the implications of this new paradigm will shift the attention to squared meter of serviced spaces and consequently increase the priority of IEQ.

3.1 Productivity and Satisfaction

In this section, we will focus on exploring the main concepts associated with well-being, productivity, and satisfaction. At the same time, we explore the definition, meaning, and importance of productivity and satisfaction.

Productivity: The term productivity tends to be used to refer to business activities and is directly affected by people's health and well-being. Staff costs typically account for 90% of business operating costs (see Fig. 5.1). Employee health and productivity can have a significant financial impact on any business. A properly assigned NZEB that takes into account the best practices of IEQ during the building design and operation can enhance occupants' productivity. A developer, owner, or tenant should aim to achieve the following for their building:

- Aim to decrease employment cost and add value
- Greater productivity of staff on core tasks

- Improve well-being, health, and sleep in households
- Reduce liability and lower the risks
- Engage occupants to become the center of the indoor environment

Satisfaction: Occupant satisfaction in the indoor environment should be taken more seriously as having a strong impact on well-being and mindset (WGBC, 2014). Satisfaction includes providing views out of windows which offer a connection to nature and natural lighting. It also entails biophilia (the presence of green space) and nature indoors and outdoors. Satisfaction includes factors related to the sleep quality of occupants in residential buildings. Satisfaction is experienced subjectively and differently by people of different cultures, genders, and ages.

3.2 Physical Environment Factors

Physical environmental factors comprise air temperature, humidity, lighting, and acoustic measures. In this section we will list and define the key factors that designers should address when designing NZEB (see Fig. 5.2).

Indoor air quality: IAQ has a significant impact on occupant health and well-being as indicated in Section 2.1. This can be quantified by low concentration of CO_2 and pollutants, and high ventilation rates. A comprehensive body of research suggests that productivity can improve between 8% and 11% as a consequence of good air quality (WGBC, 2014).

FIGURE 5.2 Health, well-being, and productivity flow chart for net zero energy buildings.

Thermal comfort: Thermal comfort has a significant impact on the health and satisfaction of building users. Depending on the type of static or adaptive comfort model used to control the thermal environment, humans will remarkably adapt to temperature. However, this temperature has to stay within a certain range. The thermal environment comprises air temperature, surface temperatures, air speed, and humidity on the one hand and the occupants metabolic rate, clothing, and personal preferences on the other hand (Attia and Carlucci, 2015). At the same time, several studies suggest that moderately high temperatures are less tolerated than low temperatures (Frontczak et al., 2012). A more recent study in a statistically controlled setting indicated a reduction in performance of 4% at cooler temperatures, and a reduction of 6% at warmers ones (Lan et al., 2011).

Daylighting and views: Our understanding of the benefits of light on health and well-being have grown in recent years. Lighting affects many aspects of well-being, including comfort, mood, safety, and health. Poor light intensity and distribution, glare, and light patches can affect productivity negatively. Visual discomfort can cause headaches, eyestrain, or disturbance of our circadian rhythm. A study by neuroscientists suggests that office workers with windows slept an average of 46 minutes more per night (Chueng, 2013). It is difficult to separate out the benefits of daylight from the benefits of views out of windows. Views are a significant factor by offering users a connection with the outdoors and nature. Daylighting in NZEB should provide maximum daylight, avoid solar gain and glare, and reduce reliance on artificial lighting. Low carbon lighting is crucial for NZEB.

Acoustic comfort: Acoustic comfort is underestimated in many high-performance buildings. Being productive in a modern knowledge and decision based office is not possible when noise presents distraction. Noise from roads affect residents' well-being and sleep quality, while noise in educational buildings like schools can lead to lower cognitive performance of students (Peacock et al., 2010). According to a study by Banbury and Berry (1998), there was up to 66% drop in performance for a "memory for prose" task when participants were exposed to different distraction noises.

Interior layout: The interior layout of office buildings can also have an effect on well-being. As mentioned earlier, the hot desking concept is centered on satisfying users' activities and tasks in the work environment. The interior office layout should allow users to have autonomy of choice between closed spaces, workstations, social spaces, or collaboration spaces. These factors influence noise, concentration, collaboration, and creativity. A heterogeneous workplace can increase productivity by making the indoor environment user-friendly and occupant-centric.

Personal control: Studies consistently show that personal control over thermal comfort, daylight, air quality, ventilation, and noise can lead to higher satisfaction and productivity (WGBC, 2014). Putting trust in occupants to make the indoor environment user-centric can reap rewards in terms of energy performance and carbon emission reduction. By providing attention in the design phase to enhance active control for users allows them to become centric in the indoor environment, bringing personalization, satisfaction, and productivity to higher levels. Occupant interaction with their surrounding environment by using various adaptive opportunities (such as opening a window) together with the effects from other environmental factors (such as visual and acoustic environment, and IAQ) can lead to significant energy saving and uplifting user satisfaction. The availability of these opportunities may make occupants feel thermally comfortable across a wider range of conditions and allow them to take responsibility to maintain the building performance and achieve the expected performance targets.

4 TARGET VALUES FOR IEQ IN NZEB

4.1 Thermal Comfort

Specifying the thermal comfort objectives of NZEB is a prerequisite for its design. Such objectives should be explicitly included as an integral part of the definition and performance thresholds of a zero energy building. These objectives need to be translated into quantifiable and measurable indicators, metrics, and assessment methods. To date, a variety of thermal comfort models exist in the literature and standardization of indoor environment.

Design Indoor Temperature

Thermal comfort is viewed as a state-of-mind where occupants are satisfied with their surrounding thermal environment and desire neither a warmer nor a cooler condition (Fanger, 1970). There are two main approaches under pinning current indoor thermal environmental standards. The first is based on a heat balance or static model calculation for occupants, while the second is known as adaptive comfort and is based on adaptive calefaction models (see Eq. 5.1). The static heat balance comfort model evolved during the 1960s and 1970s in the biometeorology research community (Höppe, 1997).

$$S = M \pm W \pm L_{cond} \pm L_{conv} \pm L_{rad} - E_{evap} - E_{res} \quad (W/m^2) \qquad (5.1)$$

With the development of air conditioning, the AEC industry has been inclined toward artificial indoor environments, sealed and pressurized buildings (CIBSE, 2007). In 1970, based on climate chamber experiments, Fanger introduced the so called Predicted Mean Vote (PMV) and Percentage People Dissatisfied (PPD) model of thermal comfort, which established a relationship between six primary factors based on a thermal balance equation developed under steady-state conditions (Fanger, 1970) (see Fig. 5.3).

Fig. 5.3 is an example which represents PPD as a function of PMV. The PMV index predicts thermal sensation as a function of metabolic rate, clothing, air temperature, mean radiant temperature, air velocity, and relative humidity. The model has been incorporated into a number of standards, simulation software, and design codes including ISO standard 7730 (1994), ASHRAE Standard 55 (2013), CEN 15251 (2012 and recently changed to CEN 16798 (2017)) and the Chinese GB/T 50785 (2012).

Today, all standards on thermal comfort basically agree with suggesting an adaptation of Fanger's model for mechanically heated and/or cooled buildings. Several international standards specify criteria for indoor environments in different classes. The higher the classification, the narrower the range of acceptable PMV values corresponding to a

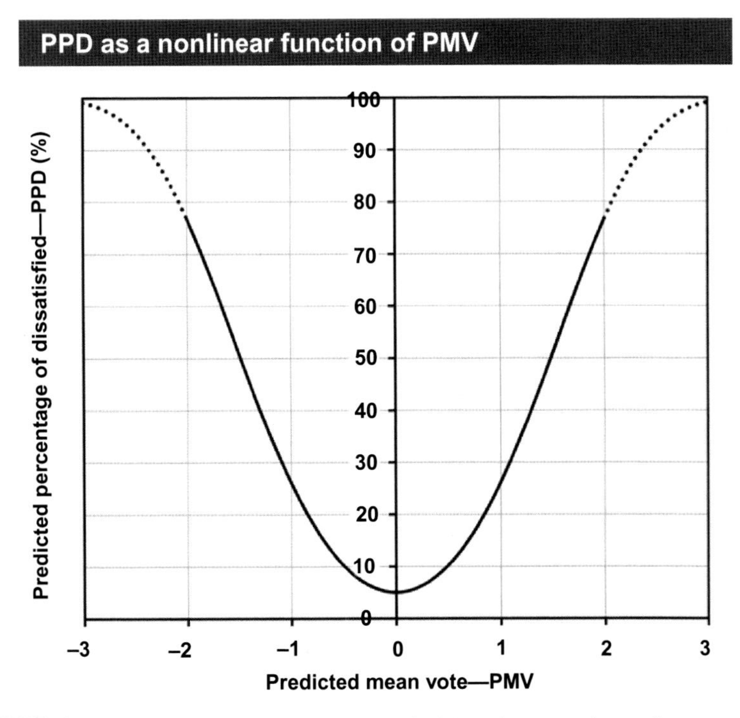

FIGURE 5.3 Predicted percentage of dissatisfied (PPD) as a nonlinear function of predicted mean vote (PMV).

TABLE 5.2 Thermal Environmental Category Label in Different Standards

	Category	PMV ranges	PPD (%)
ISO 7730	I Excellent	$-0.2 < PMV < +0.2$	<6
	II Good	$-0.5 < PMV < +0.5$	<10
	III Satisfactory	$-0.7 < PMV < +0.7$	<15
CEN 15251or CEN 16798	I Excellent	$-0.2 < PMV < +0.2$	<6
	II Good	$-0.5 < PMV < +0.5$	<10
	III Satisfactory	$-0.7 < PMV < +0.7$	<15
	IV Poor	$PMV < -0.7$ or $PMV > +0.7$	<15
ASHRAE 55	–	$-0.5 < PMV < +0.5$	<10
	I	$-0.5 < PMV < +0.5$	<10
	II	$-1 < PMV < -0.5$ or $+0.5 < PMV < 1$	10–25
	III	$PMV < -1$ or $PMV > 1$	>25

stringent comfort range variation. Table 5.2 compares these categories among ISO 7730, CEN 15251 or CEN 16798, and ASHRAE 55. So, for the CEN 15251 we added an additional category IV for existing buildings if indoor thermal conditions do not meet category III. Category IV provides less strict criteria for free-running buildings without mechanical cooling. Also Table 5.3 shows an example of different design temperature values. Those values vary depending on the activity or function, local habits, and the energy saving policy. The values in Table 5.3 are extracted from the CEN 15251 or CEN 16798, however, we advise designers to refer to values found in ASHRAE 55, ISO 7730, or CEN 15251 or CEN 16798 depending on their location for mechanically heated or cooled NZEB.

Next to the thermal environmental category of the steady-state heat balance model we find the psychrometric chart has been used to visualize the comfort zone. The psychrometric chart combines operative temperature and humidity and plots these in a chart to facilitate an understanding of human occupancy thermal comfort sensation. Fig. 5.4 shows an example of such chart.

The second fundamental approach to assess thermal comfort is known as the adaptive model. The first refined formulation of adaptive comfort for a comfort standard was done by De Dear et al. (1998). Their work is an evolution of the earlier endeavors of Olgyay and Givoni who developed bioclimatic charts (Givoni, 1992; Olgyay, 2015). The adaptive comfort approach assumes that living or working in continuously

TABLE 5.3 Examples of Recommended Design Values of the Indoor Temperature From the CEN 15251or CEN 16798

		Operative temperature (°C)	
Type of building/space	**Category**	**Minimum for heating (winter season), 1.0 clo**	**Maximum for cooling (summer season), 0.5 clo**
Residential buildings: living spaces (bedrooms, drawing room, kitchen, etc.)	I	21.0	25.5
	II	20.0	26.0
Sedentary activity 1.2 met	III	18.0	27.0
Single and open-plan offices, and spaces with similar activity (conference rooms, auditorium, cafeteria and restaurants)	I	21.0	25.5
	II	20.0	26.0
	III	19.0	27.0
Sedentary activity 1.2 met			
Classroom	I	21.0	25.0
Sedentary activity 1.2 met	II	20.0	26.0
	III	19.0	27.0

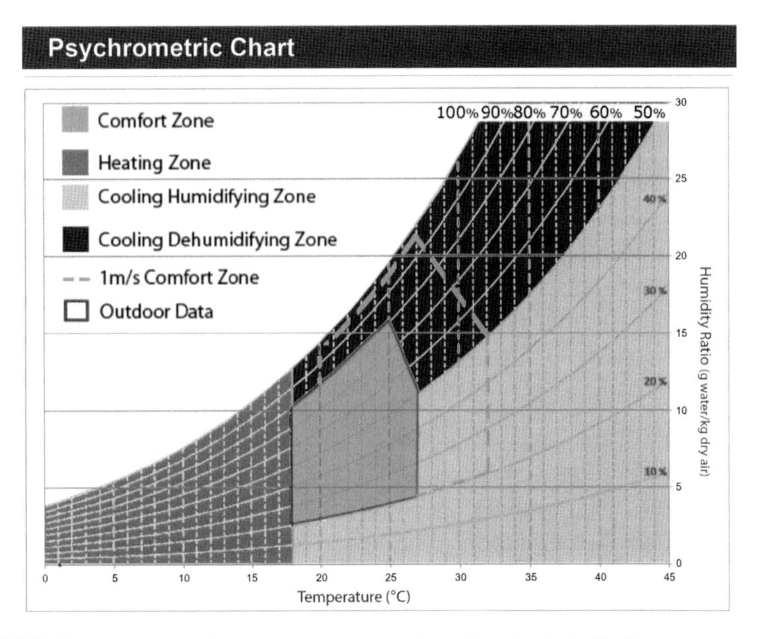

FIGURE 5.4 An example of a psychrometric chart (Lenoir et al., 2012).

stable thermal environments with tight temperature controls may cause the degeneration of the occupants' inherent ability to resist thermal stress, alter their comfort thermo regulation, and lead to higher comfort expectations and energy consumption (Zhu et al., 2016). Instead of relying on a steady-state thermal environment based on PMV, the adaptive approach aims to replace the single set point with a narrow comfort range with dynamic thermal comfort and flexibility expanding the comfort range. The advantage of this approach is that it empowers occupants and allows them to gain control over their environment, which has a strong impact on their well-being and on energy consumption and HVAC system installations impact. For NZEB, this can be an alternate path for low-tech buildings running in free mode or mixed mode (cooling or heating with natural ventilation) (Attia and Carlucci, 2015). Already ASHRAE integrated this approach in Standard 55 (ASHRAE, 2013) and the EU states incorporated it in European CEN 15251 (2012).

Generally, the implementation of the adaptive model indicates that indoor thermal comfort is achieved with a wider range of temperatures than does the implementation of the ISO 7730 model (see Fig. 5.5). Both models (CEN 15251 or CEN 16798 and ISO 7703) use statistical analysis of survey data to back up their claims with their respective areas of applicability. In some situations, it proves possible to maintain a building's interior conditions within the CEN 15251 adaptive comfort limits entirely by natural means. In these cases, there is no energy use associated with achieving indoor summer comfort.

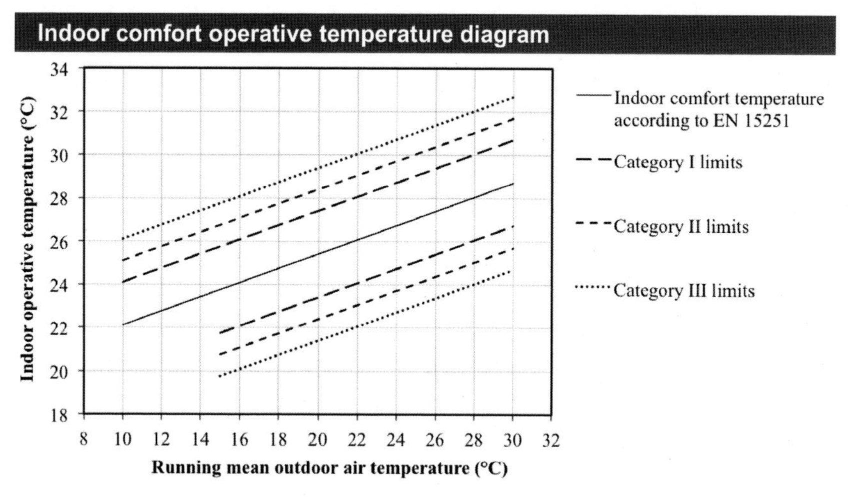

FIGURE 5.5 Indoor comfort operative temperature as a function of the monthly mean outdoor dry-bulb air temperature according to CEN 15251 or CEN 16798.

Optimal operative comfort temperature can be calculated by knowing the daily mean outdoor dry-bulb air temperature of previous days

$$T_c^{EN15251} = 0.33T_{rm} + 18.8$$
$$\text{with } T_o = (1-\alpha)\sum_{i=0}^{n}\left(\alpha^i T_e\big|_{-(1+i)}\right) \tag{5.2}$$

where T_{rm} is the exponential weighted running mean of the daily outdoor dry-bulb air temperature, $T_e|_{-(1+i)}$ is the daily mean outdoor dry-bulb air temperature of the previous $(1 + i)$ day and α is a constant included in the range [0, 1], but a recommended value is 0.8. In order to simplify calculations, the standard CEN 15251 suggests a simplified equation to calculate the exponential weighted running mean of the daily outdoor dry-bulb air temperature:

$$T_{rm} = \frac{T_e|_{-1} + 0.8T_e|_{-2} + 0.6T_e|_{-3} + 0.5T_e|_{-4} + 0.4T_e|_{-5} + 0.3T_e|_{-6} + 0.2T_e|_{-7}}{3.8} \tag{5.3}$$

The assumptions of acceptability are expressed for different categories of buildings with occupants inside a building and are expressed as symmetrical ranges around the optimal comfort temperature. Table 5.2 lists the optimal comfort temperature and the upper and lower limits of the comfort categories.

Air Velocity and Drought

Air velocity in a space influences the convective heat exchange between occupant and the indoor environment (Seppanen and Kurnitski, 2013). There are two main types of air velocity or natural air velocity. The first is the air flow caused by natural air flow, buoyancy or mainly atmospheric circulation. In contrast, the second entails air forced air velocity generated by mechanical systems including ceiling and standing fans. In general, natural air velocity provides a stronger cooling effect and better thermal comfort (Cui et al., 2013). Natural air allows a pleasant thermal sensation to control the air velocity. Table 5.4 illustrates the maximum limits of air velocity in relation to the classes of CEN 15251 (2007). The aim of these criteria is to set a limit for draft or drought complaints in buildings.

Givoni, while revising his already notable work on the building bioclimatic chart, suggested that at least air temperature, surface temperature, and air velocity should be taken in consideration (Givoni, 1992). He expanded the boundaries of the comfort zone based on the expected indoor temperatures achievable with different passive design strategies, applying a "common sense" notion that people living in unconditioned buildings become accustomed to, and grow to accept higher temperatures

TABLE 5.4 Suggested Criteria for Air Velocity and Drought Limits

		Unit	Maximum values in classes		
			S1	**S2**	**S3**
Air velocity	Winter (20°C)	m/s	0.13	0.16	0.19
	Winter (20°C)	m/s	0.14	0.17	0.20
Air velocity	Summer (24°C)	m/s	0.20	0.25	0.30
Vertical temperature difference		°C	2	3	4
Floor temperature		°C	19–29	19–29	17–31

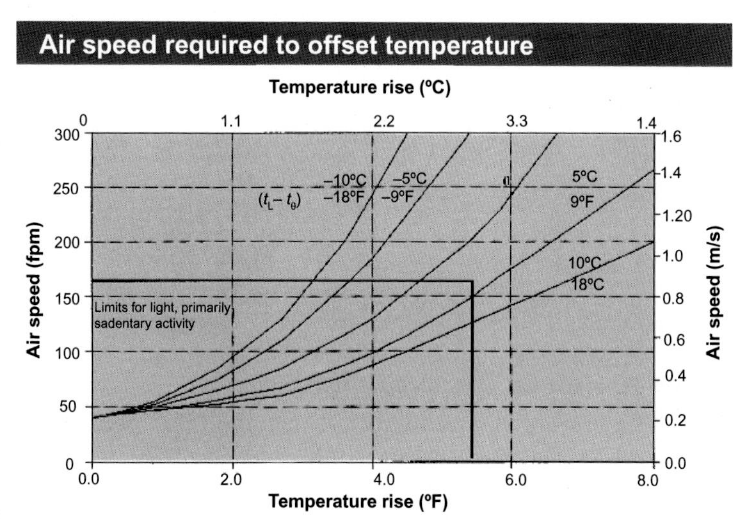

FIGURE 5.6 Air speed required for offsetting increased temperature. Source: *Reproduced from ANSI/ASHRAE 55, 2013. Thermal Environmental Conditions for Human Occupancy. American Society of Heating, Refrigerating and Air-Conditioning Engineers, Atlanta, GA, USA.*

or humidity. However, a proposed addendum to ASHRAE 55 in September 2008 suggested the use of the PMV model to air speeds below 0.20 m/s. Air speeds greater than this value may be used to increase the upper operative temperature limits of the comfort zone in certain circumstances. This could be achieved by using ceiling fans to elevate air speed to offset increased air and radiant temperatures. As shown in Fig. 5.6, elevated air speed is effective at increasing heat loss when the mean radiant temperature is high and the air temperature is low. However, if the mean radiant temperature is low or humidity is high, elevated air speed is less effective. The required air speed for light, primarily sedentary activities

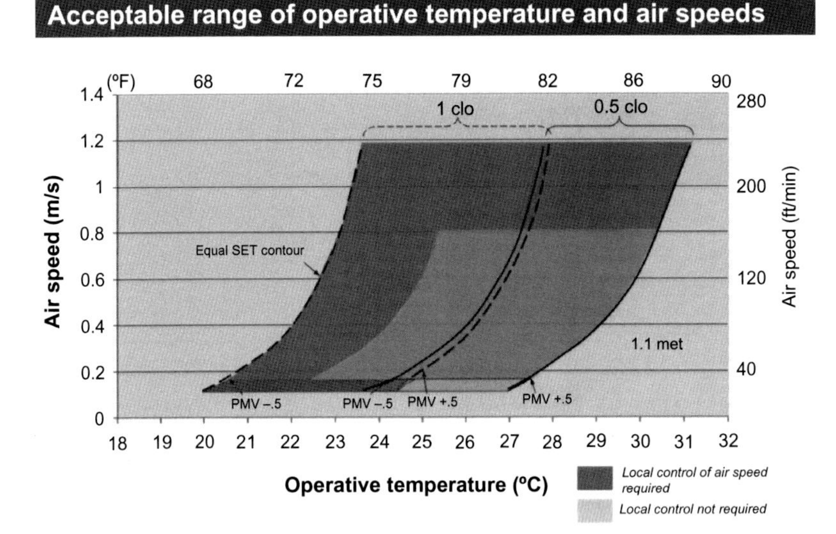

FIGURE 5.7 Acceptable ranges of operative temperature and air speeds. Source: *Reproduced from ANSI/ASHRAE 55, 2013. Thermal Environmental Conditions for Human Occupancy. American Society of Heating, Refrigerating and Air-Conditioning Engineers, Atlanta, GA, USA.*

may not be higher than 0.8 m/s. But the ceiling fans' effect cannot control humidity and depend on clothing and activity. Fig. 5.7 shows the acceptable range of operative temperature and air speed for various given clothing levels.

Vertical and Horizontal Temperature Distribution

Vertical temperature difference between ankle and neck can cause serious discomfort. The measuring height is 0.1 and 1.1 for sedentary work. Table 5.4 provides recommendations of vertical distribution limits to avoid temperature difference in system floor heating or cooling or transient air distribution. Similarly, horizontal temperature distribution aim to create homogeneous comfort conditions across spaces. Comfort conditions are different from room conditions close to the façade due to the exposure to direct solar radiation, hot glazed surfaces, glare, brightness, and draught from open windows. The role of the designer is to design a control strategy to offer the right balance between automatic, and/or manual control for occupants to stabilize the horizontal temperature variation in work spaces.

4.2 Air Quality and Ventilation

IAQ combines several indicators such as ventilation rate and carbon dioxide concentration, with emerging design strategies, including holistic emissions-based approaches (low-emitting materials), source control, and monitoring of contaminants. To achieve comfort and well-being in a NZEB, IAQ standards must be applied. To date, there is no common standard for IAQ. However, in the following paragraphs we will base our recommendations for the design target values on the WHO guidelines, ASHRAE and CEN standards, and the WELL rating system.

IAQ Ventilation Rates (Office and Residential)

By diluting pollutants created by contaminant sources and occupants in the building, ventilation contributes to occupants' comfort and well-being (USGBC, 2013).The precise correlation between ventilation rate and occupant health is still being researched. However, limit values are being used as a basis for criteria for new constructions. Table 5.5 summarizes the typical indoor air level exposure based on WHO recommendations. The table includes requirements to reduce contaminants including Volatile Organic Compound (VOC) emissions into indoor air. VOCs are off-gas elements from paints, finishes, varnishes, coatings, cleaning products, and personal care products. They are chemical compounds with a high concentration and are associated with heath disorders (IWBI, 2016).They can cause headaches, nausea, and irritation to the respiratory system, skin, and eyes, among other ailments. However, it is very difficult to verify all these contaminants during the design phase and after construction. Therefore, the common practice to provide a healthy IAQ is to focus on the design of ventilation air flows.

For office buildings, the ventilation rate is defined based on the sum of ventilation for pollution from occupancy and ventilation for the

TABLE 5.5 Typical Indoor Air Contaminants Threshold Limits According to WHO IAQ Guidelines (WHO, 2014, 2016) and WELL (IWBI, 2016)

Agent	WHO Typical ($\mu g/m^3$)	WHO Indoor source (%)	WELL
PM 2.5	10–40	up to 30	$< 15\ \mu g/m^3$
PM 10	–	–	$< 50\ \mu g/m^3$
CO	1–4	0	$< 9\ \mu g/m^3$
NO_2	10–50	up to 20	–
VOC formaldehyde	20–80	>90	< 27 ppb
VOC benzene	2–15	up to 40	Total VOC less than 500
VOC naphthalene	1–3	up to 30	
Radon	20–100	>90	< 4 pCi/L

pollution from building materials, finishes and furniture. For mechanically ventilated spaces and for mixed-mode systems when the mechanical ventilation is activated, the determination of ventilation rate should be based on ASHRAE 62.1, CEN 15251 or local equivalents (USGBC, 2013). According to CEN 15251 (2007), the two components are represented in the following equation:

$$q_{total} = n \times q_{occupancy} + A \times q_{building} \quad (5.4)$$

where q_{total} the total ventilation rate of the room, L/s; n the design value for the number of the persons in the room; $q_{occupancy}$ the ventilation rate for occupancy per person, L/s, person; A the room floor area, m^2; $q_{building}$ the ventilation rate for emissions from building, L/s, m^2.

Ventilation rates can be expressed as per square meter per floor area (L/s, m^2) or per occupant L/s per occupant. The ventilation rate for occupants ($q_{occupancy}$), and the ventilation rate ($q_{building}$) for buildings can be found in Table 5.6. However, it is more effective to calculate the ventilation rate using Eq. (5.4) with occupancy densities (floor area m^2/person) as indicated in Table 5.7.

For naturally ventilated office spaces and for mixed-mode systems when the mechanical ventilation is inactive, the determination of minimum outdoor air opening and space requirements should be based on ASHRAE Standard 62.1-2013, CEN 15251 or local equivalent, whichever is more stringent (USGBC, 2013). However, the design team must conduct an extra study to confirm that natural ventilation is an effective strategy following the diagram in the Chartered Institution of Building Services Engineers (CIBSE) Application Manual AM10, March 2005, Natural Ventilation in Nondomestic Buildings, Fig. 2.8, and meet the requirements of ASHRAE 62.1-2013, Section 4, or CEN 15251 or local equivalent, whichever is more stringent.

For residential buildings, ventilation rates and air quality are different. In residential buildings, there are specific activities such as smoking, cooking, showering, and turning on a dish washer that can cause a

TABLE 5.6 Ventilation Rates Used for Calculation in Office Areas According to CEN 15251 (2007) and CEN 16798 (2017)

	$q_{occupancy}$ **(L/s/person)**	$q_{building}$ **Very low polluting buildings (L/s, m^2)**	$q_{building}$ **Low polluting buildings (L/s, m^2)**	$q_{building}$ **None low polluting buildings (L/s, m^2)**
Category I	10	0.5	1.0	2.0
Category II	7	0.35	0.7	1.4
Category III	4	0.3	0.4	0.8

TABLE 5.7 Ventilation Rates Used for Calculation in Office Areas According to CEN 15251 (2007) and CEN 16798 (2017)

Space Type	Category	Floor area (m^2.person)	$q_{occupancy}$ (L/s/m^2)	$q_{building}$ Low polluted building (L/s, m^2)	$q_{building}$ None low polluted building (L/s, m^2)	q_{total} Total for low polluted building (L/s, m^2)	q_{total} Total for none polluted building (L/s, m^2)
Single office	I	10	1.0	1.0	2.0	2.0	3.0
	II	10	0.7	0.7	1.4	1.4	2.1
	II	10	0.4	0.4	0.8	0.8	1.2
Open space office	I	15	0.7	1.0	2.0	1.7	2.7
	II	15	0.5	0.7	1.4	1.2	1.9
	II	15	0.3	0.4	0.8	0.7	1.1
Conference room	I	2	5.0	1.0	2.0	6.0	7.0
	II	2	3.5	0.7	1.4	4.2	4.9
	II	2	2.0	0.4	0.8	2.4	2.8
Classroom	I	2	5.0	1.0	2.0	6.0	7.0
	II	2	3.5	0.7	1.4	4.2	4.9
	II	2	2.0	0.4	0.8	2.4	2.8

Adapted from Seppänen, O., & Kurnitski, J. (2013). Target Values for Indoor Environment in Energy-Efficient Design. In Cost Optimal and Nearly Zero-Energy Buildings (nZEB) (pp. 57–78). Springer London.

TABLE 5.8 Examples of Ventilation Rates for Dwellings Based on CEN 15251 or CEN 16798

	Total air change rate		Living room, bedrooms, outdoor air flow		Exhaust air flow (L/s)		
Category	L/s, m^2	ach	L/s, person	L/s, m^2	Kitchen	Bathrooms	Toilets
I	0.49	0.7	10	1.4	28	20	14
II	0.42	0.6	7	1.0	20	15	10
III	0.35	0.5	4	0.6	14	10	7

Adapted from Seppänen, O., & Kurnitski, J. (2013). Target Values for Indoor Environment in Energy-Efficient Design. In Cost Optimal and Nearly Zero-Energy Buildings (nZEB) (pp. 57-78). Springer London.

cocktail of emissions. Humidity and Particulate Matter emitted from heating systems or fire spaces are deemed the most influential parameters that can affect health in residential buildings. The ventilation rates are specified as an air change per hour, outside air supply, or required exhaust rates. Most national regulations in industrial countries force the implementation of mechanical ventilation based on the three following criteria (Seppanen and Kurnitski, 2013):

1. Exhaust of air pollution in humid rooms (kitchen, bathroom, toilet)
2. Ventilation rates for main dry rooms (bed and living rooms)
3. An overall ventilation rate of the whole dwelling volume.

Table 5.8 illustrates examples from the European Standards, ASHRAE Standard 62.2-2013 or local equivalent which can be also used in residential NZEB.

CO_2 Concentration

Carbon dioxide levels serve as a proxy for IAQ. Carbon dioxide is one of the earliest pollutants to be detected and can be used in assessing compliance with designed ventilation rates. For mechanically ventilated NZEB, a demand controlled ventilation system should keep CO_2 levels between 800 and 1000 ppm. Even for naturally ventilated spaces and mixed-mode systems when windows are operable and mechanical ventilation is inactive, the CO_2 levels should stay between 800 and 1000 ppm. Table 5.9 provides recommendations for CO_2 concentration values to be used as design target values. NZEB should comply with ASHRAE 62.1 2013 or EN 13779 requirements.

Filtration and Air Cleaning

Filtration and air cleaning is essential to reduce dust, particulate matter, and pollens. For example, carbon filters can absorb volatile

TABLE 5.9 Approximate Maximum Sedentary Carbon Dioxide Concentrations Associated With CEN 13779 Indoor Air Quality

Category	Description	Carbon Dioxide concentration in ppm	Default ppm room value	Total ppm indoor value	Fresh air flow (m^3/h person)
I	High IAQ	<400	350	700–750	>54
II	Medium IAQ	400–600	500	850–900	35–54
III	Moderate IAQ	600–1000	800	1150–1200	22–36
IV	Low IAQ	>1000	1200	1550–1600	<22

pollutants and remove the largest particles, while media filters can remove smaller particles. Mechanical filtration is efficient and requires high-performance criteria to guarantee appropriate levels of filtration and purification. Design guidelines are given in EN 13779, ASHRAE 62.1 and in REHVA Guidebook No.11. The careful position of fresh air intake is very important. Also, each ventilation system must have particle filters or air-cleaning devices that meet the CEN or ASHRAE requirements. CEN 13799: 2007 recommends installing activated carbon filters to remove NO_2 and SO_2 in polluted areas. Fine particle Filter F7 or F8 classes are recommended for locations with typical outdoor pollution. F9 classified filters are recommended to remove OM 2.5 and PM 10. A Minimum Efficiency Reporting Value (MERV) of 13 or higher is recommended by the ASHRAE Standard 52-2013. High-performance filters are associated with pressure drops, which leads to an increase of specific fan power. However, the energy cost and balance discussed in Chapter 3, NZEB Performance Indicators and Thresholds, is less of a concern if comfort, air quality compliance, and health are valued.

4.3 Moisture and Air Humidity

As mentioned in Section 3.2, moisture and air humidity can cause risks associated with the building design and use. Moisture and air humidity have an effect on thermal sensation and perceived air quality. One of the common complaints in NZEB is dryness and skin irritation especially in residential buildings located in a temperate climate. A ventilation system should maintain relative humidity between 40% and 60% at all times by adding or removing moisture from the air when necessary. Next, moisture and condensation management should be addressed by the design team and during operation. High relative humidity levels associated with oversized air conditioning systems or poor construction quality in the form of air leakage or inappropriate insulation can lead to condensation. Condensation increases microbial growth and contamination and, therefore, should be avoided through proper envelope design and quality assurance during construction.

4.4 Daylight and Visual Comfort

The intent of incorporating daylight into the space is to connect building occupants with the outdoors, reinforce circadian rhythm, and reduce the use of electrical lighting (USGBC, 2013). Daylight is variable, often unpredictable and depends on solar position and sky cover conditions. However, due to the complexity and continuous change of performance indicators for a successful daylight design, it is essential to apply climate-based daylight performance indicators and modeling during the design of NZEB. Glazing area is a sensitive design parameter that can influence the well-being and energy performance in NZEB. Based on lessons learned in Chapter 12, NZEB Roadmap and Tools, the window to wall ratio in NZEB should not exceed 50%, while the balance between the solar heat gain coefficient and visible transmittance should be optimized to the local climatic context and solar radiation intensity. Windows should be sized to fit the purpose of use and to allow functional views. The modeling should be based on typical meteorological year data, using local weather data, or TMY weather data files for the nearest city. Annual computer modeling and simulations should be made by the design team to demonstrate a spatial daylight autonomy of at least 300 Lux for 50%–90% of the regularly occupied floor area. The spatial daylighting autonomy threshold should be for the hours between 8 a.m. and 6 p.m. of local time, for the full calendar year from January 1 to December 31. Moreover, the annual sunlight exposure should not exceed 1000 Lux of direct sunlight for more than 250 hours of the hours between 8 a.m. and 6 p.m. of local time, for the full calendar year (IESNA, 2012, 2016). Glare-control devices must be provided for all transparent glazing in regularly occupied spaces regardless of whether the glazing receives direct sunlight or not. All glare-control devices must be operable by building occupants including interior window blinds, shades, curtains, movable exterior louvers, movable screens, and movable awnings.

Interior lighting: Proving high-quality lighting promotes occupants productivity, comfort and well-being. Interior lighting comprises lighting quality and light control. For lighting quality, the design team should carefully use light fixtures with proper luminance (preferably 250 Lux or more) and use light sources of color rendering index (CRI) of

TABLE 5.10 EN 12464-1 Light and Lighting of Indoor Work Places (CEN, 2011; IESNA, 2016)

Building type	Space	Luminance at working level in Lux	UGR	Ra	Working plane height (m)
Office	Single offices	300 Work	19	80	0.8
		500 Writing, typing, reading,	19	80	0.8
		data processing on PC	19	80	0.8
		750 Technical Drawing			
	Open-plan	500			
	Conference room	500			
Educational	Classroom	300	19	80	0.8
	Lecture hall	500	19	80	0.8
Circulation areas	Reception desk	300	19	80	0.8
	Corridor	100	28	40	0.1
	Stairs	150	25	40	0.1

Unified Glare Using (UGR) is used to evaluate the glare from artificial lighting.

80 or higher. For ambient light, the trend is to provide light below 300 Lux and allow users to reach the 300 to 500 Lux at work surface using task lights that are available and adaptable according to need. Task lights and individual lighting controls should enable occupants to adjust the lighting to suit their individual tasks and performance. The recommended task illuminance is defined in EN 12464-1 and IESNA (2016) as shown in Table 5.10.

Proper design of interior lighting using light fittings and ballasts should also be addressed. Flickering lighting problems are associated with risk from anxiety, migraines, headaches, and even dyslexia, and should be avoided. Avoid lighting that flickers at 100 hertz and introduce high-frequency electronic control gear and LED panels (Wilkins et al., 2010).

4.5 Acoustic Comfort

Effective acoustic design can provide occupants' satisfaction, increase their productivity through concentration, and facilitate communication in work spaces, dwellings, and classrooms. The acoustic criteria in

NZEB are focused on mitigating HVAC background levels, providing sound insulation, addressing the reverberation time, and reducing the impact noise. In the past 10 years, European building codes increased the minimum requirements for acoustic comfort in multistory buildings, schools, offices, hotels, student housing, and hospitals in response to a higher awareness of the importance of this topic.

Sound transmission: Reducing noise from adjacent spaces requires careful detailing, proper material choices, composition, and high-quality construction execution. The details include interior wall partitions and doors to reduce the sound transmission class (STCc) rating listed in Table 5.11, based on European standards or the local equivalent. Regarding exterior noise intrusion, the envelope design should prevent exterior noise. The average sound pressure level from outside noise intrusion should not exceed 50 dBA.

Sound reverberation: Reverberation time or RT60, is a metric that describes the time taken for a sound wave to decay by 60 decibels from its origin. In spaces with high reverberation times, higher levels of ambient noise cause noise and stress (ISO, 2013). Therefore, designers should limit reverberation time to maintain comfortable sound levels. Table 5.12 lists the reverberation time requirements in the performance measurement protocols for commercial buildings.

Indoor background noise: Background noise from Heating, Ventilation, and Air Conditioning (HVAC) systems can be distracting and loud enough to cause stress and loss of concentration. The best practice for mechanical systems noise control in ANSI Standard S12.60-2010 recommends that sound levels fall under the 40 dBA threshold. Table 5.12 provides indoor noise criteria according to CEN 15251 or CEN 16798 too.

TABLE 5.11 Maximum Composite Sound Transmission Class Ratings for Adjacent Spaces (CEN 15251 or CEN 16798, ANSI/ASHRAE, 2013; ASA, 2008; ANSI, 2012)

Building type	Type of space	Sound pressure level (dBA) EU	Sound pressure level (dBA) US
Residential	Living room	40	50
	Bed room	35	50
Office Building	Office space	40	45
	Conference room	40	50
	Open office	45	50
Schools	Classroom	40	45
Corridors		50	50

TABLE 5.12 Reverberation Time Requirements (CEN 15251 or CEN 16798, ASHRAE, 2007; ASA, 2008; ANSI 2012)

Building type	Type of space	T60
Residential		<0.6
Hotel	Individual room	<0.6
	Meeting room	<0.8
Office	Conference room	<0.6
Library		<1.0
Classroom		<0.6

5 QUALITY ASSURANCE

The current market trend for health and well-being within the AEC industry requires an improvement of the IEQ assessment methods across the project delivery process. Compliance is essential with the design target values and standards suggested in the previous section to achieve the full energy efficiency and carbon reduction potential of NZEB and provide the health and well-being benefits of improved IEQ. Compliance should be enforced during design, construction and operation of NZEB. There are existing products and systems that already fulfill the NZEB requirements—fans, efficient heat recovery in ventilation and air-conditioning systems, heat pumps, pumps, hydronic and air balance, room temperature and humidity control, smart controls and building automation, LED lighting, etc. (EPEE, 2016). However, the challenge remains in setting the proper IEQ criteria and following their proper execution and implementation on-site until operation. Regular inspection and continuous commissioning of technical building systems is essential to assure quality. Building construction and envelope details should be inspected during design and execution. HVAC systems should be regularly maintained and inspected. For example, proper envelope airtightness of NZEB can guarantee proper ventilation and air conditioning functioning and, as a consequence, ensure IAQ. Following the acoustic insulation details by a specialist during design and construction and providing a protocol or process supported by the specialist will ensure maintaining noise levels indoors within the required levels.

Based on lessons learned from practice in Chapter 11, NZEB Case Studies and Learned Lessons, we recommend seven steps, shown in Table 5.13, which must be followed and articulated during the project delivery process to control, validate, and follow up the designed performance requirements for thermal comfort, air quality and ventilation, moisture, air humidity, daylighting, visual comfort, and acoustic comfort.

TABLE 5.13 Quality Assurance Checklist for IEQ

1. Definition of IEQ criteria	**Design**
2. Concept design and design development	
3. RFP and final details and specifications	
4. Execution	**Construction**
5. Commissioning	
6. Monitoring + postoccupancy evaluation	**Operation**
7. Continuous commission	

These seven steps must go through three major phases. The first phase comprises the definition of IEQ criteria and integrating this in the design and concept. The criteria must be quantified and described based on standards, drawings, and specifications and must be included in the request for proposals (RFP). The second phase, namely construction, comprises the execution and commission step. Most errors associated with IEQ occur during this phase. The linear nature of project delivery process and the fragmented subcontracting culture among contractors is one of the main reasons for discrepancies between design and construction. Control, check, and validate the principle practices that should be exercised to integrate all design details material composition and systems in the new or renovated building. Chapter 9, Construction Quality and Cost, will elaborate on the importance of enabling professional builders and workers during this phase to allow them to achieve a proper envelope earthiness, thermal insulation, and acoustic insulation. The following key step in this phase is commissioning. Commissioning is the process of verifying and documenting that the building systems and assemblies are planned, designed, installed, tested, operated, and maintained to meet the owner's project requirements. Proper commissioning saves energy, reduces risk, and creates value for building operators. It also serves as a quality assurance process for enhancing the delivery of the project. The final phase of quality assurance for IEQ is monitoring postoccupant evaluation, and continuous commissioning. Chapter 10, Occupant Behavior and Performance Assurance, describes in detail the monitoring of IEQ and how it should be used to validate the defined indoor quality. Continuous commissioning goes hand-in-hand with monitoring for active systems to maintain the envisaged IEQ.

6 DISCUSSION

After exploring the concepts, design strategies, and environmental factors that influence the IEQ in NZEB, we would like to provide an informative discussion on the key challenges that face design teams. Design teams need to make important decisions related to the way people work, learn, and live in NZEB. Designers are responsible for enabling and empowering user assess and measure the outcomes of their design in relation to air quality, thermal, visual, and acoustic comfort. At the same time, NZEB built today should last for at least 60 years and allow the adaption and evolution of building use and the combination of professional and living activities while maintaining a high standard of IEQ. In this context, we share three important challenges that should be considered by design teams and addressed in any future design that aims to achieve high IEQ in high-performance buildings.

The first challenge that design teams should be aware of is health compliance versus energy compliance. Finding a balance between NZEB tightness and IEQ is not easy. NZEB tightness is associated with energy efficiency while IEQ is associated with high energy consumption. NZEB by default are tight and should avoid leaking of envelopes and ductwork to achieve energy efficiency (see Section 4.4 of Chapter 3: NZEB Performance Thresholds). Tightness is important in terms of energy use to ensure a good climate and proper building construction while controlling vapor and moisture (Carrie and Wouters, 2012). At the same time, the awareness of outdoor and indoor air pollution and the influence of fine dust (PM 2.5 and PM10) arising from automobile emissions, tire abrasion, waste incineration, and fire places is increasing (Oldenkamp et al., 2016). For example, there are more challenges and contradictions to achieve IAQ for NZEB in hot and humid climates and in densely populated cities. Cross ventilation is not easy to implement in NZEB located in hot climates (Attia, 2017). Even in temperate climates, the poor outdoor air quality requires appropriate levels of filtration, air purification, and ventilation rates indoors. The consequences of such measures are counterproductive leading to a significant increase in energy consumption and cost (see Fig. 3.3). Architects and engineers are faced with the choice of tackling air quality problems or leaving themselves open to potential liability claims. Therefore, this challenge cannot be ignored and trade-offs should be made to ensure the health and well-being in NZEB while at the same time achieving energy efficiency and construction quality.

The second challenge is to guide the decision-making in IEQ design and provide exemplary assessment and control. There are many indicators and design value targets to achieve IEQ. Integrating air quality and ventilation, thermal comfort, daylighting and views, acoustic comfort, interior design, and personal control are critical to successful IEQ. The optimization of different design strategies to maximize health, well-being, and productivity is a serious challenge. Modeling and testing design decisions with specialized building performance simulation tools is very important. Regarding the performance assessment of IEQ in NZEB, there is confusion among building professionals on the IEQ variables and methods that need to be incorporated in any impact assessment. Fig. 5.8 provides a framework for assessment of IEQ in NZEB. The focus here is not only on the classical physical indicators and metrics, but also the perceptual and functional aspects of health, well-being, satisfaction, and productivity. Based on the work of the World Green Building Council (2014), this framework suggests a relationship between these three elements. Therefore, monitoring and control of IEQ in NZEB must be planned and applied. Once design is executed, continuous monitoring and postoccupancy evaluations are the only way to ensure operational success by following up the performance and informing the building occupants and operators on the intent of IEQ, how to use personal controls, how to access and use the facility, and how users perceive thermal sensation and air quality while interacting with the building functions (see Chapter 10: Occupant Behavior and Performance Assurance). The integrative and collective decision-making of the design team should aim to create beneficial synergies and innovative solutions to achieve IEQ and provide occupant-based assessment methods at all stages of a NZEB project.

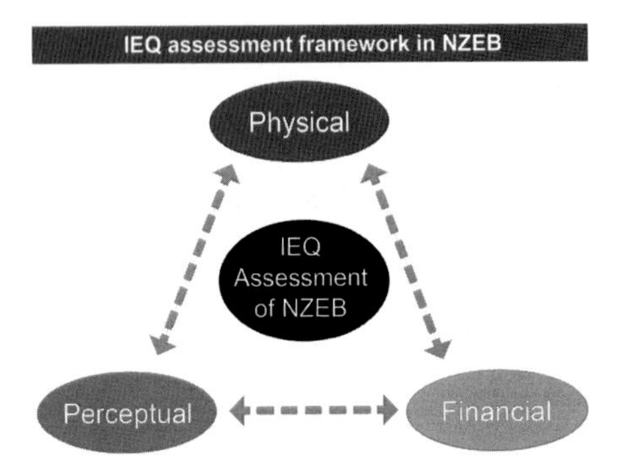

FIGURE 5.8 Indoor environmental quality assessment framework. *Adapted from World Green Building Council (WGBC), 2014. Health, Wellbeing and Productivity in Offices: The Next Chapter for Green Building. Retrieved from <http://www.worldgbc.org/news-media/health-wellbeing-and-productivity-offices-next-chapter-green-building> (accessed 21.03.17.).*

The third challenge is empowering and enabling occupants in relation to IEQ of their built environment. We should remind ourselves, again, that we build NZEB for occupants and to support their well-being and productivity. Occupant control of their indoor environment is an essential factor in the overall satisfaction and energy balance. Occupants want greater control over their desk environments by being able to control temperature or lighting. Allowing users to personalize heat and light settings empowers them and increases their productivity. Postoccupancy evaluation and measurements should be planned early on in the design process to provide evidence of end users' satisfaction with all general NZEB aspects and indoor conditions. Indoor conditions include furniture, ergonomics, and interior design as well the specific work area conditions. In order to ensure that real consumption will not exceed design targets, occupants must be educated and enabled to change and control the indoor environment and become active users. This is the only way to avoid incremental increase of energy consumption in NZEB due to the passive role of occupants. By equipping NZEB with controls and allowing users to have access to sensors information we turn NZEB into smart buildings using the latest technology. Chapter 10, Occupant Behavior and Performance Assurance, elaborates on smart control and interaction using apps for user satisfaction, well-being, and productivity. We should also foresee the changing nature of hybrid use of building spaces in relation to telework, hot desking, and the combination of professional and living activities. The creation of occupant adaptive spaces that allow concentration and contemplation, or collaboration and communication, will keep NZEB vital and useful—IEQ design and operation should allow for this adaptation.

7 LESSON LEARNED # 5

In conclusion, IEQ increases building value and reduces liability of the designer, building, and owner. The early integration of experts and articulation of the expected design target value in compliance with current standards in the request of purpose is essential. The project site and outdoor environment around the project should be analyzed in detail in relation to the selected ventilation and IAQ expectations. The majority of NZEB have a centralized, mechanical ventilation system with heat recovery. They rely on filtration and purification systems and continuous IEQ monitoring. Mixed mode buildings rely more on adaptive comfort models and hybrid solutions (a combination of mechanical and natural ventilation), when the outdoor conditions can provide fresh and clean air and noise levels are low. In hot climates, adaptive comfort models and adaptive occupant behavior can have a significant impact on the proliferation NZEB in the future. Overheating and comfort problems remain a high risk in NZEB. Therefore, glazing selection, window design, shading devices and systems sizing must be optimized using BPS tools. Quality assurance is a very important measure to ensure that the installed systems and building components operate as planned. Quality assurance during the construction phase, after completion of the construction, and during the operation of the building is essential. Commissioning of all the technical installations is the success key for IEQ. Lessons learned from practice show that the use of third-party rating systems, certification or standards can increase trust, accelerate, and guarantee a good IEQ. Continuous monitoring and postoccupancy evaluation during operation in NZEB is as important as the design phase. Enabling occupants and allowing them to control their environment is fundamental to make NZEB smart, maintain the energy budget, energy balance, and achieve maximum satisfaction and productivity.

References

ACAS, 2014. Promoting Positive Mental Health at Work. Retrieved from <http://www.acas.org.uk/media/pdf/1/a/Promoting_positive_mental_health_at_work(SEPT2014).pdf> (accessed 16.03.17.).

Akbari, H., Pomerantz, M., Taha, H., 2001. Cool surfaces and shade trees to reduce energy use and improve air quality in urban areas. Sol. Energy 70 (3), 295–310.

ANSI/ASHRAE 55, 2013. Thermal Environmental Conditions for Human Occupancy. American Society of Heating, Refrigerating and Air-Conditioning Engineers, Atlanta, GA, USA.

ANSI, 2012. Standard for Sound Quality Evaluation Procedures for Air-Conditioning and Refrigeration Equipment. AHRI, November 2012.

ANSI/ASHRAE, 2013. ANSI/ASHRAE Standard 62.1-2013 Ventilation for Acceptable Indoor Air Quality. American Society of Heating, Refrigerating and Air-Conditioning Engineers, Inc., Atlanta, GA.

ASA, 2008. Criteria for Evaluating Room Noise. American National Standards Institute, USA.

Attia, S., 2017. Best practices for NZEB design in mixed & cooling dominated climates. In: Symposium for Improving Energy Efficiency in Buildings Project, 13–14 February, Ankara, Turkey.

Attia, S., Eleftheriou, P., Xeni,F., Morlot, R., Ménézo, C., Kostopoulos, V., et. al., 2016. Overview of challenges of residential nearly Zero Energy Buildings (nZEB) in Southern Europe, Sustainable Buildings Design Lab, Technical Report, Liege, Belgium, 9782930909059.

Attia, S., Bashandy, H., 2016. Evaluation of Adaptive facades: the case study of agc headquarter in Belgium. In: Belis, J.L.I.F., Bos, F.P., Louter, Ch. (Eds.), ChallengingGlass 5—Conference on Architectural and Structural Applications of Glass. Ghent University, Belgium, ISBN 978-90-825-2680-6.

Attia, S., Carlucci, S., 2015. Impact of different thermal comfort models on zero energy residential buildings in hot climate. Energy Build. 102, 117–128.

Banbury, S., Berry, D.C., 1998. Disruption of office-related tasks by speech and office noise. Br. J. Psychol. 89 (3), 499–517.

Barthelmes, V.M., Becchio, C., Corgnati, S.P., 2016. Occupant behavior lifestyles in a residential nearly zero energy building: effect on energy use and thermal comfort. Sci. Technol. Built Environ. 22 (7), 960–975.

Carrie, F.R., Wouters, P., 2012. Building Airtightness: A Critical Review of Testing, Reporting and Quality Schemes in 10 Countries. TightVent Report, 4.

CEN 13779, 2007. Ventilation for Non-Residential Buildings—Performance Requirements for Ventilation and Room-Conditioning Systems. 2007. EN 13779. European Committee for Standardization, Brussels, Belgium.

CEN 12464, 2011. 12464-1: 2011 Light and Lighting. Lighting of Work Places Part, 1.

CEN 15251, 2012. Indoor Environmental Input Parameters for Design and Assessment of Energy Performance of Buildings Addressing Indoor Air Quality, Thermal Environment, Lighting and Acoustics. European Committee for Standardization, Brussels, Belgium.

CEN 16798, 2017. Energy Performance of Buildings, Parts 1–7. European Committee for Standardization, Brussels, Belgium.

Cheng, W.C., Liu, C.H., Leung, D.Y.C., 2009. On the Correlation of air and pollutant exchange for street Canyons in combined wind-buoyancy-driven flow. Atmos. Environ. 43, 3682–3690. Available from: https://doi.org/10.1016/j.atmosenv.2009.04.054.

Chueng, I., 2013. Impact of workplace daylight exposure on sleep, physical activity, and quality of life. Am. Acad. Sleep Med. 36, 30.

Cui, W., Cao, G., Ouyang, Q., Zhu, Y., 2013. Influence of dynamic environment with different airflows on human performance. Build. Environ. 62, 124–132.

De Dear, R.J., Brager, G.S., Reardon, J., Nicol, F., 1998. Developing an adaptive model of thermal comfort and preference/discussion. ASHRAE Trans. 104, 145.

EPEE, 2016. Strengthening Indoor Environment Quality (IEQ) in the Revised Energy Performance of Buildings Directive, Position of Industry and Professional Associations, European Partnership for Energy and the Environment.

Eurostat, 2018. General and regional statistics 2018. European Union ISBN 978-92-79-49273-0, Eurostat Statistical Books <http://ec.europa.eu/eurostat/documents/3217494/7018888/KS-HA-15-001-EN-N.pdf>.

Fanger, P.O., 1970. Thermal comfort. Analysis and applications in environmental engineering. McGraw-Hill Book Company, New York.

Frontczak, M., Schiavon, S., Goins, J., Arens, E., Zhang, H., Wargocki, P., 2012. Quantitative relationships between occupant satisfaction and satisfaction aspects of indoor environmental quality and building design. Indoor Air 22 (2), 119–131.

GB/T, 2012. National Standard China GB/T 50785: Evaluation Standard for Indoor Thermal Environment in Civil Buildings. Architecture and Building Press, Beijing, China.

Givoni, B., 1992. Comfort, climate analysis and building design guidelines. Energy Build. 18 (1), 11–23.

Hart, M.A., Sailor, D.J., 2009. Quantifying the influence of land-use and surface characteristics on spatial variability in the urban heat island. Theor. Appl. Climatol. 95 (3–4), 397–406.

Hermelink, A., 2016. Societal impact of unhealthy homes energy efficiency as a key to create healthier homes. ECOFYS, presentation available from <http://www.buildup.eu/sites/default/a/content/presentations_all_breakfast_debate_20042.pdf> (accessed 13.01.18).

Hermelink, A., Ashok, J., 2016. The Relation Between Quality of Dwelling, Socio-economic Data and Well-Being in EU28 and Its Member States—Initial Results, ECOFYS 15.07.16.

Höppe, P., 1997. Aspects of human biometerology in past, present and future. Int. J. Biometeorol. 40 (1), 19–23.

IESNA Lighting Measurements Committee, 2012. IES LM-83-12, IES Approved Method for Measuring Spatial Daylight Autonomy (sDA) and Annual Sunlight Exposure (ASE). Illuminating Engineering Society of North America, New York.

IESNA, 2016. The Lighting Handbook, 10th edition Illuminating Engineering Society of North America, New York.

ISO 7730, 1994. Determination of the PMV and PPD Indices and Specification of the Conditions for Thermal Comfort, Moderate Thermal Environments. International Organization for Standards, Geneva, Switzerland.

ISO, E. 717-1, 2013. Acoustics—Rating of Sound Insulation in Buildings and of Building Elements—Part, 1, 31-032.

IWBI, 2016. WELL: The WELL Building Standard v1. Delos Living LCC, New York.

Jantunen, M., Oliveira Fernandes, E., Carrer, P., Kephalopoulos, S., 2011. Promoting actions for healthy indoor air (IAIAQ). European Commission Directorate General for Health and Consumers, Luxembourg.

Judd, K.S., Zalensny, M., Sanquist, T., & Fernandez, N., 2013. The Role of Occupant Behavior in Achieving Net Zero Energy: A Demonstration Project at Fort Carson. Pacific Northwest National Laboratory Report, PNNL, 22824.

Klingenberg, K., Kernagis, M., Knezovich, M., 2016. Zero energy and carbon buildings based on climate-specific passive building standards for North America. J. Build. Phys. 39 (6), 503–521.

Lan, L., Wargocki, P., Wyon, D.P., Lian, Z., 2011. Effects of thermal discomfort in an office on perceived air quality, SBS symptoms, physiological responses, and human performance. Indoor Air 21 (5), 376–390.

Leung, D.Y., 2015. Outdoor-indoor air pollution in urban environment: challenges and opportunity. Front. Environ. Sci. 2, 69.

Lenoir, A., Baird, G., Garde, F., 2012. Post-occupancy evaluation and experimental feedback of a net zero-energy building in a tropical climate. Archit. Sci. Rev. 55 (3), 156–168.

Li, X.X., Liu, C.H., Leung, D.Y.C., 2009. Numerical investigation of pollutant transport characteristics inside deep urban street canyons. Atmos. Environ. 43, 2410–2418. Available from: https://doi.org/10.1016/j.atmosenv.2009.02.022.

Li, X.X., Britter, R.F., Koh, T.Y., Norford, L.K., Liu, C.H., Entekhabi, D., et al., 2010. Large-eddy simulation of flow and pollutant transport in urban street canyons with ground heating. Bound. Layer Meteorol. 137, 187–204. Available from: https://doi.org/10.1007/s10546-010-9534-8.

Oldenkamp, R., van Zelm, R., Huijbregts, M.A., 2016. Valuing the human health damage caused by the fraud of Volkswagen. Environ. Pollut. 212, 121–127.

Olgyay, V., 2015. Design With Climate: Bioclimatic Approach to Architectural Regionalism. Princeton University Press, Princeton.

Peacock, A.D., Jenkins, D.P., Kane, D., 2010. Investigating the potential of overheating in UK dwellings as a consequence of extant climate change. Energy Policy 38 (7), 3277–3288.

Pope III, C.A., Ezzati, M., Dockery, D.W., 2009. Fine-particulate air pollution and life expectancy in the United States. N. Engl. J. Med. 2009 (360), 376–386.

REHVA, 2012. Guidebook No.11, Air Filtration in HVAC Systems. Federation of European Heating and Air-Conditioning Associations, Brussels.

Roshan, Gh.R., Ghanghermeh, A., Attia, S., 2016. Determining new threshold temperatures for cooling and heating degree day index of different climatic zones of Iran. Renew. Energy 101, 156–167, ISSN 0960-1481, 10.1016/j.renene.2016.08.053.

Sailor, D.J., 2014. Risks of summertime extreme thermal conditions in buildings as a result of climate change and exacerbation of urban heat islands. Build. Environ. 78, 81–88.

Santamouris, M., 2013. Environmental Design of Urban Buildings: An Integrated Approach. Routledge, London.

Scott, D., Amelung, B., Becken, S., Ceron, J.P., Dubois, G., Gössling, S., et al., 2008. Climate Change and Tourism: Responding to Global Challenges. World Tourism Organization, Madrid, p. 230.

Seppänen, O., Kurnitski, J., 2013. Target values for indoor environment in energy-efficient design. In Cost Optimal and Nearly Zero-Energy Buildings (nZEB). Springer, London, pp. 57–78.

Shepherd, D., Welch, D., Dirks, K.N., McBride, D., 2013. Do quiet areas afford greater health-related quality of life than noisy areas? Int. J. Environ. Res. Public Health 10 (4), 1284–1303.

Synnefa, A., Santamouris, M., Akbari, H., 2007. Estimating the effect of using cool coatings on energy loads and thermal comfort in residential buildings in various climatic conditions. Energy Build. 39 (11), 1167–1174.

Tofield, A., 2017. Air pollution and traffic noise effect on blood pressure. Eur. Heart J. 38 (2), 71.

US DOL, 2016. Absences From Work of Employed Full-Time Wage and Salary workers by Occupation and Industry. Retrieved from <https://www.bls.gov/cps/cpsaat47.htm> (accessed 16.03.17.).

USGBC, 2013. LEED v4: Reference Guide for Building Design and Construction (v4). US Green Building Council, Washington, D.C.

Wilkins, A., Veitch, J., Lehman, B., September 2010. LED lighting flicker and potential health concerns: IEEE standard PAR1789 update. In: 2010 IEEE Energy Conversion Congress and Exposition (ECCE). IEEE, New York, pp. 171–178.

World Green Building Council (WGBC), (2014), Health, Wellbeing and Productivity in Offices: The Next Chapter for Green Building. Retrieved from <http://www.worldgbc.org/news-media/health-wellbeing-and-productivity-offices-next-chapter-green-building> (accessed 21.03.17.).

World Health Organization, 2016. Air Pollution Levels Rising in Many of the World's Poorest Cities.

World Health Organization, 2014. Burden of Disease From Household Air Pollution for 2012. World Health Organization, Geneva.

WMO, 1991. International Meteorological Vocabulary, W.M.O, No. 182, TP 91, P.116.

Yanase, T., 1998. A study on the physiological and psychological comfort of residential conditions. J. Home Econ. Jpn. 49 (9), 975–984.

Zhu, Y., Ouyang, Q., Cao, B., Zhou, X., Yu, J., 2016. Dynamic thermal environment and thermal comfort. Indoor Air 26 (1), 125–137.

Further Reading

ANSI/ASHRAE, 2009. Indoor Air Quality Guide. Best Practices for Design, Construction, and Commissioning. American Society of Heating, Refrigerating and Air-Conditioning Engineers, Atlanta, GA.

Brandemuehl, M., Field, K.M., November 2011. Effects of variations of occupant behavior on residential building net zero energy performance. In: 12th Conference of International Building Performance Simulation Association, Sydney, pp. 14–16.

CIBSE, E.D., 1999. CIBSE Guide A. The Chartered Institution of Building Services Engineers, London.

GB/T5701, 2008. Thermal Environmental Conditions for Human Occupancy, National Standard, Standardization Administration of the People's Republic of China, SAC, General Administration of Quality Supervision, Inspection and Quarantine of the People's Republic of China. AQSIQ, Beijing, China (in Chinese).

CHAPTER

6

Materials and Environmental Impact Assessment

ABBREVIATIONS

ADP abiotic depletion potential
AEC Architectural, Engineering, and Construction
BAMB building materials banks
C2C cradle to cradle
EIA environmental impact assessment
EPC Energy Performance Certificate
EPD environmental product declaration
EPBD Energy Performance of Buildings Directive
EPS extruded polystyrene
EU European Union
FSC Forest Stewardship Council
GHG greenhouse gas
GWP global warming potential
HPD Health Product Declaration
HVAC Heating, Ventilation, and Air Conditioning
ISO International Standardization Organization
LC life cycle
LCA life cycle assessment
NRE nonrenewable energy
nZEB nearly Zero Energy Buildings
NZEB Net Zero Energy Buildings
PCR Product Category Rules
PEFC Program for the Endorsement of Forest Certification
PH Passive House
PIR polyisocyanurate
PPM particle per million
VOC Volatile Organic Compound
XPS expanded polyester

Net Zero Energy Buildings (NZEB)
DOI: https://doi.org/10.1016/B978-0-12-812461-1.00006-X

1 INTRODUCTION

Worldwide there is an intensified competition for raw materials (Wiebe et al., 2012). At the same time, there is a growing demand for ecological and sustainable buildings materials. The demand includes responsibly harvested wood, recycled content materials, bio-based materials, and building solutions designed for disassembly. In Northern countries, the most common approach for Net Zero Energy Buildings (NZEB) is to comply with the locally modified version of the German Passive House (PH) Standard that requires a very low conductivity of exterior walls. The conventional PH constructions are dominated by building materials with high environmental impact including concrete blocks, firebrick and petrochemical insulation materials that produce a great amount of greenhouse gases (GHGs). Moreover, there are very few studies that assessed the holistic environmental impact of construction systems and material compositions against ecological and sustainable NZEB. Therefore, we should not only seek to lower the operation energy through better energy insulation and by generation of energy on-site, but we should also lower the embodied energy consumed upfront in making NZEB. The environmental impact of the used building materials arises during various stages of building life cycle (LC). Most buildings go through three main phases along their lifespan, construction, operation, and demolition. This means that we should not only think about energy efficiency, but we should also consider overall resource efficiency.

This chapter is dedicated to stakeholders who should be concerned and responsible concerning materials selection based on their multiple attributes and purchasing including construction estimators, owners, manufacturers, contractors, specification writers, interior designers, construction purchasing agents, architects, and project managers. In this chapter, we aim to understand the environmental implications of building materials and their complex and multi-criteria attributes. This includes products such as concrete, masonry, wood, plastics, composites, thermal and moisture protection, openings, finishes, equipment, furnishing, plumbing, and electrical wiring. Also, we explore the policies, recent trends, and best practices of building structural systems and ecological materials. Finally, the chapter provides a critical discussion on the environmental consequences of building materials used for NZEB.

2 BUILDING MATERIALS ENVIRONMENTAL IMPACT

NZEB are realized by lowering the energy demand through passive design and energy conservation measures, and by generating energy that meets this demand on-site. Both strategies have implications on the building construction system, technology and materials selection for the envelope, systems, and finishes. Worldwide, 50% if all extracted materials are delivered to serve the built environment (CESBA, 2014). At the same time, the consumption of material resources is increasing significantly to meet humans' living standards and the development of modern societies. There is sufficient empirical evidence that the use of building materials has a negative environmental impact associated with extraction, processing, transport, maintenance, and disposal. In this section, we will briefly present the environmental and health impact of building materials in relation to the linear exploitation and disposal process.

2.1 Environmental Impact

Around 50% of all extracted materials are used by the building industry. Building construction and operation contribute greatly to resource depletion and consumption. The increased level of building material consumption worldwide leads to land use, land deforestation, soil erosion and degradation, and more mined minerals. The materials used for building construction represent a significant share of extracted materials and include steel, copper, and aluminum. For example, more than 30%–50% of total material use in Europe goes to housing, and mainly consists of iron, aluminum, copper, clay sand, gravel, limestone, wood, and stone (EEA, 2010). Minerals have the highest share of all building materials, around 65% of total aggregates (sand, gravel, and crushed rock), and approximately 20% of total metals are used by the construction sector. The depletion or degradation of the earth's natural resources is exhausting the earth's ability to sustainably provide the resources to meet the consumption level.

The depletion and degradation of natural capital includes fresh water soil, forest land, rivers shores, wetlands, and biodiversity. For example, sand and gravel account for approximately 75% of the 60 billion tons of material we pull out of the ground every year (Peduzzi, 2014). Sand and gravel are being consumed at a rapid and growing rate, though we hardly ever hear alarm bells sounding over a global sand shortage (Floyd, 2017). As sand from land quarries and rivers runs out, the industry is moving out to sea. Desert sand, due to its shape after heavy weathering, is not suitable for most applications, which is why Saudi Arabia buys sand from Australia, and Dubai imports sand from Belgium. Another example is timber and its consumption in the construction industry and its use as a heat-energy source. Forests cover about 31% of the Earth's land and accommodate as much as 80% of all fauna and flora of the world. Deforestation or degradation of forests is happening fast. On average, about 17 football fields disappear naturally from forests every minute. This does not only endanger the animals that live in the forests, but also humans and even the entire planet. Worldwide, there is a loss of forest areas for the construction of agricultural land, but also for wood production. Even the labels such as the Forest Stewardship Council (FSC) or the Program for the Endorsement of Forest Certification (PEFC), which aim to protect and manage our forests, are not effective. The FSC certification is almost 13% and the PEFC is almost 9% of the world's total forest area we are logging (PEFC, 2017; FSC, 2017). In fact, demand for timber resources is on the rise for fuel and construction and consequently deforestation is widespread.

Another significant environmental impact of building construction materials' is the carbon emissions associated with the use of nonrenewable energy (NRE) resources during the extraction, transportation, use, and disposal (see Fig. 6.1). In the past 30 years, the amount of energy necessary to process raw materials and modify materials for buildings significantly increased due to the use of insulation materials and installations (Dixit et al., 2010). Unfortunately, the energy use and carbon emissions are seldom evaluated. According to the resource efficiency report published by ECORYS in 2014, the embodied energy in building products was around 1.9 million TJ in 2011. Steel and aluminum together are responsible for approximately 20% of the total embodied energy in building materials.

Water is another significant resource that is used in association with construction. The concept of quantifying the amount of water used throughout the supply chain of building materials was introduced a

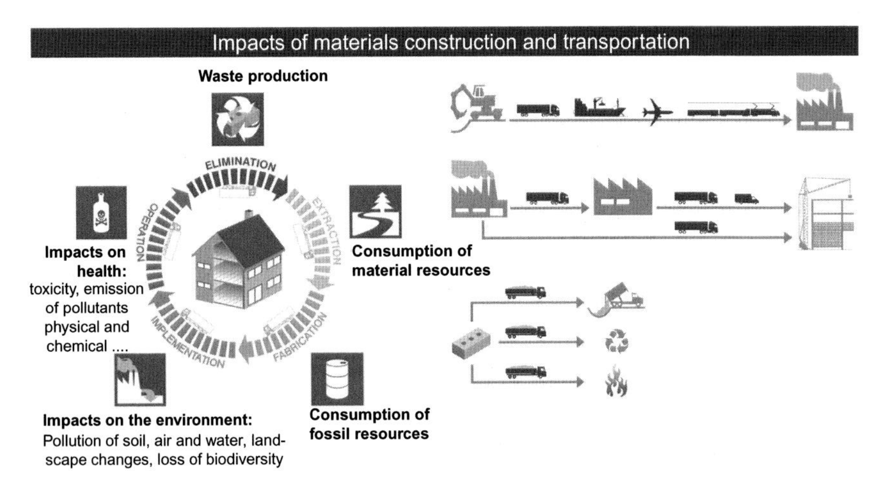

FIGURE 6.1 Impact of material construction and transportation (Architecture et climat).

decade ago. Water use can be classified as Green Water (rainwater evaporated or incorporated into a building material), Blue Water (surface water or groundwater incorporated into a building), or Gray Water (polluted water). The classification and quantification of water associated with building materials became essential to assess quarried products such as sand, gravel, marble, and granite that use huge amounts of water during the extraction phase. Also, products that use huge amounts of water during the production phase such as wood, ceramics, tiles, stones, or other flooring materials need to be assessed. There are some recent methodologies and metrics that assess the water footprint of building materials. The key message for building suppliers and designers is that they need to be aware of the water foot print or impact of building materials on water use.

Construction and demolition waste contribute about 25% of the total waste stream in the European Union and about 40% of the total solid waste stream in the United States (EU, 2011a). In general, the building construction industry follows a linear pattern of consumption of resources resulting in landfill disposal without qualitative and quantitative recycling or reuse (Leduc et al., 2009). Manufacturing, implementation, and disposal inevitably lead to demolition waste. Construction and demolition waste is one the heaviest and most voluminous waste streams that have a high potential of recycling and reuse. However, separation technology and recovery of construction and building material waste was not always practiced. As a consequence, back filling or incineration of construction and demolition waste, which includes hazardous wastes, pose particular risks to soil and underground water while hampering the reuse and recycling of those materials.

2.2 Health Impact

The extraction and processing of building materials is associated with toxicity. Pollution includes the toxicity of substances that impact freshwater, terrestrial ecosystems, and humans, covering a number of emissions such as NOx, HCB, Pb, Hg, Ni, Cu, Cd, Zn, Sox, NH3, Cr and PM. According to the study of ECORYS (2014), the production of materials used for building in the EU-27 is responsible for two percent of the global emissions with toxicity potential. There is a high environmental burden associated with building material extraction and processing associated with the most impacting materials namely copper, steel, aluminum, and concrete. The construction industry is responsible for around four percent of particulate emissions, water pollution, and noise pollution. The clearing of land causes soil erosion, silt runoff, and sediment pollution. Also, sites include oil, paints, solvents, and other harmful chemicals that run into natural waterways, groundwater or public sewage.

Moreover, construction sites produce noise that can disturb cycles of animals and eliminate the habitat for several species. In other words, construction can cause pollution risks and, thus, requires serious measures to prevent pollution. The respiratory system is the primary target of air pollutants and airborne dust that are consumed through the air we breathe. The effects can range from temporary disorders of respiratory dysfunction and permanent or chronic diseases.

Another health concern related to materials is the presence of hazardous chemicals that influences building occupants and builders through exposure. Numerous studies have demonstrated the presence hazardous chemicals in occupants' tissue (MBDC, 2012). Chemical composition data for building materials down to the 100 ppm level shows not only the environmental hazards and risks associated with building materials, but also almost 24 human and health hazards. The presence of hazardous substances in building materials and the Volatile Organic Compounds (VOCs) that spread indoors through off-gassing is not sufficiently investigated worldwide. However, the selection of materials such as paints, floorings, adhesives, sprays, and furniture must be made to avoid hazardous health effects especially for surfaces in the indoor environment of NZEB. Material finishing is more complex and involves many components and activities. The pollutants can have a negative effect on the quality of indoor air, well-being, and health (allergies, irritation of the eyes and throat, headaches etc.) In addition, these emissions can be amplified by climate conditions including sunshine temperature increase, humidity etc.

2.3 Materials Life Cycle and Circularity

Resources such as land, water, fossil energy, materials, and air are in decline because they are being used, exhausted, or damaged faster than nature can regenerate them. At the same time, the demand for these resources is growing. The previously mentioned negative impacts associated with building materials are a result of a linear process of using materials. This process is considered linear because it starts with building materials extraction, processing of materials, and the use of materials until they end up in landfills. The linear approach of building materials uses leads to the use, depletion and down-cycling of materials. This will definitely lead to a deficit of resources and an increase in pollution and health problems. Therefore, a recent design paradigm named "circularity" aims to adapt the linear process of the current use of material resources by mitigating the negative impact into a circular one that is considered healthy and is operated within a closed loop (Attia, 2016, 2018; van Dijk et al., 2014).

3 MATERIALS ENVIRONMENTAL IMPACT ASSESSMENT APPROACHES

Definitions, concepts, and approaches of materials environmental impact assessment (EIA) vary between industry, academia, and standardization organizations. In this section, we explain the main terms and definitions related to EIA and the approaches to measure the impact of construction materials and their implications for NZEB.

3.1 Definitions and Concepts

Resource efficiency: Resource efficiency promotes creating buildings with less, delivering greater value with less input, using resources in a sustainable way, and reducing their impact on the environment. The environmental impact of building materials should be described based on life cycle analysis information, complex modeling for the climate change impact of GHG and using CO_2 equivalent emissions indicators.

Building life cycle: The linear life of the building and its material components from extraction, processing, transportation, mounting of building material, use, demolition and waste treatment.

Cradle to gate: Cradle to gate is an assessment of the partial impact of a material or product LC. The assessment starts from extraction and manufacture (cradle) to the factory gate. The cradle to gate is the basis for environmental product declaration.

Cradle to grave: Cradle to grave is a concept that is an assessment of the entire life of a building material or product to the point of disposal.

Cradle to cradle (*C2C*): Cradle to cradle is a concept that advocates the elimination of waste by recycling a material or product into a new or similar product at the end of its intended use, rather than disposing of it.

Life cycle analysis or assessment (*LCA*): LCA is based on the principles of sustainable development by providing the means to assess the environmental impacts (including health) of a product, service, or process. The LCA aims to minimize the impact on resources and the environment throughout the LC. The LCA is normalized with standard ISO 14040 for environmental management.

Building carbon footprint: The building carbon footprint derived from the carbon dioxide emission resulting from the building materials' manufacture and carbon dioxide storage.

Global warming potential (*GWP*): A metric used to relate a compound to the CO_2 equivalents to measure the potential heating effects on the atmosphere. The greenhouse gas effect is presented in kgCO_2 equivalent per kilogram of produced material.

Abiotic depletion potential (*ADP*): ADP describes the decrease of availability of total reserve functions of a resource. The quantities of raw materials used are given in kg antimony equivalents per kg produced material.

Embodied energy: Embodied energy is the energy consumed by all of the processes associated with the production of a building, from the mining and processing of natural resources, to manufacturing, transport, and product delivery. Embodied energy does not include the operation and disposal of the building materials, which would be considered in a life cycle approach. Embodied energy is the "upstream" or "front-end" component of the life cycle impact of homes (Dixit et al., 2010, 2012; Verbeeck and Hens, 2010; Thormark, 2002).

Embodied carbon emissions: Carbon emissions are the side-effects from fossil energy use of building materials. Embodied carbon emission is the carbon dioxide produced in the making of a building upfront. However, it is only a portion of the carbon emissions produced during the full LC of a building.

Nonrenewable energy (*NRE*): NRE come from sources that will run out or will not be renewed. Most nonrenewable energy sources are fossil fuels: coal, petroleum, and natural gases.

Material efficiency index: Material efficiency can be specified by using the carbon emission value represented in kilograms of CO_2-emissions per floor area. This indicator and metric corresponds to the carbon dioxide emissions released during the extraction, processing, transportation, and mounting of building materials.

Environmental product declaration (*EPD*): EPDs are statements prepared under the responsibility of manufacturers and third-party assessors according to ISO 14025. The purpose of EPD is to compare products with the same functional unit. They are based on fact sheets verified by an independent third party, in which the manufacturer provides, based on a LCA, quantitative data on the impact of the product on the environment for a cradle to gate or cradle to grave process.

Recyclable: Building materials that can technically be recycled at least once after its initial use phase.

Recycling potential: The recycling potential is the ability to upcycle or downcycle a material. For example, the upcycling of crushed concrete can happen by reusing it as an aggregate in the production of new concrete. On the other hand, down-cycling of crushed concrete can happen by using the crushed concrete in the foundations of asphalt roads.

Biodegradable: The process by which a building material is broken down or decomposed by microorganisms and reduced to organic or inorganic molecules which can be further utilized by living systems.

Readily disassembled: Building materials or components that are capable of being deconstructed with the use of common hand tools.

C2C Certification: A scientifically based process and system of certification that established specific criteria to assess the environmental attributes of inputs and outputs used in manufactured goods.

Life Cycle NZEB: LC NZEB is a definition of NZEB that accounts for the embodied energy of building components together with energy use in operation (Attia, 2010; Attia and De Herde, 2011). According to Hernandez and Kenny (2010), a LC-ZEB is defined here as a building whose primary energy use in operation plus the energy embedded in materials and systems over the life of the building is equal or less than the energy produced by renewable energy systems within the building.

3.2 Approaches

Measuring the impact of construction materials is complicated and can be based on different approaches. For example, 1 m^3 of concrete (equivalent to 2400 kg) requires 200 L of water, 100 kg of reinforced steel, 1800 kg of aggregates, and 350 kg of cement. To assess the impact of this block we will need to break it down and trace each component and define the type of resources included in this cubic meter of concrete. Each component should be traced and classified according to the list below:

- Materials from recycling
- Natural materials
- Synthetic materials
- Renewable materials
- Nonrenewable materials
- Rare or limited materials
- Abundant materials

Then, an EIA needs to take place investigating the consumption of fossil resources associated with the transport, the different stages of industrial transformation, the implementation, and at the end of life. This includes the embodied energy consumption at each processing stage and the total energy contained in the product (Moncaster and

Symons, 2013). Embodied energy is calculated as primary energy and presented as MJ/kg of produced material:

$$\text{Embodied Primary Energy} = \sum_{n=1}^{n} \frac{\text{Primary Energy per product}}{\text{mass of product}} \tag{6.1}$$

Then the embodied carbon emission can be calculated as follows:

$$\begin{aligned}\text{Embodied Carbon Emissions} = \sum_{n=1}^{n} & \text{ECE building element} \\ & + \text{ECE transp} + \text{ECE Construction} \\ & + \text{ECE Mainenance} + \text{ECE endlife} \\ & - \text{ECE reover}\end{aligned} \tag{6.2}$$

Therefore, in order to assess the impact of building materials and products, we should understand that most of the time materials are comprised from different resources. Depending on the type of resources of construction materials we can figure out the most suitable approach to evaluate their impact. In general, the characterization of building materials is based on the ADP factor, which quantities raw materials used in kg antimony equivalents per kg produced material. However, the most expected measure of the environmental impact of building materials is LCA. Most of the time the LCA results are presented as follows:

- Conventional air pollution or GHG emissions namely, CO, CO_2, NO_2, SO_2, Particulate, and VOC.
- Global warming potential (GWP) or the greenhouse gas effect is presented in kgCO_2 equivalent/kg produced material.
- Acidification of Air, or the acidification potential (AP), is presented in kgSO_2 equivalent/kg produced material.
- Tropospheric ozone formation, or the potential photochemical ozone creation (POCP), is presented in kgC_2H_2 equivalent/kg produced material.

To come back to our example of one cubic meter of concrete block (2400 kg/m^3), its environmental impact for cradle to gate LCA is:

- GWP: 0.12 kgCO_2 eq./kg
- AP: 0.00028 kgSO_2 eq./kg
- POCP: 0.00001 kgC_2H_2 eq./kg

However, to complete the LCA we should take into account the effect of the production of the concrete block on landscape and biodiversity, and the health impacts indoors and finally the waste associated with the production and disposal of the cubic concrete block.

Today, there are several approaches to address material resource efficiency for NZEB. Building and material reuse is the most effective strategy. Building material reuse avoids the environmental risks and burdens of the manufacturing process. The reuse of materials can happen off-site or on-site, or as part of a material purchase strategy. Another effective approach is recycling. Recycling is effective in diverting waste from landfill and saves land, soil, and groundwater pollution as well as waste transportation. However, in building any project the owner can make a choice between three options:

1. *Build a new construction*: Purchase virgin land and build a new NZEB.
2. *Reconstruct*: Purchase an existing building to demolish and reconstruct a new NZEB
3. *Renovate*: Purchase an existing building to execute a renovation with major energy efficiency improvements and generation capacity to reach NZEB performance.

In this context, the ownership duration (or period) and the building function plays a major role to make a decision. For the renovation option (3), the owner and design team must be sure that they can reach zero energy performance targets when the building gets upgraded. In many cases, the renovation decision will not be sufficient to achieve the ambitions of a NZEB (Dubois and Allacker, 2015). Therefore, we would like to identify four key factors that influence the resources and material efficiency approach for a NZEB:

1. *Linear or circular life cycle assessment?*
   The AEC is on the verge of a market transformation of building products. LCA is a key assessment methodology of building materials' environmental impact that is relatively recent. Its application depends on many factors and can function during the partial or full LC of a product. The classical LCA follows a linear approach that is based on a linear cradle to grave examination. The circular LCA follows a linear approach that is based on a circular C2C examination. It is therefore, crucial to decide at the beginning of the design process whether the comparison of the environmental impact of potential design solutions regarding construction systems and building materials selection will be based on a cradle to grave approach, or a C2C approach (Bor et al., 2011).
2. *Environmental or health risks, or both?*
   Moreover, the owner and the design team should set up a scope for the environmental assessment boundaries of building materials. LCA of building materials is tedious and can include many variables related to the materials' environmental risks (atmospheric pollution, ecosystems modifications, and waste production), or materials

resource depletion NRE consumption, nonrenewable material consumption) or health risks (allergies, irritation of the eyes and throat, etc.). Therefore, the decision maker should clearly define the scope of the LCA and the criteria that will be addressed. Similarly, qualifying products and exclusions must be done. Examples include structure and enclosure elements, furniture, installed finishes, framing, interior walls, cabinets, doors, windows, and roofs. The inclusion or not of piping, ducts, pipe and ducts insulation, conduit, plumbing fixtures, faucets, showerheads, and lamps must also be made. Defining the LCA scope and assessment criteria next to the products that will be assessed is an essential step in any EIA approach.

3. *Functional unit of analysis*

   The third factor that influences any LCA approach is setting the functional unit for calculation. The functional unit defines precisely the metric that will be used to quantify the impact or service delivered by the building material. Defining the functional unit is important to enable and compare alternative sustainable building materials. We advise to use the method specified in EN 15978 (2011) that shows how many kilograms of CO_2-emissions per floor area ($kgCO_2/m^2$) are released during the production of construction materials. This method became more commonly used in Europe in recent years and allows comparing the embodied carbon emissions of building materials, similar to the energy use intensity indicator.

4. *Duration of analysis*

   The last factor of LCA is to define the calculation period based on the expected occupancy or use of a building. Some LCA studies go for 60 years or 30 years, however the longer the study period the more we can get a sustainable understanding on the impact of material choices. By dividing the period or duration of the LCA for each material annually, the design team can use metrics such as $kgCO_2/m^2$ per annum.

Around these four factors, building assessment systems are crucial tools in driving NZEB toward sustainability. There is a wide range of LCA tools and LCA-based building performance simulation tools that have been developed which calculate the environmental impact of the building as a whole. There are currently more than 50 building assessment systems in Europe (EU, 2013). Examples of EIAs include ECOBAU-KBOB and eLCA in Germany, ecoinvent in Switzerland, Greencalc + in the Netherlands, SimaPro in the United States and Elodie in France. Life cycle assessment (LCA) is a widely known methodology for "cradle to grave" investigation of the environmental impacts of NZEB. However, the LCA methodology has not yet been broadly used among practitioners of the building sector (Asdrubali et al., 2013). It is, therefore, important to use LCA and the abovementioned tools for decision support and assessment on construction systems and building materials in the early design phases.

4 POLICIES AND BEST PRACTICES

In the past years, the market has witnessed an increase of awareness and transformation of building products by creating a cycle of consumer demand and industry delivery of environmentally preferable products (USGBC, 2013).The green movement worldwide has created a demand for sustainable products, and suppliers, designers, and manufacturers are acting. Green building practices have become foundational in building codes. The efforts to reduce global climate change, enhance individual human health, protect and restore water resources, protect and enhance biodiversity, and ecosystem services are stimulating the creation of green NZEB. Promoting sustainable and regenerative material cycles is leading a market transformation. Recently announced policies in North America, Asia, and Europe are seeking to improve the environmental and health performance of building material section. Europe's 2020 program proposed a flagship initiative, "Roadmap to resource efficient Europe," setting milestones and actions for resource efficiency in the building sector by 2020 (EU, 2011b).

4.1 Policies

Sustainable construction legislation, regulation, and drivers have a significant influence on the building design and construction of NZEB. Despite the abundant number of initiatives worldwide relating to sustainable construction and material selection, there are very few countries that have applied mandatory requirements to drive the AEC industry. The EPBD requires all new buildings to be nearly zero energy by 2020 in EU-27. However, it is mainly focused on operational energy and carbon without any ambition or legally binding target regarding material use. Even the Energy Performance Certificates (EPCs) are mainly focused on operational energy. In fact, reporting and quantifying the sustainability of materials is a work in progress worldwide. We still cannot find LCA, chemical hazard assessment, and supply chain transparency as part of the common practice for manufacturers in the construction materials sector. In Europe, the European Committee for Standardization has published EN 15804, a common Product Category Rules (PCR) for EPD development in the construction sector. Other complementary standards, for example for environmental building assessment (EN 15978) were also published by this technical committee.

In this context, we believe that greater attention is being paid to sustainability attributes of building materials used to construct NZEB. NZEB require high volumes of insulation, heavy triple gazing window frames, complex wall compositions, and significant quantities of adherents, seams, fasteners, and tape. Those materials are already used on a large scale and several measures (technical, environmental and health) of the used materials within the concept of NZEB must be investigated. We provide four major measures that strongly influence the regulatory landscape of current construction materials and those used in the near future:

Firstly, policy is encouraging the use of LCA beyond single attributes. The EPD became a critical and unique label that helps forensic examination of the environmental impact of energy components that make a building product. The EPD is based on a LCA following internationally agreed standards of ISO. After the publication of EN 15804, a large number of EPDs is available on the market. An EU framework published in 2015 requires EPDs to be phased in overtime before being mandatory by 2020. Fig. 6.2 shows an example of an EPD with six LCA impact measures. The EPD helps design teams in better material and product selection and rewards manufacturers to reduce the environmental impact of their products.

Secondly, policy is encouraging the disclosure and transparency of building components and materials. As part of the European Commission road map to a resource-efficient Europe, several standards and protocols are encouraging transparency and reporting for building materials. For example, an EU advisory panel recommends that building materials manufactured and sold in Europe should have a product

EPD label for a building product

Amount per unit

LCA impact measures	Total
Primary energy (MJ)	12.4
Global warming potential (kgCO_2 eq.)	0.96
Ozone depletion (kgCFC 11 eq.)	1.80E–08
Acidification potential (mol H^+ eq.)	0.93
Eutrophication potential (kgN eq.)	6.43E–04
Photo-oxidant creation potential (kg O_3 eq.)	0.121

FIGURE 6.2 Example of a third party certified product and the EPD.

passport—a declaration on the materials that are used and their potential for reuse at the end of the product life. In 2015, the EU funded the building materials banks (BAMB) project to transform the construction industry, where buildings are designed as material banks for the future (EU, 2017).

Thirdly, policy is encouraging the use of higher standards to assess building materials by third party programs and certificates. For example, EN 15978 is a standard that allows calculating material volumes including load-bearing structures and building envelope following ISO 14025. Another example of a fast-growing certificate is the C2C certificate. From a health point of view, the use of petrochemical materials, such as extruded polystyrene (EPS), expanded polyester (XPS) and polyisocyanurate (PIR) or even bio-based materials such as hemp or flax wool mixed with polyester support fibers is raising many concerns. The use of formaldehyde-based binders used in mineral wool or wood panels and the use of fire retardants such hexabromocyclododecane used in EPS can cause serious health risks. With the high airtightness of NZEB and with the raised awareness of indoor air quality, the developments for the use of third party certificates, such as REACH, Green Screen, Health Product Declaration (HPD), are already in progress and in an advanced phase.

Fourthly, policy is encouraging setting threshold limits for the environmental impact of building materials. France is one of the leading countries to set a new regulation seeking low carbon in nZEB and NZEB. For example, the proposal for the new thermal regulation in France (Règlementation Thermique 2018) envisages low-embodied carbon for free standing houses up to 800 $kgCO_2/m^2$, apartments up to 950 $kgCO_2/m^2$, and office buildings up to 1300 $kgCO_2/m^2$. If this regulation is passed, we will have a carbon and energy consumption limit for NZEB in Europe for the first time. Even though no concrete targets (on European level) are set yet, legal requirements on sustainable material use and on the environmental impact thresholds of new NZEB is expected in the near future.

4.2 Best Practices

The selection of low-impact materials is a complex and multifaceted task. Material purchasing agents, architects, and contractors should seek reducing environmental effects during the initial project decision-making regarding the construction structure, technology, and building materials. As sustainability goals become more prominent, best practices can guide the design team, ensure compliance, cost effectiveness, and decrease liability (Kibert, 2016). To guide project teams in selecting a construction structural system, envelope, products, and materials we list eight criteria that need to be addressed when designing a low-impact NZEB. The list is extracted from several case studies and aims to guide project teams (Attia, 2011, 2016).

1. Design for flexibility and disassembly allowing for extended producer responsibility.
2. Select the structural construction system with low-embodied energy and environmental impact.
3. Select the envelope construction system and cladding with low-embodied energy, environmental impact, and health risks.
4. Specify and select compliant and regional materials and products.
5. Reduce the amount to raw materials and opt for reused materials.
6. Use renewable and bio-based materials and product.
7. Use responsibly sourced and extracted materials.
8. Track purchasing throughout construction.

Table 6.1 lists the key decisions that need to be addressed when designing NZEB. Together with Table 6.2, design teams can rely on either tables or other materials with EPDs. Until this moment, the calculation of carbon emissions associated with NZEB is calculated in relation to the squared area of buildings. By observing Table 6.2, we can realize that bio-based materials have a high chance to lead the construction industry toward an energy and low carbon transition. Bio-based materials are identified as materials with harvesting maturity within 10 years. They require less land and examples include bamboo, hemp, wood fibers, and cork. Timber will be a key resource in the future due to the ability of forests to get restored or grow the wood and manage the harvesting process in a sustainable way. In fact, multi-story timber buildings can store carbon in their construction systems, which can turn NZEB into carbon sinks. However, bio-based materials are still under testing until they can prove their robust resistance to moisture, fire, and thermites.

In parallel to the abovementioned recommendations, best practice recommends complying with rating systems and environmental labels. Rating systems such as LEED and BREEAM—voluntary building rating systems that succeeded to implement more holistic sustainability reporting and apply LCA, chemical hazard assessment, and supply chain transparency for sustainable NZEB. Also, several labels on material sustainability exist worldwide. Environmental labels are considered as helpful indicators to guide the decision makers. There are three different types of labels including official labels, private labels, and privately controlled labels. Environmental labels should respect the ISO 14024 requirements and follow transparent evaluation criteria and methodology. They must also be controlled by independent and accredited authorities. Examples of widely recognized labels include PEFC, FSC, C2C, NaturePlus, and the EU Ecolabel. In general, rating systems mainly promote the use of sustainability and Eco labels. For example, C2C certificates rate the products that strive

TABLE 6.1 Key Decisions Related to Material Selection for NZEB

Design stage	Decisions	Expertise	Example
Predesign	Massing	LCA expert	Select structural system based on its flexibility and environmental impact
	Structural system	Structural engineer	
	Construction	Architect	
	Technology		
Schematic	Envelope system	Mechanical engineer	Comparison of environmental cost and benefits of different envelope technologies.
	Insulation materials	Structural engineer	
	Façade disassembly	LCA expert	Comparison of different insulation materials
	Cladding	Building Physics Expert	
Design Development	Interior finishes	Specifications consultant	Choose low-impact interior finishes
	Furniture		Investigate the available labeled and certified materials
	Material specification	Purchasing agent	
	Products with EPD and other labels		Perform an LCA
Construction	Purchase tracking	General contractor	Report and track purchases throughout construction
		Façade contractor	
		Purchasing agent	Track substitutions and change orders

to be environmentally positive. There are three grades of certification namely silver, gold, and platinum. The certification includes social and ethical criteria, is managed by McDonough Braungart Design Chemistry (MBDC), and is not subject to external control. The C2C certificate is based

TABLE 6.2 Specific Emissions of Different Building Materials Based on ENC 15804 (2012)

Material	Mass per volume (kg/m^3)	GHG emissions (gCO_2 eq./kg)	Carbon storage (gCO_2 eq./kg)
Steel reinforcement	7850	440	
K35 ready-mixed concrete	2400	140	
K90 ready-mixed concrete	2400	200	
K60 multicore concrete slab	2400	170	
Steel pipe	7950	1090	
Aluminum profile (85 percent)	2700	3640	
Fired Clay Brick	1300	220	
Lightweight concrete Block	700	330	
Mortar	880	270	
Lime plaster	1800	130	
Concrete roof tile	44	140	
Glued laminated timber	440	330	1600
Sawn Lumber	480	70	1600
Glued laminated beam	500	230	1600
Hardboard	950	130	1600
Plywood Conifer	450	680	1600
Plywood Birch	660	720	1600
Gypsum Board	10	390	
Polyethylene (HD)	940	2410	
Polypropylene	905	2100	
PUR, polyurethane	Variant	4230	
Polystyrene (floor board)	15	3600	
Compressed polystyrene	38	3600	

(*Continued*)

TABLE 6.2 (Continued)

Material	Mass per volume (kg/m^3)	GHG emissions (gCO_2 eq./kg)	Carbon storage (gCO_2 eq./kg)
Bitumen—polymer sheeting	–	1120	
Cellulose (paper wool)	30–70	?	
Flax (flax wool)	28	100	1600
Hemp (hemp wool)	40	200	1600
Wood Fiber (wood wool)	55–140	70	1600

on five criteria including water stewardship, renewable energy, material reutilization, materials, as well as health and social responsibility.

Finally, Eco labels and rating systems are not known by most professionals. There is a plethora of different sustainability schemes worldwide and the European Commission issued a mandate to the European Committee for Standardization (CEN) to develop horizontal standardized methods for the assessment of building materials. The importance of European Standards is that EU Member States have to follow best practices as part of the new norm for manufacturers and building owners.

5 CONSTRUCTION SYSTEMS AND MATERIALS

A sustainable construction of NZEB embraces the choice of material or construction technology in a responsible way, taking into account the environmental performance of the latter and its entire life. There are, in general, two aspects of building technology: (1) conventional construction methods and (2) modern or industrialized methods. Conventional construction methods allow smaller types of structures such as domestic dwellings with one or two stories built by traditional methods. Generally it is cheaper to construct this type of building and the owner can get involved in a self-building process. By modern industrial methods we mean those which are mainly composed of factory produced components to a module or standard increment. In both cases, NZEB rely on advanced building technologies to guarantee envelope insulation, airtightness, structural stability, and durable cladding. It is essential that all building professionals involved in the construction to have a good knowledge of construction technology and ecological materials.

Ecological and sustainable building materials include responsibly harvested wood, recycled content materials, bio-based materials and building solutions designed for disassembly. In the next section, we briefly explore different construction technologies and multi-criteria decision-making to select suitable materials for NZEB.

5.1 Building Construction Systems and Materials

The built environment on the world scale is diverse regarding different living standards (influencing the space area per capita), climatic conditions (influencing the cooling and heating demand, and envelope design), and construction technology (influencing the material choice, construction system, and fabrication). The choice of construction systems, construction technique, building components and materials is usually based on a multi-criteria approach and should cover different aspects, including:

- Functionality
- Technical Performance
- Architectural Aesthetics
- Economic Cost
- Sustainability

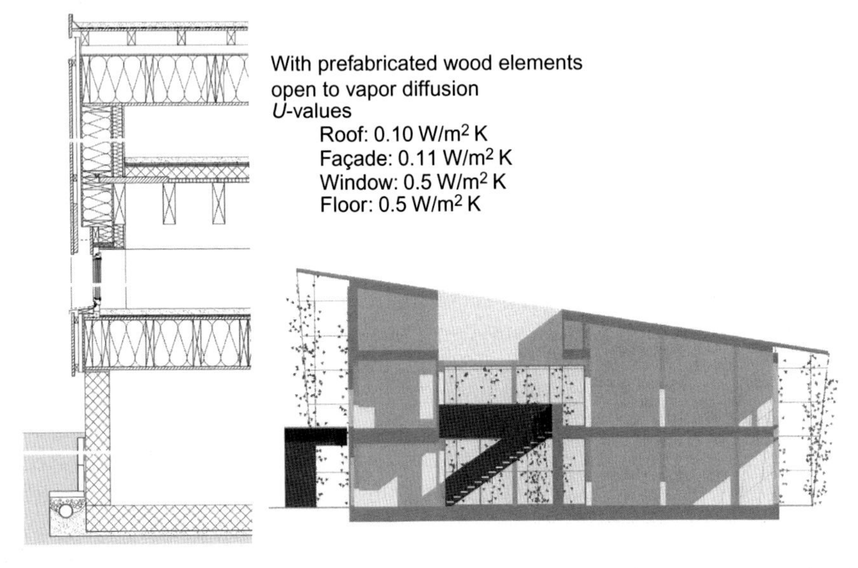

FIGURE 6.3 Example for a timber structural system and envelope construction detail (Green Offices by Conrad Lutz).

- Durability and Maintenance

The large number of materials in the selection parameters often makes it difficult and tedious to choose a given component (see Fig. 6.3). Gauvreau-Lemelin and Attia (2017) proved that very few studies have assessed the holistic environmental impact of NZEB. Therefore, the first question that needs to be answered when designing NZEB is:

- What is the most suitable building construction system for a NZEB?

Worldwide, there are several types of building construction systems shown below. For each of those building construction systems, different combinations of building materials can be used, including brick, concrete, steel, aluminum, copper, plastic, glass, ceramics, and plaster.

- Timber Construction (framed or post and column)
- Load-Bearing Masonry Wall Construction
- Steel Frame Structure
- Precast Concrete Construction
- Earthen Construction

Ideally, the question needs to be answered during the early design stages. During the first design iteration, priority should be given to components with clear ingredient and environmental impact documentation. During early design stages, it is possible to find sufficient materials and components that are readily available, minimally processed, nontoxic, and renewable or recyclable. The focus should be on the structural system in relation to the fabrication and construction on-site or off-site. The optimal sustainable and cost-effective structural system should be selected first. This can be done by eliminating undesirable substances, where energy-intensive and polluting materials, products, and processes are used only if no other better equivalent product is available. Undesirable materials and products typically relate to building construction systems and components which imply a large consumption of fossil-based resources and those which are known to contain toxic substances. The use of products that are coming from very far and transported over thousands of kilometers is undesirable too. A positive list of potential materials can be created by emphasizing renewable materials like timber and avoiding fossil fuel based products, especially insulation materials. The positive list should include solar panels.

After deciding on the most suitable structural and construction system, the envelope construction should be addressed. The choice of a low-carbon envelope system encourages reducing the quantities of concrete, aluminum, and steel as well as the overall embodied energy of the facade. This step involves the selection of the insulation material in combination with the envelope and construction technology. Locally harvested or produced materials, recycled content materials, bio-based materials like earth, hemp, bamboo, and timber are smart choices (Attia et al., 2013). Natural and recycled materials can be used in NZEB construction including curved glulam beams and recycled steel. Concrete mixed with volcanic ash requires lower temperatures to mix it and has an expected longer lifespan. However, we should not underestimate the importance of other performance criteria including fire safety, hygrothermal efficiency, material durability, water proofing, and thermal mass. Between a market-driven logic of architects, contractors, construction purchasing agents, and the sustainability consultant the design team must find a market-driven logic that embraces sustainability and decreases the embodied energy, increases operational efficiencies of the buildings, and reduces carbon emissions to save costs. Finally, the design team should prepare their NZEB solutions to be ready for disassembly. NZEB construction systems and material assemblies are mostly hybrid and therefore they require to be easily disassembled for reuse or recycling. As mentioned in Section 4.1, the European Union aims to transform the construction industry to design NZEB as material banks for the future through the BAMB project.

5.2 Ecological Building Construction and Materials

The use of ecological building materials and products represents one important strategy in the design of NZEB. The challenge in the professional world is to define the characteristics of ecological building construction and not ecological material. Building professionals are faced with strong marketing from different industrial material manufacturers and construction lobbies that make it difficult to select ecological building material. For example, materials such as concrete and timber have the lowest embodied energy intensities, but are consumed in very large quantities, whereas materials with high energy content such as stainless steel and aluminum are used in much smaller amounts. Thus, the greatest amount of embodied energy in a building is often in concrete and steel (see Fig. 6.4).

Another example is the reuse of building materials that saves about 95% of embodied energy. However, the savings from recycling of materials for reprocessing varies considerably, with savings of up to 95% for aluminum, but only 20% for glass. Even aluminum that is recycled up to 95% is the heaviest impacting material when it comes to energy and CO_2. According to Fig. 6.4, the embodied energy for aluminum is around 220 MJ/kg, which is considerably high when compared to timber (around 10 MJ/kg). This means that the recycled content or recycling ability of a material is not a distinctive sustainable characteristic.

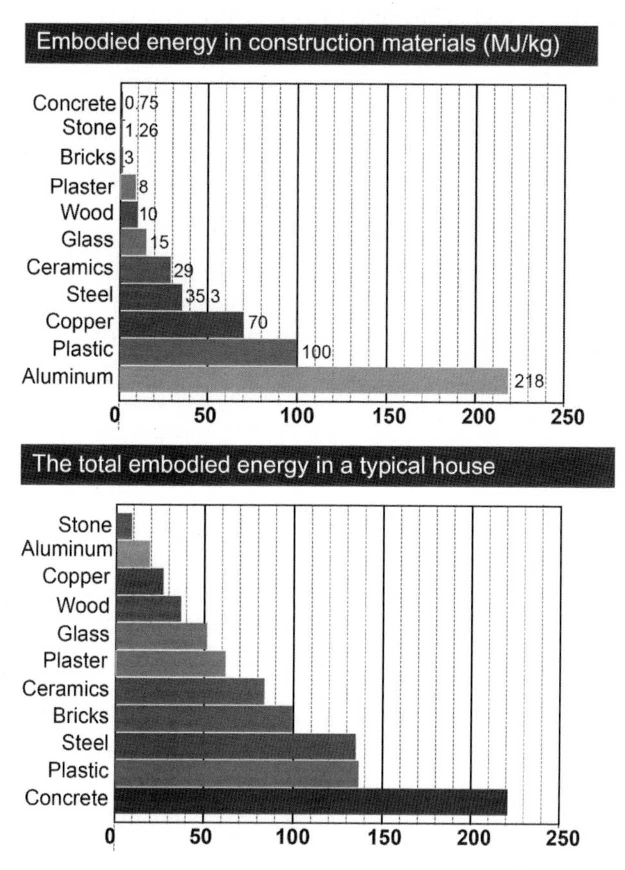

FIGURE 6.4 Estimates the global embodied carbon of construction materials (Hammond and Jones, 2015).

Even the CEN 350 Technical committee, working on the development of horizontal standardized methods for the assessment of the integrated environmental performance of buildings, has advised not to include recycling in the function system boundaries, known as module D. At the same time, we need to remind the reader that if timber is not harvested from a sustainably managed forest, the ecological impact of deforestation can be worse than the impact of timber. Therefore, we should avoid misleading single attribute indicators (embodied energy, carbon emissions, etc.) and opt for multi-criteria assessments based on third party verified EPDs when selecting ecological materials. Falling into the trap of single criteria comparisons and selections disqualifies materials by being labeled ecological.

In this context, we compared two NZEB by conducting a LCA and weighed the results of different environmental impact categories and indicators (Attia, 2018). The first case study is the Green Offices high-performance building in Switzerland mainly built from bio-based materials. The second case study is the Research Support Facility (RSF) of the National Renewable Energy Lab (NREL) in Golden Colorado, mainly built from concrete and steel. See Chapter 11, NZEB Case Studies and Learned Lessons, for further descriptions of these case studies.

Materials used in the RSF contain recycled content, rapidly renewable products, regional procured products within a 500-mile radius of Golden (DOE, 2012). The precast panels that make up the exterior walls of the RSF consist of 2 in. of expanded polystyrene of rigid insulation (R-14) sandwiched between 3 in. of precast concrete on the outside and 6 in. of concrete on the inside. The panels, which were fabricated in Denver using concrete and aggregate from Colorado sources, constitute the finished surface on both the inside and outside of the wall except that the interior is primed and painted. Wood originates from pine trees killed by beetles used for the lobby entry. Recycled runway materials from Denver's closed Stapleton Airport were used for aggregate in foundations and slabs. Reclaimed steel gas piping was used as structural columns. About 75% of construction waste materials have been diverted from landfills (DOE, 2012). Table 6.3 summaries the mid-point environmental indicators relevant to the LC of the RSF. Preuse and maintenance impacts are higher than those relevant to the use phase.

For Green Offices, the architect used wood as the raw material as it is widely available and with the least possible impact on the environment in Switzerland. Some 450 m^3 of wood were transported from a forest

TABLE 6.3 Mid-Point Environmental Indicators Relevant to the Life Cycle of the RSF and Green Offices Buildings (Attia 2018)

RSF	**Carbon sequest.**	**Preuse**	**Operation**	**End of life**	**Life cycle**
PE MJ/m^2 per annum	/	274 (23%)	785 (66%)	131 (11%)	1190
NRE MJ/m^2 per annum	/	274	785	131	880
GWP kgCO_2 eq./m^2 per annum	/	56 (20%)	197 (71%)	25 (9%)	275
Green Offices		**Preuse**	**Operation**	**End of life**	**Life cycle**
PE MJ/m^2 per annum	/	27 (10%)	168 (86%)	−14 (4%)	181
NRE MJ/m^2 per annum	/	27	56	5	88
GWP kgCO_2 eq./m^2 per annum	−5.9 (−13%)	6.5 (14%)	40 (90%)	3.4 (7%)	44

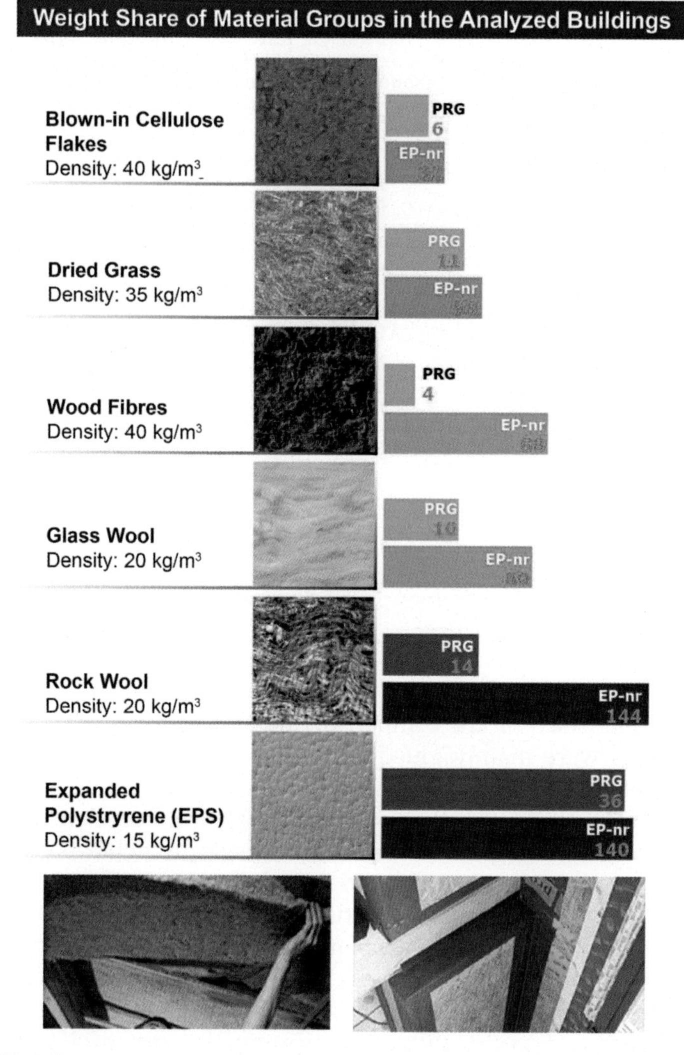

FIGURE 6.5 Example of different insulation material and their environmental impact (Attia 2018; Lutz 2012).

20 km away. The forest wood was sustainably managed and each tree was selected explicitly with lower possible moisture content to reduce the energy of the wood kiln. As shown in Table 6.3, the use of wood resulted in a carbon negative footprint. By carbon negative we mean a negative outcome of the carbon footprint of wood—i.e., when carbon credits through carbon sequestration and energy production at the end of life phase are higher than the emissions caused by production and transport. The architect designed prefabricated wooden panels filled with wood

TABLE 6.4 Weight Share of Material Groups in the Analyzed Buildings

	Building 2 RSF		Building 1 green offices	
Building material category	**Amount (kg)**	**Share (%)**	**Amount (kg)**	**Share (%)**
Concrete	32,500,000	79	788,650	73.4
Brick	–	–	10,890	1
Gravel	6,000,000	14.6	50,000	4.6
Ceramics	84,000	0.2	–	0
Mineral binding materials	82,600	0.2	2000	0.1
Wood and wood based materials	10,000	0.2	144,200	13.5
Insulation materials	110,000	0.3	55,000	5.1
Metals	1,904,762	4.7	12,600	1
Glass	53,460	0. 1	5680	0.05
Paints and preservatives	48,240	0. 1	1340	0.1
Plaster and Gypsum board	229,680	0.5	2000	0.1

fiber insulation based on the comparison illustrated in Fig. 6.5. The structural elements were mainly glued laminated timber trusses and beams. The whole construction was designed to be easily dismantled and with materials that could be, for the most part, reused or recycled. The compactness of the building space not only strategically achieved heat loss reduction, but also reduces the material total quantity while reducing the embodied energy of building materials. MINERGIE-ECO required the use of an exclusion list that prevents materials that end up in the landfill and are not compatible with a healthy indoor environment. Concrete was used in the foundation from a cement factory 100 km away and other materials were transported from a maximum distance of 1000 km. All materials from a distance less than 500 km were transported with 3.5–20 t trucks, materials transported from further away came on 32 t trucks. Table 6.4 shows the important contribution of the building LC which corresponds to the two different NZEB designs.

Table 6.4 shows primary energy and carbon emissions calculations and provides a new perspective for the overall LCA of both buildings. While both buildings succeeded to achieve the zero-energy annual balance, the carbon emissions associated with the generation and importing of energy were positive. The main reason is due to carbon emissions associated with the energy imported from the grid. The use of cellulose insulation will require two times replacement, which increased the operational energy. Even the use of bio-based construction materials like wood or wood fibers was not enough (−13 percent) to create a carbon negative outcome. However, if we take into account the biogenic CO_2 captured in wood and wood fibers and make sure to have a zero-carbon operational energy we might reach a total negative balance of carbon.

On the other hand, the life cycle analysis of the RSF building shows carbon emissions intensity constituting about 275 kgCO_2 eq./m^2 per annum and about 44 kgCO_2 eq./m^2 per annum for Green Offices buildings. Attention should be paid to the fact that the use of concrete and steel has a very high environmental impact on carbon emissions. Even in Green Offices building, which has very low emission value, the impact of foundations and concrete walls (average 1400 kg/m^2) has been high (73% according to Table 6.4). However, reaching 44 kgCO_2 eq./m^2 per annum for the Green Offices is a very good record, because if the emissions associated with operation could be neutralized through a greener grid energy mix where the total carbon emissions might reach a negative balance taking into account the carbon sequestration or bio generation of wood. The role of reaching a negative CO_2 balance over the whole building LC should become increasingly prominent. Reducing the embodied energy and carbon of NZEB of Green Offices was achieved in this project in a cost-neutral manner starting with simple cost-effective reduction measures.

Finally, we should recognize that ecological materials are environmentally friendly materials that maintain accountable performance when used in NZEB. Ecological and regenerative materials have different properties including energy and resource saving ability, reusability, structural reliability, recyclability, and (most importantly) are used without causing negative effects to environment (Attia 2018). There is a wide choice of eco-materials including wood fibers, cellulose, hemp wool, and compressed straw. These materials can be used in different formats such as panels, frames, or even blocks. Before the arrival of petrochemical construction materials, eco-materials were used widely to insulate buildings or improve the hygrothermal properties of building envelopes. The regulation of moisture and the ability of absorbing and restoring water, or allowing water vapor to pass outward, makes those materials very effective. When mixed with clay, it can stabilize the indoor temperature and provide a healthy indoor atmosphere without chemical compounds or harmful substances. Even the fire resistance properties improve without emitting toxic gases. The use of eco-materials as local and sustainable building materials that respect and protect the environment will play a major role in NZEB in the near future.

6 DISCUSSION

Meaningful progress has already been made by the construction sector in order to achieve better economic and environmental sustainability of high-performance buildings. The selection of sustainable and environmentally friendly materials remains a challenge because it happens with single attributes for individual building elements and products. NZEB must be viewed holistically and examined in the context of their impact on the whole environment and across the LC. In fact, addressing the embodied energy and the environmental impact of building materials will change the way NZEB are designed. Material selection and the choice of construction technology will serve as a corner stone to achieve the energy performance gains of NZEB. At the same time, the environmental impact of materials and resources should be taken into account. Until now, our policy review showed that there are no binding regulations on carbon emissions on environmental or health impacts associated with the use of building materials. New constructions should maximize the use of materials with EPDs. However, in the near future, we expect to have stringent regulations regarding the material used and its embodied energy in NZEB. Next, we will discuss some important considerations that need to be addressed in relation to building materials to raise the awareness of professionals and prepare for current and future challenges.

NZEB require high quantities and volumes of materials. Construction technology is advanced requiring controlling the hygrothermal performance of the envelope, maintaining airtightness, and avoiding thermal bridges. As a consequence, this requires volumes of thermal insulation, thermal-performing glazing, air and moisture barriers beside low-impact structural systems, envelope, and cladding systems. In this context, professionals have to make sure that the embodied energy and emissions associated with those materials are low. Studies show that NZEB might need more materials and related CO_2 input (Attia et al., 2013; Rovers et al., 2017). A rooftop of a NZEB with solar panels is energy intensive. However, if the solar panels are manufactured using entirely on-site renewables, within 4–6 years of the solar panels, they generate more energy than it takes to manufacture them (Attia et al., 2013). Therefore, we recommend the use of the maximum possible amount of bio-based construction materials that are sustainably harvested and have EPDs—in particular for insulation materials. Bio-based materials are of high density and great insulating capacity. For bio-based materials there a chance to restore or grow the material and manage the harvesting in a sustainable way when associated with certificates such as the PEFC and FSC certificates. However, this is more difficult with metals which just get depleted. Multi-story timber buildings can store carbon in their construction systems, which can turn NZEB into carbon sinks.

Another important issue we should be aware of in relation to material quantities is the rebound effect of growing building space and material needs in relation to NZEB. The growing size of new homes and the complexity of construction technology can cancel out the gains from lower impact material choices. The key message here is that we should always embrace resource-centered thinking and take into account the ecological impact of the amount of building materials and the associated construction technology.

Secondly, we should not underestimate the complexity and difficulty to assess the durability and sustainability of buildings in a holistic way. This does not only include environmental attributes of materials and resources, but also functional attributes such as cost effectiveness, durability, fire resistance, moisture resistance, thermite resistance, ease of installation, and maintenance. The best way to breakthrough this complexity is to embrace resource-centered thinking. Resource-centered thinking is an effective approach to close the resource cycles, explore and analyze the paradigm of circular economy, explore the consequences of the Paris Climate agreement, and look for strategies and solutions that can achieve this approach. It involves inquiring about the traceability of materials and assessing the materials from a multi-attribute approach. Materials should be traceable. As a consequence, the importance of local materials for sustainable NZEB increased in recent years due to its ability to strengthen local economy and valorize natural resources for eco-materials.

During early design stages, stakeholders should explore the different attributes of building materials and collaborate to find a sustainability balance. Therefore, materials should be considered early on in the design process and assigned to a responsible professional who will set up a sustainability purchasing plan and inventory. With the help of LCA experts, stakeholders should identify the GHG emissions, LC costing, resource efficiency, water efficiency, health, comfort and resilience of their potentially used material, land use changes associated with materials harvesting or extraction, resources, and products for the NZEB. The LCA calculations should be transparent and should communicate the environmental damage using universal metric to quantify the damage without involving any environmental costing transformation. The idea of transforming the environmental damage into cost damage based on the willingness of the polluter to pay is flawed and counter productive. On the other hand, decision makers should be aware of the "claim effects" that material manufacturers use to sell their products. There is very strong competition in the construction material supply sector and marketing is fierce to influence client choices based on single quality attributes. Some manufactures focus only on one quality of their products such as recyclability, low-embodied energy, low-embodied carbon, or regionality and claim it as the most important factor to market their products. This approach is misleading because sustainability of buildings is holistic and multi factorial. EPDs are a good start so far, but the building professionals' community should learn more about the impact of materials beyond short tractability claims and single attributes. We should integrate embodied energy calculations for material choices and land use changes in the calculation of CO_2 emissions associated with NZEB construction and operation. We are reaching the limit of single attributes materials and the focus has shifted to whole building sustainability assessment.

The third consideration we would like to point out in our discussion is the importance of design for circularity and disassembly (Rau and Oberhuber, 2017). Clients may not be longer be the owners of a number of building components, but instead leases them from the manufacturer through the use of long-term service contracts. This requires optimizing the reuse and recycling of components and materials within the construction industry. The principle of circular economic development is to preserve components and materials within closed loops of either biological or technical nutrients, while maximizing the conserved value for any particular component. Instead of becoming waste, buildings must function as banks of valuable materials, slowing down the usage of resources to a rate that meets the capacity of the planet (EU, 2017). The key message here is that renovation will be the future challenge in many industrial countries and we need to be prepared for this systematic shift that will reflect on the building construction technology and the construction manufacturing industry.

Finally, a harmonization of building assessment systems is needed in order to make the evaluation of NZEB' environmental impact comparable and enable building professionals to better select sustainable materials and lead the development and innovation of the AEC industry.

7 LESSON LEARNED # 6

We face an increasingly resource-constrained future. EIAs and EPDs can contribute to the sustainability of NZEB and can become a part of the roadmap of their implementation. Professionals should prepare to develop energy-neutral and resource-efficient buildings embracing resource-centered thinking for NZEB. This chapter makes it clear that addressing the materials' environmental impact including carbon reduces health risks and cost. Reducing the embodied energy and carbon of

NZEB can make a valuable contribution to reducing global greenhouse gas emissions. It is often possible to reduce the embodied carbon of a building by around 20% in a cost-neutral manner starting with simple cost-effective reduction measures. It is part and parcel of saving materials, reducing energy demand, and delivering a healthy indoor environment. Pursuing extremely positive carbon NZEB is not yet mandatory, but will be soon. European legislation is very close to limiting the environmental impact of new NZEB using LCA to quantify the energy and resources used in the processes and waste, emissions and GHGs produced per unit mass of building materials and products. This will be supported by a compilation of the hazards and health risks associated with the chemicals used in building materials. Building professionals should use that target to stimulate innovation and prepare the leadership in redefining how NZEB are designed and built to achieve a low-carbon future. The value of conducting LCAs, tracing materials, using EPDs, and setting up a sustainability purchasing program early on in NZEB design development and construction is that it provides professionals with an understanding of where to focus on future developments. It also provides a better understanding of the absolute carbon (operational and embodied) impact over the life of the NZEB.

References

Asdrubali, F., Baldassarri, C., Fthenakis, V., 2013. Life cycle analysis in the construction sector: Guiding the optimization of conventional Italian buildings. Energy Build. 64, 73–89.

Attia, S., 2010. Towards 0-Impact Buildings and Built Environments. Techne Press, Amsterdam, The Netherlands, ISBN: 978-9085940289.

Attia, S., De Herde, A., July 2011. Defining Zero Energy Buildings from a Cradle to Cradle Approach, Passive and Low Energy Architecture. Louvain La Neuve, Belgium.

Attia, S., 2011. A case study for a zero impact building in Belgium: Mondo Solar-2002. J. Sustain. Build. Technol. Urban Dev. 2 (2), 137–142.

Attia, S., Beney, J.F., Andersen, M., 2013. Application of the Cradle to Cradle Paradigm to a Housing Unit in Switzerland: Findings From a Prototype Design. PLEA, Munich, Germany.

Attia, S., 2016. Towards regenerative and positive impact architecture: A comparison of two net zero energy buildings. Sustain. Cities Soc. 26, 393–406. Available from: http://dx.doi.org/10.1016/j.scs.2016.04.017. ISSN 2210-6707.

Attia, S., 2018. Regenerative and Positive Impact Architecture: Learning from Case Studies. Springer International Publishing, London, UK978-3-319-66717-1.

Bor, A.M., Hansen, K., Goedkoop, M., Rivière, A., Alvarado, C., van den Wittenboer, W., 2011. Usability of Life Cycle Assessment for Cradle to Cradle Purposes. NL Agency, Utrecht.

CESBA, 2014. Initiative Policy Paper Towards a Common Sustainable Building Assessment in Europe. Available from <http://www.fedarene.org/wp-content/uploads/2014/09/CESBA-Policy_paper-final.pdf> (accessed 20.05.17.).

Dixit, M.K., Fernández-Solís, J.L., Lavy, S., Culp, C.H., 2010. Identification of parameters for embodied energy measurement: a literature review. Energy Build. 42 (8), 1238–1247.

Dixit, M.K., Fernández-Solís, J.L., Lavy, S., Culp, C.H., 2012. Need for an embodied energy measurement protocol for buildings: a review paper. Renew. Sustain. Energy Rev. 16 (6), 3730–3743.

DOE, 2012. The Design-Build Process for the Research Support Facility, p. 60. Retrieved from <http://www.nrel.gov/docs/fy12osti/51387.pdf>.

Dubois, M., Allacker, K., 2015. Energy savings from housing: Ineffective renovation subsidies vs efficient demolition and reconstruction incentives. Energy Policy 86, 697–704.

ECORYS, 2014. Resource Efficiency in the Building Sector, Final Report DG Environment Rotterdam, May 23, 2014. Available from <http://ec.europa.eu/environment/eussd/pdf/Resource%20efficiency%20in%20the%20building%20sector.pdf> (accessed 25.05.17.).

EEA, 2010. SOER 2010, Material Resources and Waste—SOER 2010 Thematic Assessment. Available from <https://www.eea.europa.eu/soer/europe/material-resources-and-waste> (accessed 20.05.17.).

EN, B., 2011. 15978: 2011 Sustainability of Construction Works. Assessment of Environmental Performance of Buildings. Calculation Method. BSI, London.

EN, C., 2012. 15804 (2012): Sustainability of Construction Works. Environmental Product Declarations. Core Rules for the Product Category of Construction Products. CEN, Brussels.

EU, 2011a. European Commission Service Contract on Management of Construction and Demolition Waster, Final Report. Available from <http://www.eu-smr.eu/cdw/docs/BIO-construction%20and%Demolition%20Waste-Finl%report-09022011.pdf> (accessed 20.05.17.).

EU, 2011b. The Roadmap to a Resource Efficient Europe, The Council, The European Economic and Social Committee and the Committee of the Regions.

EU, 2013. Open House Assessment Guideline, Version 1.2 New Office Buildings. Available from <http://www.openhouse-fp7.eu/assets/files/OPEN_HOUSE_AG1.2.pdf> (accessed 16.05.17.).

EU, 2017. BAMB Project—Buildings As Material Banks. Available from <http://www.bamb2020.eu/about-bamb/> (accessed 20.05.17.).

Floyd, D., 2017. This Material is 85% of Everything We Mine, and It's Running Out, Investopedia, April 19, 2017.

FSC, 2017. Facts & Figures. Available from <https://ic.fsc.org/en/facts-and-figures> (accessed 20.05.17.).

Gauvreau-Lemelin, C., Attia, S., July 2017. Benchmarking the Environmental Impact of Green and Traditional Masonry Wall Constructions, Passive and Low Energy Architecture, Edinburgh, UK.

Hammond, G., Jones, C., 2015. Inventory of Carbon and Energy (ICE). Sustainable Energy Research Team, Dept. of Mechanical Engineering, University of Bath, Bath, UK.

Hernandez, P., Kenny, P., 2010. From net energy to zero energy buildings: defining life cycle zero energy buildings (LC-ZEB). Energy Build. 42 (6), 815–821.

Kibert, C.J., 2016. Sustainable Construction: Green Building Design and Delivery. John Wiley & Sons, New York.

Leduc et al., 2009. Expanding the Exergy Concept to the Urban Water Cycle, SASBE 2009, Delft, NL.

Lutz C., 2012. Construction, écologie et impact environnemental: Exemple du Green Offices, 1er bâtiment administratif Minergie-P-Eco de Suisse, EPFL, Switzerland.

MBDC, 2012. Material Health Assessment Methodology, Cradle to Cradle Certified Product Standard, Version 3,0.

Moncaster, A.M., Symons, K.E., 2013. A method and tool for 'cradle to grave'embodied carbon and energy impacts of UK buildings in compliance with the new TC350 standards. Energy Build. 66, 514–523.

Peduzzi, P., 2014. Sand, rarer than one thinks. Environ. Dev. 11, 208–218.

PEFC, 2017. Facts & Figures. Available from <https://www.pefc.org/about-pefc/who-we-are/facts-a-figures> (accessed 15.01.18.).
Rau, T., Oberhuber, S., 2017. Material Matters, het alternatief voor onze roofbouw-maatschappij, Bertram en de Leeuw Uitgevers, ISBN: 9789461562258.
Rovers, R., Lützkendorf, T., Habert, G., 2017. A Near CO_2 Neutral Built Environment: iiSBE Expert Explorations, v1.0 April 2017.
Thormark, C., 2002. A low energy building in a life cycle—its embodied energy, energy need for operation and recycling potential. Build. Environ. 37 (4), 429–435.
USGBC, 2013. LEED v4: Reference Guide for Building Design and Construction (v4). US Green Building Council, Washington, D.C.
van Dijk, S., Tenpierik, M., van den Dobbelsteen, A., 2014. Continuing the building's cycles: a literature review and analysis of current systems theories in comparison with the theory of Cradle to Cradle. Resour. Conserv. Recycl. 82, 21–34.
Verbeeck, G., Hens, H., 2010. Life cycle inventory of buildings: a contribution analysis. Build. Environ. 45 (4), 964–967.
Wiebe, K.S., Bruckner, M., Giljum, S., Lutz, C., Polzin, C., 2012. Carbon and materials embodied in the international trade of emerging economies. J. Ind. Ecol. 16 (4), 636–646.

CHAPTER

7

Energy Systems and Loads Operation

ABBREVIATIONS

AEC Architectural, Engineering, and Construction
EPBD Energy Performance of Buildings Directive
BIM Building Information Modeling
BMS Building Managements Systems
BPS Building Performance Simulation
CDD cooling degree day
CxA Commissioning Authority
DX direct expansion
DHW domestic hot water
DOAS dedicated outdoor air system
EV electric vehicle
FCU fan-coil unit
GHG greenhouse gas
GSHP ground source heat pump
HDDs heating degree days
HVAC Heating, Ventilation, and Air Conditioning
HVACR Heating, Ventilation, Air Conditioning, and Refrigeration
HVLS High-Volume Low Speed
IAQ indoor air quality
IDP integrative design process
IEQ indoor environmental quality
LED Light Emitting Diode
M&V measurement and verification
MEP Mechanical, Electrical, and Plumbing
MERV minimum efficiency reporting value
MVHR mechanical ventilation heat recovery
NREL National Renewable Energy Lab
NZEB Net Zero Energy Buildings
O&M operation and maintenance
RES renewable energy systems

Net Zero Energy Buildings (NZEB)
DOI: https://doi.org/10.1016/B978-0-12-812461-1.00007-1

TABS thermally activated building systems
TASB thermal active storage banks
VFD variable frequency drive
VRF variable refrigerant flow

1 INTRODUCTION

In this chapter, we will summarize the technologies and systems available to reduce energy use and achieve indoor environmental quality (IEQ) in Net Zero Energy Buildings (NZEB). We are reaching the technological limits of increasing energy efficiency of building services. For NZEB, the focus has shifted from equipment efficiency to whole building performance efficiency and controls. Therefore, highly efficient energy systems and optimal loads operation should rely on properly sized and installed HVAC components and renewable energy systems (RES) together with integrated automated control systems. From a strategical point of view, designers should enable passive design strategies, improve the envelope efficiency, reduce plug loads, reduce lighting loads, and then design the HVAC system to match the remaining loads. Therefore, this chapter is dedicated to Mechanical, Electrical, and Plumbing (MEP) engineers, MEP consultants, MEP contractors, trade-specific MEP, firefighting engineers and subcontractors, fabricators, and installation and maintenance specialists including facility managers.

In the first section, we will discuss the importance of energy systems integration to manage complex cascades of interactions between energy systems. The importance of fit-to-purpose design, integrative design, and commissioning are summarized. Then (in Section 3), we summarize different Heating, Ventilation, and Air Conditioning (HVAC) technologies and renewable systems that have been used successfully in several countries. We present some techniques, such as energy recovery ventilators, liquid desiccant dehumidification systems for hot-humid climates, building integrated photovoltaic (BIPV), solar assisted heating, solar driven cooling systems, and ground source heat pumps (GSHPs) that require further development to bring down cost and improve performance. The review includes hydronic heating systems with low temperature radiant floors and panel radiators for temperate and cold climates. Coupling hydronic heating systems and GSHPs to photovoltaic systems is a reasonable and cost-effective choice when it can be done on large-scale projects.

In Section 4, we address plug loads and electricity as well as the importance of reducing the energy consumption of devices and appliances when they are operational, or when they are off and not being used. In Section 5, we discuss how controls are critical to NZEB. NZEB require controls expertise that handles the complexity of controls in quantity, quality, and derived data means. Finally, Sections 6 and 7 discuss the implications and recommendations of energy systems and loads operation on overall building performance and occupants. We discuss how Central and Northern European Countries, under the Passive House Standard, couple energy recovery ventilators to hydronic heating systems, while in the United States and Canada forced air furnaces are the norm in relation to ASHRAE 90.1 and 90.2, which are used more often in high-performance buildings. We also present lessons from practice suggesting simplifying system designs, loads operation, and controls automation while suggesting energy system integration, performance optimization, third party commissioning, extensive monitoring, and user-centered operation approaches.

2 ENERGY SYSTEMS INTEGRATION

In this section, we discuss the role of energy systems integration as well as the role of energy modeling and commissioning. NZEB employ the latest systems, controls, and technologies in order to minimize energy consumption and produce as much, or more, energy than the building consumes. They require highly efficient products and systems to meet the high-performance building goals. Smarter controls and automated demand response are playing a large role in ensuring the operation of NZEB. Additionally, a renewed focus on IEQ in combination with overall building performance efficiency is forcing professionals to adopt an integrated project delivery approach. We are reaching the limits of the linear design approach and with the emergence of Building Information Modeling (BIM) energy systems, integration became essential. Fig. 7.1 illustrates the different aspects that need to be addressed regarding systems and loads operation of NZEB.

Energy systems integration requires following a holistic approach on energy systems levels and how they connect together. By energy systems we mean HVAC systems, renewable systems, energy storage, electric vehicles (EVs), batteries, electric lighting, controls, and IT including data centers and plug loads. MEP engineers are responsible for ensuring the integration of those systems and the links between them. NZEB are complex and rely on renewable supply with emerging smart devices including electric cars. Therefore, they must be integrated in an effective way through an intelligent system that achieves indoor environmental qualities in an affordable way. MEP coordination service is part of the overall energy systems integration that is critical to guarantee optimal use. By bringing all systems components together for NZEB in a cohesive way, the MEP team can avoid clashes, system errors, and conflicts.

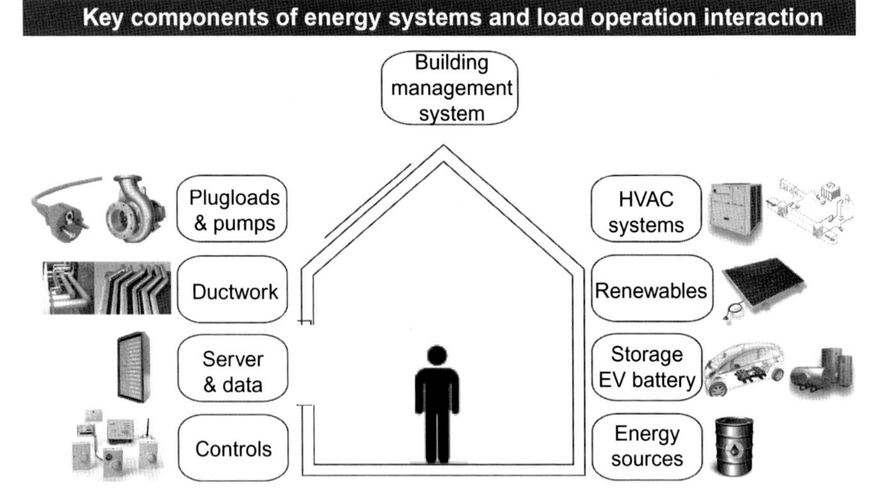

FIGURE 7.1 Illustration of the key components of energy systems and load operation interaction.

The only way to manage these complex cascades of interaction in NZEB is to follow an integrative and iterative approach design process, involving all the team members and following best practices of Building Performance Simulation (BPS). Together with BIM, building energy modeling plays a major role to assure an integrated design process (IDP). Mechanical HVAC systems must be integrated with passive heating, passive cooling, and passive ventilation so that passive strategies and measures can be used whenever conditions permit.

As part of the IDP, the Owner's Project Requirements (OPRs) documents must be completed during the predesign phase to plan for commissioning and measurement and verification (M&V) of energy systems and load operation (McFarlane, 2013). A Commissioning Authority (CxA) should be part of the OPR development and Basis of Design (BOD). Including a CxA early on in the design process is the only guarantee that the MEP engineers will design HVAC systems and controls in harmony and compliance with performance requirements.

3 HVAC SYSTEMS AND RENEWABLE ENERGY SYSTEMS

In this section, we will discuss mechanical ventilation, heating systems, cooling and air conditioning systems, and renewable systems. The HVAC system size and energy consumption directly relates to key components of the building design, such as orientation, building envelope, heated/cooled volume, lighting system, and appliances. Mechanical system designers strive to reduce the cooling and heating loads these elements impose in order to minimize the energy consumption of the HVAC systems and then overlay renewable energy sources to achieve a net-zero-energy balance. The largest energy consumer in a building is the heating, air conditioning, and ventilation systems, therefore, technological improvements in these categories have contributed toward the NZEB. For heating-dominated NZEB, the design of HVAC systems should be based on eliminating cooling loads with minimum extra energy consumption—preferably by using free or renewable cooling sources. On the other hand, for cooling-dominated NZEB, the design of HVAC systems should be based on eliminating heating loads with minimum extra energy consumption—preferably using passive or renewable heating sources. In cooling- and heating-dominated NZEB, heat pumps coupled to radiant floors can be a good choice if the soil conditions and the budget allow for this.

A high-efficiency HVAC system is a critical part of an NZEB, and depends on two strategies. The first strategy is thermal zoning. Thermal zoning allows both spatially and temporally space condition control. Studies have indicated that at least 5%−15% of waste is due to unoccupied spaces being conditioned (Pitts and Potentia, 2007; Vargas et al., 2017). Most of the energy wasted by HVAC systems goes toward heating or cooling unoccupied spaces during long periods while people only use a small fraction of a building. Mechanical systems designers should minimize this waste to enable thermal zoning where each space is conditioned based on its functional occupancy. This will allow only occupied spaces to be conditioned while saving the energy used to condition unoccupied spaces. The volume of the building has a significant impact on thermal zoning. One of the influential strategies to reduce the space condition energy demand in NZEB is limiting the airtight internal air volume within the building. Using the building's net heated or cooled internal air volume (m^3) rather than its floor area reduces the heating and cooling needs while increasing comfort. Dividing building thermal zones that are controlled by separate thermostats or excluded from HVAC system so that occupants can schedule each zone to be heated or cooled separately and scheduled with a low frequency will reduce the energy needs.

The second strategy is to use BPS for load calculation. HVAC load calculations require time and attention to detail, so most HVAC contractors rely on rule-of-thumb to determine the size of the heating or cooling systems they install. For NZEB, this process is not possible. As mentioned in Section 2, HVAC load calculation requires the use of accurate dynamic BPS. It's vitally important to have all the specifications and plans for the building and a local weather file prior to designing a

HVAC system for a NZEB. The load calculations are used to find the right equipment size. However, NZEB are ultra-low energy buildings and more complicated than standard HVAC buildings. HVAC systems for NZEB push the limits of cooling tower systems, making cold water at night, using warmer temperature water through chilled beams, radiant slabs, or displacement systems. The use of BPS tools is the only way to make sure that the HVAC products are fully integrated to provide comfort and significant savings. By reducing the space condition demand to the effectively used areas only, and performance accurate load calculations using dynamic modeling and BPS, mechanical engineers can achieve significant energy saving and assure comfort.

3.1 Mechanical Ventilation

Mechanical ventilation systems for NZEB serve many functions including air filtration, dilution, and removal of indoor pollutants through code-mandated air change rates of supply and outdoor air, as well as the precise control of space temperature, humidity, and pressurization. In this section, we will discuss the distribution of heat or coldness throughout a building and the concurrent supply of ventilation air. Mechanical ventilation is obligatory in any NZEB according to European and US regulations. Therefore, ventilation systems selection and sizing should take place as early as possible during the design process. Mechanical ventilation does not function alone and should be integrated with other functions and equipment, such as hot water services, heating, cooling humidification, allergy control, and so on. In Chapter 5, Occupants Well-Being and Indoor Environmental Quality, we discussed why we should ventilate and the benefit or consequences associated with poorly or well ventilated buildings. In this section, we will discuss how to ventilate and the best practices of ventilation for NZEB. In practice, mechanical ventilation is associated with all types of NZEB. Mechanical ventilation should be used to guarantee good indoor air quality and air distribution.

Along with the insulation, sealed envelope and solar protected windows (highly insulated), heat recovery ventilation (HRV) is the most important active feature in any NZEB. To provide maximum efficiency in NZEB, the use of air to heat or cool the indoor spaces via air supply is not recommended. However, we can use the hygienic air and recover heat from it. An HRV system or mechanical ventilation heat recovery (MVHR) system can do this efficiently and effectively. In essence, it is a ventilation system with a heat exchanger, so that in winter, when cold air is drawn into the building, it is warmed with the heat from the stale outgoing air (and vice versa in summer). For individual housing

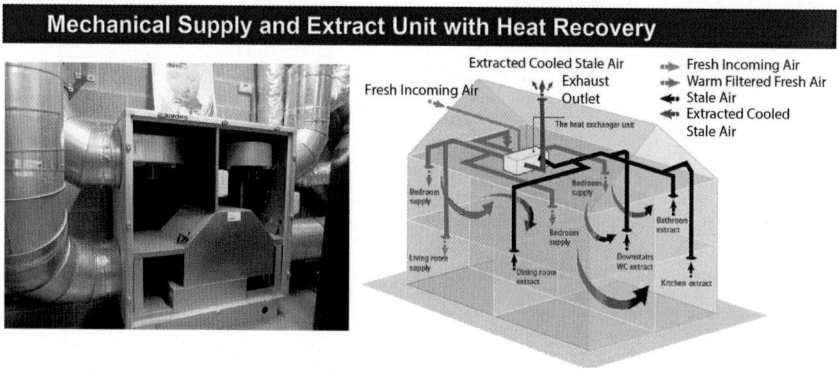

FIGURE 7.2 Mechanical supply and extract unit with heat recovery.

and small apartments, designers rely on small DC powered HRV (see Fig. 7.2). For small buildings, fresh air needs to distributed around the building. For middle-sized and large-scale buildings, packaged dedicated outdoor air systems (DOASs) with direct expansion (DX) and HRV are recommended (see DOAS Section for further explanation). In general, the HRV or MVHR ensures a uniform temperature throughout the building and is mainly used for ventilation.

Heat Recovery Ventilation (HRV)

In residential buildings, the HRV supplies dry spaces (bedrooms, offices, living rooms) with fresh air and extracts air from wet spaces (bathroom, kitchen, laundry room). A heat exchanger is coupled to the system to preheat the fresh air while crossing it with exhaust air as shown in Fig. 7.2. Openings in the form of vents integrate the doors, or lifted doors, guarantee the air movement between the dry and wet spaces before being extracted as shown in Fig. 7.3.

The efficiency of HRV depends on the airtightness of the buildings. If the envelope airtightness is poor, then the ventilation system has a partially short circuit. A volume of the heated air will infiltrate with passing by the exchanger and the left volume of air will enter the building directly without any heat exchange. Therefore, to guarantee the effectiveness of HRV, a good envelope airtightness is necessary ($n50 < 0{,}6$ [1/h]). Moreover, the efficiency of the mechanical ventilation system is determined by:

- The efficiency of the heat recovery system.
- The efficiency of the ventilation fans.
- The heat losses of loads to be compensated (distribution network design).

HVR Unit Mounted within the Ceiling Void Space or on the Wall

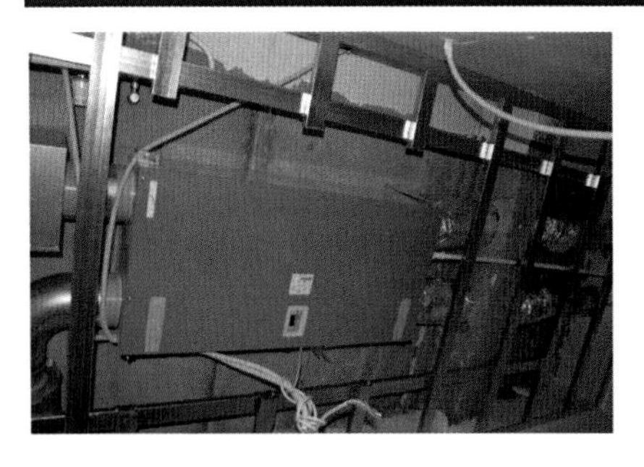

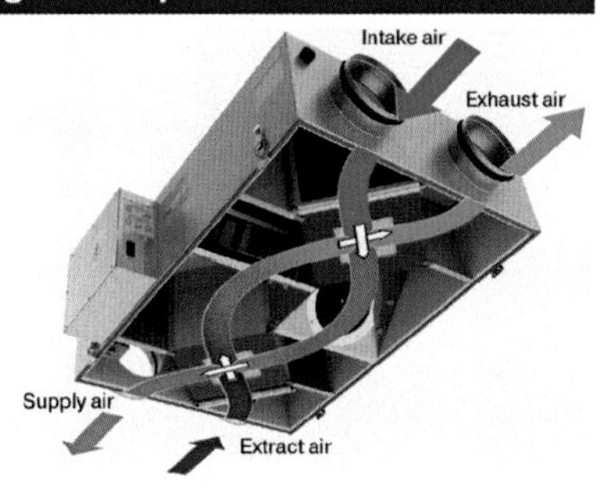

FIGURE 7.3 HVR unit mounted within the ceiling void space or on the wall (left: Vents US)

Some further guidelines:

- The efficiency of a heat recovery exchanger of good quality is higher than 85% complying with quality standards.
- The global thermal efficiency of the system: 93%–97% (> 85% due to the thermal losses of the fan engine that participate in preheating the supply air).

Moreover, the difference between the global efficiency and the efficiency of the heat recovery exchangers should be minimal, despite the gained thermal kWh. They are considered as electric loss in relation to the fan engine. Manufacturers mostly only reveal the global efficiency.

- Electric consumption of the mechanical ventilation system depends on the pressure drop of fan quality and supply flow. For example, a system with airflow of 325 m^3/h on maximum power, has power of approximately 243 W, while on minimal power (absence, night), it has a power of 17 W. In this case, the electric consumption is largely compensated by the energy needs reduction due to heating.

> The global efficiency of the heat recovery system plaque reaches 95%. This is valid for an external temperature of 0°C and interior temperature of 20°C, the supply air is preheated at 19°C by contacting only the exhaust air.

Dedicated Outdoor Air System (DOAS)

For commercial NZEB, a DOAS with HRV is essential. A dedicated outdoor air system is a type of HVAC system that consists of two parallel systems: A dedicated system for delivering outdoor air ventilation that handles both the latent and sensible loads of conditioning the ventilation air, and a parallel system to handle the (mostly sensible heat) loads generated by indoor/process sources and those that pass through the building enclosure. DOASs provide fresh air, balance moisture, and provide neutral air. By separating ventilation from air conditioning and applying heat recovery technology, significant savings can be achieved. DOAS that combines ventilation and heat recovery with a high level of efficiency is optimal for offices, schools, restaurants, multifamily, and assembly spaces. It is ruggedly engineered for commercial rooftop application and provides fresh air to the building. Finally, it is important to design the full ductwork in time to integrate with architectural design. For roof-mounted HVAC units, designers should make sure to integrate them on the roof which might cause conflict with photovoltaic panels.

3.2 Heating Systems

By reducing the heating requirements through a highly insulated envelope and the maximum use of passive solar heating, the mechanical ventilation system designer can estimate the amount of energy required to heat the NZEB. In this section, we will discuss the efficiency and characteristics of various heating systems and the different heat distribution solutions.

Electric Resistance Heating

Currently, electricity is generated from fossil fuels in most parts of the world. Although the power sector is decarbonizing, we are still far away from recommending electric resistance heating. Despite its efficiency of 100%, the amount of primary energy used to deliver end-use energy is three times as high. Electric resistance heating contributes to peak demand and is associated with high CO_2 emissions. We only recommend electric heating in small NZEB with good thermal mass and very small peak- and overall-heating loads. At the same time, the building should be preferably charging an electric vehicle's battery or more. This could be a reasonable choice when heat pumps are not practical and renewable energy sources can supply the peak electric loads.

Air Source, Ground Source, and Water Heat Pumps

Heat pumps can be used for heating, cooling, and generation of hot water. Residential and commercial air conditioners operate on the same principles as heat pumps, however, they operate in one direction only. In this section, we focus mainly on reversible heat pumps. By reversing the direction of the flow of working fluid, the evaporator works as a condenser and the condenser serves as an evaporator. Thus, heat is transferred in the opposite direction—a heat pump can act as a heater in winter and air conditioner in summer. The critical parameter measuring the performance of a heat pump is the coefficient of performance (COP). The COP of a heat pump depends on the difference between the outside and inside air temperature. The keys to improve the COP of a heat pump are (Harvey, 2012):

- To improve the efficiency of the compressor.
- To make use of variable-speed compressors of part-load operation.
- To minimize the difference in temperature required between the evaporator and condenser.

A heat pump is an efficient heating and cooling system that transfers heat from one way to another. A typical air source heat pump will have an outdoor unit containing a compressor and condenser working in conjunction with the building indoor handling unit. In the summer, the heat pump pulls the heat from inside the house to the outside, resulting in cooling the home like an air conditioner. During extreme summer days, the heat pump will intensively consume electricity and will have difficulty to transfer heat in an efficient way. In the winter, it reverses this process and extracts the heat from the outside into the home. On extremely cold days, the air handling unit should have an electric resistance to provide auxiliary heat. This means that *air source heat pumps* have to lower the efficiency in extreme climates. Electric air source heat pumps can barely achieve a (COP) of 3.0. Therefore, it is recommended for small buildings in moderate and warm climates.

For *ground/water source heat pumps*, the use of horizontal ground collectors or vertical borehole pipes to extract heat from the ground depends on a water loop. This water loop is then coupled to a distribution system for underfloor heating or radiators. A ground source heat pump (GSHP) requires heating exchange tubing underground or inserted in a water body. The effectiveness of this technology depends on the temperature difference between air and soil. In other words, it depends on the volumetric heat capacity of soil, of the project site, variability of speed compressor, the cost of drilling or excavation for underground boreholes (vertical), or underground tubing (horizontal) and the building scale or conditioned space area. Typical tubing laid horizontally in the ground need five to ten times the footprint of a house. Drilling boreholes under the building foundations is not recommended. A typical GSHP will deliver a COP about 3.5–4.5. The European performance of hydronic heating systems with low temperature radiators or radiant floors makes GSHP a fit-to-purpose technology for NZEB when it can be done cost effectively. Due to the high initial investment cost of GDHPs, it is recommended to use them for large projects. The lifespan of the piping is usually set at about 25 years. There is recent guidance on the design of GSHP provided by ASHRAE (Kavanaugh and Rafferty, 2014) and the International Ground Source Heat Pump Association.

Wood-Burning Boilers

Wood-burning boilers are a reliable technology if they are sourced from sustainability managed wood wastes. Hot water can be heated from the same boiler and used to distribute heat through a radiant floor or a thermally activated building systems (TABS). Wood-pellet boilers have an efficiency reaching 95% with automatic ignition, electronic throttling down to 30% of peak output, and automatic pellet feeding and metering (see Fig. 7.4). The pellets are made from wood waste (such as sawmill dust) and compressed yielding a density of at least 650 kg/m^3. The energy density is 3.3–4.2 kWh/L or 4.905.4 kWh/kg. Wood-burning boilers, and in particular pellet heating systems, provide a natural complement to active solar heating systems as the primary reliance on solar energy (Harvey, 2012).

Natural Gas and Petroleum Fuels

Natural gas and petroleum fuels might be a practical solution in countries with a developed natural gas infrastructure. However, there are two major problems associated with the use of natural gas in NZEB. The first problem is related to the nature of natural gas—it is a fossil fuel that releases carbon dioxide when burned. Natural gas contributes to greenhouse gas emission, and consequently climate change. At the same time, it complicates the net-metering process and discourages users to install on-site solar systems that generate more electricity than

FIGURE 7.4 Wood-pellet heating system with up to 56 kW coupled to a storage unit that monitors the pellet consumption.

is used in the building. Technically speaking, the efficiency of natural gas is around 95% which is not nearly as efficient as electric heat pumps for space and water heating which range between 300% and 400%. The second problem of natural gas is related to the low peak load and demand of NZEB. The majority of available gas boilers on market are oversized for the heat requirements of a single family NZEB. Installation of a heat buffer is necessary to avoid the frequent on/off cycle of the boiler (Georges et al., 2012). The low heat demand of NZEB and the high investment cost makes this solution unfavorable and not recommended for NZEB (Hopfe and McLeod, 2015).

District Heating and Cooling

District heating provides large energy savings through water-based distribution networks, known as district energy systems. Currently, district heating has an average only 9% market share in the heat market in the EU. This is well below the feasible market share, which has been estimated to be between 60% and 80% of the heat market in various countries (Connolly et al., 2013). The main advantage of district heating is that it is provided by dedicated centralized boilers that can operate more efficiently than individual boilers in each building (Lund et al., 2010). A district heating system consists of a network of insulated underground pipes carrying hot water or steam. Similarly, chilled water supplied to a district cooling network can be produced through a centralized chilling plant. District cooling is a relatively newly applied solution, but built on the experience acquired in the district heating field. District heating and cooling can make it easier to shift to renewables energy at feasible cost. In combination with heat storage they can stabilize the energy supply to NZEB. However, they require upfront capital investment and involve a large number of institutional, technological, legal, and financial planning and issues (Harvey, 2012).

Thermally Activated Building Systems

Thermally activated building systems (TABSs) are becoming a key component of HVAC systems designed for NZEB. Fig. 7.5 illustrates an example of a floor heating system in a residential building. To provide maximum efficiency in NZEB, the use of air to heat or cool indoor spaces is not efficient. TABSs can provide warmer distribution temperature (16°C) compared to air-forced cooling systems (10°C). Instead of using air to condition spaces, hydronic radiant heating and cooling systems are effective and efficient in NZEB. TABSs actively incorporate thermal storage in the building structure including walls, ceilings, and floors. TABS require low distribution temperatures and can improve the COP of a heat pump by 1.0 or more because a smaller temperature lift is required. Also, a TABS provides thermal storage in the structure

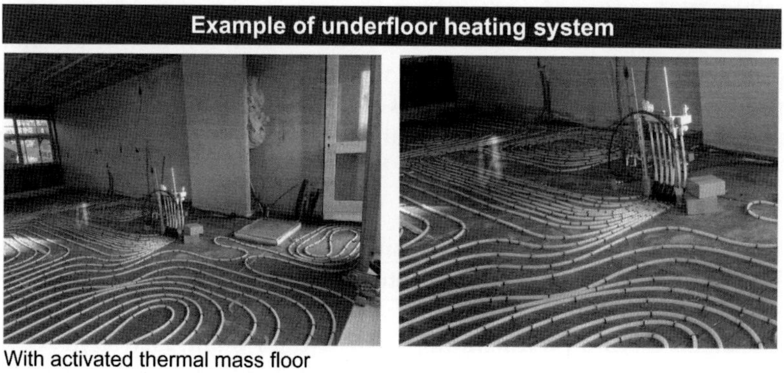

Example of underfloor heating system

With activated thermal mass floor

FIGURE 7.5 An example of underfloor heating system with activated thermal mass floor.

itself, so pumps do not need to run all the time. TABSs are not advisable in buildings with low thermal mass or storage capacity, or for spaces that require individual or precise control to achieve a target space temperature (Hopfe and McLeod, 2015; Levihn, 2017).

3.3 Cooling and Air Conditioning

Several tendencies indicate that the need for cooling in residential buildings is generally increasing. This is partly because (1) we are experiencing more extreme weather types with warmer summers, because (2) people's requirements for indoor comfort are rising and last, but not least, because of (3) low-energy building codes with stricter requirements for tightness and insulation have brought about significant cooling loads during the warmer seasons. One of the options for cooling in residential NZEB is to use a heat pump and recover coldness from the air, or make use of earth pipe loops to precool the ventilation air. Variable refrigerant flow (VRF) systems and air-to-air heat pumps can provide cooling and dehumidification in summer for cooling-dominated residential buildings. A VRF is an HVAC technology of ductless mini-splits. VRFs use refrigerant as the cooling and heating medium. This refrigerant is conditioned by a single outdoor condensing unit, and is circulated within the building to multiple fan-coil units (FCUs).

For commercial buildings, there is a significant energy saving benefit of distributing coldness with water rather than air, with airflow restricted to that needed for hygienic ventilation purposes. The dehumidification and cooling functions should be separated by using solid or liquid desiccants for dehumidification, with chillers for removal of sensible heat only. Mechanical system designers should use chilled TABSs with separate dehumidification of ventilation air. They should supply ventilation air and chilling water at the warmest possible temperatures, as this will increase the COP of mechanical chillers and will extend the range of outside temperate conditions where free cooling can be used. The following systems can be used for medium and large commercial NZEB:

- Packaged DOAS with direct expansion or VRF system and HRV can be a good option to balance moisture and provide fresh air for small commercial buildings.
- Packaged medium-temperature chilled-water plant. A system packaged and optimized to make 12°C chilled water with integrated economizer, full part-load compressor system, on board controls to manage sequences.
- Packaged ambient central plant—air to water heat pumps, 12°C chilled-water, 45°C hot water with booster helical coils (add on) with packaged controls to be stand alone.

Indoor Fans

Air movement is an energy-efficient alterative to air cooling in NZEB (Garde and Donn, 2014). Air speeds greater than 0.20 m/s may be used to increase the upper operative temperature limits of the comfort zone under certain circumstances. This could be achieved by using ceiling, desk, or pedestal fans to elevate air speed to offset increased air and radiant temperatures. As shown in Figs. 5.6 and 5.7, elevated air speed is effective at increasing heat loss when the mean radiant temperature is high and the air temperature is low. However, if the mean radiant temperature is low or humidity is high, elevated air speed is less effective. The required air speed for light, primarily sedentary activities may not be higher than 0.8 m/s. But, the ceiling fans effect cannot control humidity and depends on clothing and activity (ASHRAE 2013). Figs. 5.6 and 5.7 show the acceptable range of operative temperature and air speed for a given clothing level. When air speed is under individual control, it can maintain comfort for occupants in NZEB. The air speed should be adjustable continuously or in maximum steps of 0.25 m/s as measured at the occupant's location (Attia and Carlucci, 2015).

High-performance ceiling fans ensure consistent silent operation with the practicality of air movement that ensures energy savings of existing heating and cooling energy needs. High-performance ceiling fans are powered by energy saving drag reduction brushless motors, and self-balancing rotor coupling systems. Thus, it is able to eliminate unnecessary vibrations and wobbling by keeping their evenly weighed form in consistent aerodynamic equilibrium. This minimizes the ratio of power and torque loss, increasing both efficiency and long-term reliability without relying on external balancing kits, improving the smooth delivery of a natural wind momentum thus achieving silent efficiency within the living or working space. High-performance ceiling fans minimize vortexes and wind noise—these winglets optimize structural flexibility while promoting the flow of natural cross ventilation patterns during operation across their individual stepped-speeds of control (Attia, 2012).

Another successful alternative for indoor fans in NZEB are High-Volume, Low Speed (HVLS) fans. HVLS fans use variable frequency drives (VFDs) to help with destratification, increase comfort and reduce the need for heating and cooling. A high-volume, large-diameter ceiling fan is a ceiling fan greater than 2.1 m in diameter (US DOE, 2015). HVLS fans can create a light 2–3 mph breeze that produces an evaporative cooling effect and reduces the effective temperature by 1–3°C. By moving large volumes of air in an area up to 2000 m^2, a single HVLS fan can replace as many as 10–20 traditional high-speed floor fans. In air conditioned facilities (mainly office buildings) the breeze from a HVLS fan typically allows up to a 3°C increase in thermostat setting with no change in comfort.

3.4 Renewable Energy Systems

Today, we have the know-how, experience and technology to tap into the abundantly available renewable energy sources (RESs) and supply the energy in an efficient manner. Solar energy is one of the most important renewable sources of energy in NZEB. Another popular choice for RES in NZEB is geothermal energy. Geothermal energy is a renewable energy source that helps to significantly lower heating and cooling costs. However, we elected to discuss geothermal energy systems in the previous section under the topic of heat pumps. In this section, we focus mainly on active solar energy systems represented by solar electric and solar thermal panels.

A renewable power system comprises a PV unit, inverters, an electricity meter, a mounting system, and electrical cables. By default, NZEB are grid-connected and do not rely on storage batteries. When the renewable energy system produces more electricity than the building requires, the inverter exports the excess to the grid. The energy meter tracks the energy flow and is used to check the annual energy balance. Solar cells can be made from monocrystalline silicon, poly crystals, amorphous (noncrystalline) silicon, or monocrystalline dye cells. The best efficiency obtained for monocrystalline cells of commercial modules is 10%–18%. Poly-crystalline silicon consists of numerous individual grains made from lower-grade silica, which impedes electron conduction leading to lower efficiencies ranging between 10% and 15%. Thin film cells or amorphous cells are made using amorphous (noncrystalline) silicon with efficiencies ranging between 5% and 6%. For crystalline silicon modules, the efficiency in generating electrify decreases with the increase of temperature. Therefore, it is recommended to ventilate the PV panels to guarantee optimal electrical output.

PV panels involve building integration, require support structures, and conversion to AC electricity with appropriate frequency and voltage. Solar modules can be mounted on roofs and envelope surfaces, or can be a part of building cladding units—roof cladding, wall siding, curtain walls, skylights, windows, or even roof shingles. Modules that are part of the building envelope are referred to as BIPV, but are restricted to the building geometry and orientation leading to lower efficiency. Modules can be attached to clamps fixed to the roofing tiles with screws or attached to specially built tiles. An air gap underneath the PV panels can limit the temperature increase. Another option is to integrate them by replacing conventional roofing or cladding which tend to be more aesthetically appealing than modules mounted onto the roof. A third option is to mount the PV modules tilted on a flat roof. This requires a robust support structure, but also provides shading to the roof and airflow underneath. The need of space between the modules to reduce shading of modules by adjacent modules makes this option unfavorable. Horizontal modules can be better used on flat roofs and elevated slightly for ventilative cooling.

The optimal interaction of PV modules for maximum electric generation is related to the panel orientation mounting slope. The orientation is to the south in the Northern Hemisphere and to the north in the Southern Hemisphere. The tilt angle should match the latitude of the project location as close as possible. The solar access should be properly simulated and the solar rights associated with urban planning and surrounding building heights and obstructions must be studied beforehand to avoid any shade cast on the PV panels during the building operation. PV technology form a significant part of a NZEB project budget. Therefore, the PV panel sizing and integration in NZEB should be accomplished after achieving the maximum energy efficiency and loads reduction. Every watt that gets saved in HVAC, lighting, and plug loads is a watt that does not need to be generated by renewable energy or sourced from the grid.

The second most effective renewable technology for NZEB is solar thermal energy. Solar thermal energy for heating and hot water can be collected, stored in hot water tanks, used for hot water services, and provide space heating. In residential NZEB, water uses more energy than heating. Mechanical engineers should pay attention to design an efficient system. Similarly to PV panels, the solar thermal collectors should be mounted on the roof. The main types of solar collectors are (1) evacuated-tube thermal collector and (2) solar flat collectors. A solar collector absorbs solar energy and heats up. Heat losses through emission of informed radiation thorough convection offsets some of the absorption of solar energy, thereby limiting the net efficiency of the collector. Typical flat-plate collector efficiency will be 0.5, and for evacuated tubes 0.6. The collector design should include antireflective coating on the lags cover, while the temperature of the water that returns to the collector inlet should be minimized.

Solar thermal energy can be directly used for cooling and dehumidification. A design handbook developed as part of IEA Task 25 provides guidelines to use solar energy for solar assisted building air conditioning (Henning, 2004). Heat from solar thermal collectors can be used to drive a variety of different heat-driven chilling systems: Absorption chillers, adsorption chillers, ejector chillers, both solid and liquid desiccant dehumidification, and cooling systems. The solar coefficient of performance (solar COP) of these systems tend to be rather low in the order of 0.2 for adsorption chillers, 0.3 for ejector systems, 0.7 for desiccant systems, and up to 0.9 for double-effect absorption chiller systems. An advantage of solar thermal air conditioning is that the same solar collectors can be used for heating in winter.

Finally, integrated PV and thermal collectors provide useful heat or generate electricity. The sizing, mounting, orientation, design, and integration of those technologies must be accurately done. At the early stage of the design, a simplified software tool such as RETScreen, PVSyst, PVWatts, or the System Advisor Model may provide enough accuracy to size a BIPV or a solar thermal system as it provides monthly estimates of energy generated. Renewable energy technologies require estimation of the heat recovered and how it can potentially be used—to heat ventilation air, to heat water, or space heating (directly or through a heat pump) (Henning, 2004). To properly simulate these systems, there is a need for tools characterized by a high integrity representation of the dynamic and connected processes (Athienitis et al., 2010). Designers should also allow for some margin of error and oversized PV systems because they never put out power exactly as calculated. Maintenance is very important to clear any dust or obstructions from the panels' glazing. Airborne particles and their accumulation on solar cells cut energy output by more than 25% in certain parts of the world (Paudyal and Shakya, 2016). The use of these systems as basic design elements of NZEB opens new architectural and technological possibilities. The growth of building integrated renewable systems is evident leading to a decrease of primary energy use and its future looks fastly assured.

4 PLUG LOADS AND ELECTRIC LIGHTING

Plug loads and electric lighting can reach up to 40% in NZEB (Kaneda et al., 2010). This makes them a significant part of energy needs that is dependent on occupant's behavior and preferences, Over long periods of use and under the rebound effect, occupants can increase plug loads and electric loads due to the incremental proliferation of appliances, delivery and lighting fixtures (Attia, 2016). It is imperative to consider them during NZEB design and operation.

4.1 Plug Loads

The NZEB definition includes the plug loads from the balance boundary as described in Chapter 2, Evolution of Definitions and Approaches. NZEB must have low plug and process loads, when compared with conventional buildings, for home appliances such as refrigerators, cooking devices, office equipment, computers, and other electrical equipment to achieve zero. Case studies confirm the plug and process loads are one of the largest energy end-uses in NZEB (Kaneda et al., 2010). Therefore, every watt counts and for every watt saved the design team can avoid PV needed to offset that watt.

Lessons learned from practice indicate that reducing plug loads and process load energy use can only be achieved by submetering the building energy down to the circuit level. Occupancy sensors and load shedding devices should be installed in most building outlets, including wireless controls. Advanced power strips are convenient and low-cost solutions that can help reduce the electricity wasted when these devices are idle (Sheppy and Lobato, 2011). They are designed where there are many consumer electronics plugged in. They work by preventing electronics from drawing power when electronics are off or not being used (US DOE, 2017). Fig. 7.6 illustrates the case study of the Research Support Facility (RSF) of the US National Renewable Energy Lab (NREL) in Golden, Colorado (see Chapter 11: NZEB Case Studies and Learned Lessons for case studies). For this NZEB fluorescent task light, personal space heaters, desktop computers, and desktop printers are replaced with sensor-controlled Light Emitting Diode (LED) task lights (15 W), LCD energy efficiency monitors (24″ screens of 25 W), laptops and thin client computers (35 W). Telephones have been changed from standard units that consume approximately 15 W each to voice over internet protocol units that consume 4 W each. Even data projectors (300 W) are replaced with LCD (140 W) screens (Torcellini et al., 2010).

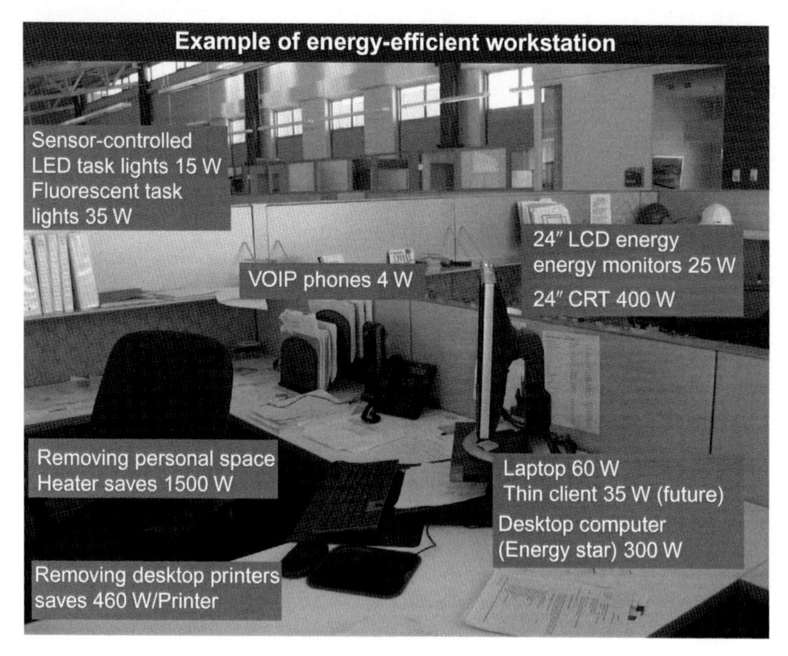

FIGURE 7.6 Energy-efficient workstation in the RSF Building, NREL, Golden, Colorado.

To ensure that plug loads and process loads are low, we list the key strategies that need to be addressed by the design team and owner team:

- Plug loads should be addressed and regularly monitored in NZEB projects. This includes plug load management systems that provide easy switching and monitoring-based feedback. As well as submetering at the circuit level. The building system performance should be separated from individual devices and equipment.
- Night plug loads energy use contributes to a large fraction of wasted energy. Therefore, computers and unused office space equipment must be truly turned off at night and when not in use.
- Postoccupancy evaluation should consider occupants' needs and revise the operation and control strategies to better meet occupant needs and reduce the plug loads.

4.2 Electric Lighting

Smart and energy efficiency lighting is an important step to reduce NZEB energy demand. In Chapter 5, Occupants Well-Being and Indoor Environmental Quality, we discussed the best practices for daylight and visual comfort. In this section, we focus mainly on electric lighting.

Depending on the brightness that can be achieved by properly designing the openings in a NZEB, electric lighting will make up the missing part. Lighting efficacy is measured by calculating the lumens of lights delivered per watt of electricity. This includes both light bulbs and fixtures that distribute and direct the light. For NZEB, smart controls are essential to detect users' movements and deliver light depending on daylighting intensity. Separating the task and ambient lighting is an effective energy efficiency measure because it shifts the control concept from universal to individual control. Instead of using T-8 electrical lighting (25 W), which is designed at a weighted average of 7 W/m^2, and provides 300 lux at the desktop, the lighting control concept can be developed to be as simple as possible and still respond to available daylight. Manual switches, photosensors, and occupancy sensors ensure electric lighting is provided only when daylighting cannot supply adequate light.

Open offices can be controlled with local daylight dimming sensors and a building occupancy schedule. LED task lighting provides localized electric lighting control to office spaces—each consumes only 13 W. For corridors and enclosed spaces, vacancy sensors can be used to encourage appropriate occupant interaction with the lighting controls. The control concept is to encourage occupants to turn their lights on if daylighting is not sufficient, and to turn them off when they leave the space (Torcellini et al., 2010). Vacancy sensors will turn the lights off if the occupants are not present. Finally, LEDs and ultra-efficient lighting solutions lesson total energy consumption and coupled to daylighting strategies should be used in balance while maintaining optimal visual comfort. Energy saving control solutions and daylight control algorithms can help to reduce the electrical power share of lighting. Wireless switches, dimmers, and controls are essential in NZEB.

5 CONTROLS AND LOADS OPERATION

Controls and loads operation became more critical in achieving and maintaining energy and operation expectations in NZEB. A control system in the building is designed to serve as a platform for monitoring and controls in the building. Controls are programmable devices that enable high resolution monitoring of HVAC, lighting, and metering systems. They include smart power strips, outlet level controls, and centralized power management—all incorporated with energy-use dashboards and occupant feedback. Loads operation is essential to balance and stabilize the building performance regarding HVAC equipment loads, plug loads, and lighting loads. Controls and load operation provide feedback to maintain comfort and energy operation in a robust way. Therefore, we briefly discuss the controls and loads operation of NZEB next.

5.1 Controls

Controls are playing a large part in ensuring a NZEB is operating as efficiently as possible. For example, occupants' sensors guarantee optimal comfort without scarifying energy efficiency by automatically adjusting thermostats and directing the air flow louvers depending on occupants' needs. Equipment control solutions (or intelligent equipment) also offer facility managers direct access to data points to help achieve a more accurate power measurement at the equipment level. Intelligent equipment offers automated demand response and predictive control to shift and shed load to lower-priced time, or avoid energy use during peak demand periods. Controls that are coupled to HVAC, lighting, and plugs can reach up to 15% savings of the total annual energy demand of NZEB (Higgins et al., 2015).

To manage and control the use of the HVAC systems, control technology is necessary to react depending on climate conditions, energy availability, types of energy supplied, and energy costs. Building control and management systems increase the ability of building managers and operators to integrate many desperate building systems into a functional high-performance building. Building Management Systems (BMS) help owners and facility managers analyze energy performance and operate HVACR and lighting systems at optimum levels and discover potential system failures and errors. Based on the study of Higgins et al. (2015) there are three best practices to apply in NZEB:

- Use integrated and low-cost controls systems. This includes integrating end-uses, and wired and wireless connectivity between sensors and controls (e.g., light levels, occupancy, CO_2, and temperature).
- Use standardized protocol for network entries and automation of the control management.
- Use adaptive controls that can learn and respond to occupant-based needs and preferences.

5.2 Part-Load Operation

Building control systems continuously monitor installed components and use efficient algorithms to optimize the load operation. Load operation incorporates sophisticated controls strategy such as demand ventilation, economizer based on enthalpy, chilled water reset, heat recovery wheel optimization algorithm, and chiller performance optimization.

NZEB rely on variable-speed technologies to ensure only the amount of heating or cooling that is needed will be delivered. This includes drive-motors solutions for air handlers, as well as roof top units and pumps. Also, for variable-speed compressors an inverter or variable frequency drive (VFD) runs the motor at different speeds, thereby modulating refrigerant flow and cooling output. Speed reductions that precisely match the load help reduce energy consumption up to 50% or more, depending on the application (Georges et al., 2017). Less energy is wasted since the variable-speed compressor delivers precisely the capacity required by speeding the motor up or slowing it down. The VFDs can drive elevator systems as well as geothermal systems.

Load operation and management can configure various indoor comfort condition algorithms such as operation based on schedule, custom adaptive comfort equation, the equation of outdoor/indoor conditions, or the predicted mean vote algorithm. Key energy and operational parameters should be continuously displayed on a display screen. Based on the study of Higgins et al. (2015), there are three recommendations to apply in NZEB:

- Integrate the controls subcontractor during the early design phases to become part of the design team through to occupancy.
- Increase operator training and support for improved engagement and control access.
- Open the team's perspective to emerging trends and awareness of (1) integrated, wireless and adaptive controls, (2) user feedback and dashboards, and (3) DC systems and renewable integration.

6 DISCUSSION

The design of simple and effective energy systems and the control and operation of loads in NZEB requires the adoption of a holistic and integrative approach. Adopting an integrative holistic building approach to building technology helps improve the chances of meeting the technical design challenge of NZEB. Lessons learned from practice indicate that NZEB owners and operators can get easily overwhelmed by the flood of information, wide range of technologies, and available technical equipment and solutions. Combining the personal preferences of stakeholders and varied expectations on the particular project can lead to complex and unfit-to-purpose planning concepts and realization. For NZEB, designers must simplify the system design and technologies. There is a high risk associated with not simplifying the heating, cooling, and ventilation systems together with renewables which can lead to challenging load operation and loss of control effectiveness and consequently jeopardizing comfort. Moreover, in many NZEB users suffer during the first two years of occupation and complain about discomfort. Cumbersome learning curves to optimize the systems operation, comfort issues, and load management challenges are often related to energy systems calibration and load management. Soft-landing, monitoring, and postoccupancy evaluation are very important during the early years of occupation. Chapter 10, Occupant Behavior and Performance Assurance, elaborates on this issue and provides recommendations on the role of continuous commissioning and controls adaptive adjustment.

> It is important to simplify the building services and building controls to better guarantee high comfort and user satisfaction.

The use of BPS and BIM tools can help in selecting highly efficient heating and cooling plants with a significantly reduced auxiliary energy while tracing the ventilation system components and modeling expected performance. The role of MEP contractors is to do the detailing, fabrication, spatial coordination, and installation using BIM models to make sure the system is optimized and complete before installation—and performs as designed after operation. Bringing the operation team or facility manager on board early into the design process is another key to success. Operators involved in all sequences development efforts including the commissioning process and continuous monitoring is a best practice for NZEB. The holistic and integrative design approach will create synergies and can lead to the design, installation, and operation of simple and effective HVAC systems and integrated renewables while managing and controlling building demand and services.

The second important lesson learned is that the selection of energy systems for NZEB is climate- and user-dependent. In Central and Northern European countries, energy recovery ventilators are coupled to hydronic heating systems, while in the United States and Canada, forced air furnaces is the norm and used more commonly in high-performance buildings (Klingenberg et al., 2016). Therefore, MEP designers should seek to achieve a global optimum of thermal comfort, air quality, and primary energy use and avoid overheating risks. The reduction of auxiliary energy use for heating, cooling, energy generation, distribution, and delivery must be optimized and coupled to renewable energy from environmental heat and electric sources, including solar power, geothermal energy, and biomass. Electric loads and plug loads should be low and rely on high-efficiency electrical appliances. The optimal HVAC and renewable system should be carefully coordinated with optimal solar protection, thermal insulation, thermal storage capacity, and airtight building envelope. For NZEB, it is important to follow the recommendations below:

- HRV is essential in NZEB energy systems. Select and size the HRV systems properly.
- Separate heating and cooling from mechanical ventilation to guarantee a simple operation and control by users for their indoor environment. Based on lessons learned in Chapter 10, Occupant Behavior and Performance Assurance, NZEB users and operators can be easily dissatisfied in fully automated systems that do not allow them to interact and feel they control the HVAC system.
- Select a heating system that covers hot water or domestic hot water (DHW) energy demands at approximately the same temperature, or even lower than hot water temperature. The heating system should primarily provide hot water plus heating as and when needed.
- Select a solar thermal system (collectors or evacuated tubes) with a large buffer storage and significantly oversize the area of collectors to cover the annual demand for both space heating and hot water. Depending on the climate, a back-up system of biomass (pellet, wood or other) heating system, or electric (direct electric heater or heat pump) should be combined with the system to meet the hot water demand during extremely cold periods.
- Electrification of NZEB is becoming a trend and is expected to increase in the future. It is worth it to take this into account when designing building services of NZEB. Heat pumps become a particularly important technology in this scenario. NZEB are expected to switch from fossils fuels to electricity with the integration of heat pumps. Even heating radiators are expected to be replaced by underground heating systems. Heat pumps can provide the heat required for space heating and hot water—including DHW in residential buildings. At the same time, heat pumps can be used in a reversible mode to provide active cooling in the summer. The performance of heat pumps can be coupled to PV systems to compensate the electricity consumption. The use of storage tanks or thermo-active storage banks (TASB), electric bikes, and electric cars can balance the demand and production curves as well as balance the daily electric loads mismatch.
- Hydronic radiant heating and cooling systems including TABS, radiators, and cooling panels suspended from the ceiling are very suitable for NZEB. Air cooling or heating is not recommended in NZEB. Hygienic air should be delivered to the chilled beams or underground heating slabs, and do not recirculate air.

For cooling-dominated commercial NZEB, zones should be cooled using cold ceiling chilled beams while outside air is strictly controlled and mainly used to meet the minimum of air exchange rate per hour indicated in Chapter 5, Occupants Well-Being and Indoor Environmental Quality.

- Personal HVAC is becoming essential to increase occupants' satisfaction and achieve, at the same time, the expected high energy savings. The AEC industry is increasingly looking at personal, or portable, HVAC technologies that are closely connected to individuals in a commercial NZEB or home. In this context, cooling and heating is focused on an individual or groups of individuals instead of the space or room itself. These energy control technologies will be capable of optimizing the energy performance of NZEB.

Third, controls and load operation are critical and will become significantly important to achieve NZEB and net zero energy communities. Smart controls will be essential under the NZEB electrification scenario. In this context, electric cars and bikes are expected to get charged by the PV generated electricity when electricity is at its cheapest or when neighbors have excessive production. Smart control and operation will play key roles to manage electricity use and exchange. Smart dishwashers or washing machines will run around noon with energy generated by neighbors.

Controls and loads operation might appear simple, however, with the increasingly user-centered NZEB design, the occupant becomes empowered as an active operator. The level of sophistication to control and integrate the HVAC, renewables, and lighting systems in harmony is high. The management of building services including presence sensors, lighting sensors, CO_2 sensor, thermostat, dampers, and pumps depends on automated and centralized operation. With the diversity of these elements, the BMS can easily become too complex and the users and facility managers lose their control over the building. This complexity can hinder the easy access of users and operators to guarantee comfort and optimize energy consumption and production. The management optimization of complex controls systems in NZEB to meet comfort requirements and increase energy savings is not always easy. It is recommended, therefore, to simplify the management, control, and load operation as much as possible to empower users.

Fourth, commissioning and system control checks should be planned and executed to ensure that the system is operating as designed (NEBB, 2014). The OPRs are vital to energy systems commissioning (Building Commissioning Association (BCxA), 2016). The goal of commissioning is to have a set of building systems which work together smoothly and reliably as the designers and owners intended. For example, MVHR needs to be ducted correctly and airflow rate balance needs to be verified. Controls must be checked for function and activation. Noise control and fire prevention should be checked to undercut air and noise transfer. Grills, fans, and ducts should comply with national fire regulations. MVHR filters should be changed after 6 months and a full maintenance plan held once the building is commissioned. Special attention should be made to the ventilation system design. The MVHR system should be easily monitored and should achieve the required ventilation in the most energy efficient and fit-to-purpose manner. It is vital to avoid overheating and prepare the NZEB for heat waves. Removal of heat using ventilation requires high levels of air exchange and night cooling or purging might not be effective. Based on lessons learned from Chapter 11, NZEB Case Studies and Learned Lessons, delivering a high air flow rate in NZEB is (in many cases) blocked by ductwork and supply and extract values. Therefore, very serious consideration must be given to providing active cooling. As mentioned in Chapter 4, Integrative Project Delivery and Team Roles, commissioning should be included in the OPR and BOD. The Commissioning Authority should be part of the predesign phase to formulate the commissioning criteria and commission process in the OPR (Table 7.1).

7 LESSON LEARNED # 7

The high-performance envelopes of NZEB lead to significant energy savings and, more importantly, downsizing of mechanical systems (cooling and heating plants, ductworks, floor heating) and associated electrical and operation systems. However, for NZEB lessons learned indicate that the reduction of heating and cooling loads and downsizing of mechanical systems is associated with a risk that the comfort of occupants is neglected. In order to realize high energy savings and provide optimal IEQ for NZEB, a fully integrated and iterative design process is essential. The MEP design team, Commissioning Authority, and operator must be bought into the design process as early as possible. Operators should be involved in all sequence development efforts. The commissioning process should start during the predesign phase and defined and documented in the OPRs and BOD.

Many high-efficiency design strategies involving HVAC systems (such as heat exchangers, thermally activated building systems, heat pumps and high-volume, low speed fans) allow downsizing of HVAC systems, equipment, and fans. Investing in a high performing envelope can lead to low or negligible need for heating systems and heating demand. However, HRV remains crucial for NZEB and poses the most challenge of integration with the heating and cooling systems as well as the load operation. The HVAC systems in NZEB tend toward radiant

TABLE 7.1 Example of Compliance Form Within the OPR

Owner's Project Requirements (OPR)

Example of Compliance Form

COMPLETE AND INCORPORATE THIS FORM INTO THE PLANS

Project Address: ____________________ Permit Number: _________-_________-________

Item #	OPR items	Page Number in OPR Document
PROJECT PROGRAM		
1	General building information (size, stories, construction type, occupancy type and number)	
2	Intended uses and schedules	
3	Future expandability and flexibility of spaces	
4	Quality and/or durability of materials and desired building lifespan	
5	Budget or operation constraints	
ENVIRONMENTAL AND SUSTAINABILITY GOALS		
6	Level of compliance with the Building Standard: Mandatory	
7	Specific environmental or sustainability goals (e.g. water efficiency, water reuse, CO_2, monitoring, xeriscaping, etc.)	
ENERGY EFFICIENCY GOALS		
8	Overall efficiency of building: meet Local Energy Code or exceed by (%)	
9	Lighting system efficiency: meet Local Energy Code or exceed by (%)	
10	HVAC equipment efficiency and characteristics	
11	Other measures affecting energy efficiency desired by owner (e.g. Building orientation, shading, daylighting, natural ventilation, renewable power, etc.)	
INDOOR ENVIRONMENTAL QUALITY REQUIREMENTS		
12	Lighting	
13	Temperature and Humidity	
14	Acoustics	
15	Air quality, ventilation, and filtration	
16	Desired adjustability of system controls	
17	Accommodations for after-hours use	
18	Other owner requirements (e.g., natural ventilation, daylight, views, etc.)	

References: ANSI (2014), McFarlane (2013), BCxA (2016), NEBB (2014).

heating/cooling ground and air source heat pumps and variable refrigeration flow systems. Cost, primary energy, and carbon emissions are also important to determine the fit-to-purpose heating systems. Also, we would like to highlight the importance of systems selection and accurate load calculations when considering mechanical solutions. Promising low-energy cooling techniques, such as evaporative cooling with solid or liquid desiccants, are good energy-efficient cooling systems. For heating-dominated NZEB, low temperature water-based heating systems are a key element. Hydronic water systems such as GSHPs or water-based TABS remain difficult to install properly. Finding qualified installers for hydronic systems, especially outside Europe, remains challenging.

At the same time, solar energy systems are becoming dramatically cheaper year after year, which can accelerate the process of loads electrification and the reliance on air and water source heat pumps. Electricity from PV and BPIV in sunny locations is essential for NZEB. PV energy systems should be oversized to avoid the margin of error. PV energy system operation and maintenance should consider the full performance of the system typically after about 25 years (NREL, 2017). Solar thermal energy with seasonal storage is already a robust technology. District energy and heat storage can be combined as a backbone for future fully RES and NZEB electrification.

Ultra-efficient appliances and advanced lighting systems involving day lighting controlled dimming remain essential for NZEB. However, we should not forget that sophisticated BMSs and controls remain expensive and require skilled designers and installers. Commissioning and continuous monitoring are the only way to assure that energy systems and controls operate in harmony. For NZEB, HVAC and renewable systems contractors have no room to put together systems and components that do not match the intended design or fail to provide comfort, efficiency, and reliability. Energy modeling is the best tool that MEP engineers can use to design high performance and cost-effective HVAC systems for NZEB. Finally, empowering users with maximum interaction with HVAC systems and controls is the key to achieve expected energy savings. NZEB occupants are more aware and interested in the benefits of higher efficiency HVAC systems, app-driven systems, and appliances than those that complement today's connected homes and workplaces.

References

ASHRAE 55, 2013. Thermal Environmental Conditions for Human Occupancy. American Society of Heating, Refrigeration and Air-Conditioning Engineers, Atlanta, USA.

Athienitis, A., Torcellini, P., Hirsch, A., O'Brien, W., Cellura, M., Klein, R., Delisle, V., Attia, S., Bourdoukan, P., Carlucci, S., 2010. Strategic design, optimization, and modelling issues of net-zero energy solar buildings. In: Proceedings of Eurosun 2010.

Attia, S., 2012. A Tool for Design Decision Making-Zero Energy Residential Buildings in Hot Humid Climates (Ph.D. thesis). UCL, Diffusion universitaire CIACO, Louvain La Neuve, ISBN 978-2-87558-059-7.

Attia, S., Carlucci, S., September 2015. Impact of different thermal comfort models on zero energy residential buildings in hot climate. Energy Build. 102, 117–128, ISSN 0378-7788, http://dx.doi.org/10.1016/j.enbuild.2015.05.017.

Attia, S., 2016. Towards regenerative and positive impact architecture: a comparison of two net zero energy buildings. Sustain. Cities Soc. 26, 393–406.

Building Commissioning Association (BCxA), 2016. New Construction Building Commissioning Best Practice. Building Commissioning Association, Beaverton, OR.

Connolly, D., Mathiesen, B.V., Østergaard, P.A., Möller, B., Nielsen, S., Lund, H., Persson, U., Werner, S., 2013. Heat Roadmap Europe 2050. Second Pre-Study for the EU27, 236 pp.

Georges, E., Cornélusse, B., Ernst, D., Lemort, V., Mathieu, S., 2017. Residential heat pump as flexible load for direct control service with parametrized duration and rebound effect. Appl. Energy 187, 140–153.

Garde, F., Donn, M., 2014. Solution Sets and Net Zero Energy Buildings: A Review of 30 Net ZEBs Case Studies Worldwide, Technical Report of Subtask C – DC.TR1.

Georges, L., Massart, C., Van Moeseke, G., De Herde, A., 2012. Environmental and economic performance of heating systems for energy-efficient dwellings: case of passive and low-energy single-family houses. Energy Policy 40, 452–464.

Harvey, L.D., 2012. A Handbook on Low-Energy Buildings and District-Energy Systems: Fundamentals, Techniques and Examples. Routledge, London.

Henning, H.M. (Ed.), 2004. Solar-Assisted Air-Conditioning in Buildings: A Handbook for Planners. Springer Vienna Architecture, Vienna.

Higgins, C., Miller, A., Lyles, M., 2015. Zero Net Energy Building Controls: Characteristics, Energy Impacts and Lessons. NBI, New Buildings Institute, Continental Automated Buildings Association, Ottowa.

Hopfe, C.J., McLeod, R.S. (Eds.), 2015. The Passivhaus Designer's Manual: A Technical Guide to Low and Zero Energy Buildings. Routledge, New York.

Kaneda, D., Jacobson, B., Rumsey, P., Engineers, R., 2010. Plug load reduction: the next big hurdle for net zero energy building design. In: ACEEE Summer Study on Energy Efficient Buildings, pp. 9–120.

Kavanaugh, S.P., Rafferty, K.D., 2014. Geothermal Heating and Cooling: Design of Ground-Source Heat Pump Systems. ASHRAE, Atlanta, USA.

Klingenberg, K., Kernagis, M., Knezovich, M., 2016. Zero energy and carbon buildings based on climate-specific passive building standards for North America. J. Build. Phys. 39 (6), 503–521.

Levihn, F., 2017. CHP and heat pumps to balance renewable power production: Lessons from the district heating network in Stockholm. Energy . Available from: https://doi.org/10.1016/j.energy.2017.01.118 Retrieved 15.03.15.

Lund, H., Möller, B., Mathiesen, B.V., Dyrelund, A., 2010. The role of district heating in future renewable energy systems. Energy 35 (3), 1381–1390.

McFarlane, D., 2013. Technical vs. Process Commissioning: Owner's Project Requirements. ASHRAE J. 55 (8), 32.

NEBB, 2014. Procedural Standard for Whole Building Systems Technical Commissioning for New Construction, fourth ed. National Environmental Balancing Bureau, Gaithersburg, Maryland, USA.

NREL, 2017. New Best-Practices Guide for Photovoltaic System Operation and Maintenance, NREL/Sandia/Sunspec Alliance SuNLaMP. *PV* O&M Working Group, second ed. National Renewable Energy Laboratory, Denver, USA.

Paudyal, B.R., Shakya, S.R., 2016. Dust accumulation effects on efficiency of solar PV modules for off grid purpose: a case study of Kathmandu. Sol. Energy 135, 103–110.

Pitts, A., Saleh, J.B., 2007. Potential for energy saving in building transition spaces. Energy Build. 39 (7), 815–822.

Sheppy, M., Lobato, C., 2011. Assessing and Reducing Plug and Process Loads in Commercial Office and Retail Buildings. NREL is a national laboratory of the U.S. Department of Energy, Denver, USA.

Torcellini, P., Pless, S., Lobato, C., Hootman, T., 2010. Main street net-zero energy buildings: the zero energy method in concept and practice. ASME 2010 4th International Conference on Energy Sustainability. American Society of Mechanical Engineers, Phoenix, AZ, pp. 1009–1017.

Vargas, G., Lawrence, R., Stevenson, F., 2017. The role of lobbies: short-term thermal transitions. Build. Res. Inf. 1–24.

US DOE, 2015. Department of Energy 10 CFR Parts 429 and 430 (PDF). Energy.gov. U.S. Department of Energy. Retrieved 15.01.18.

US DOE, 2017. Plug Load Strategies for Zero Energy Buildings. Better Buildings Solution Center, USA.

Further Reading

Vents US, 2017. Available from <http://vents-us.com/images/image/Frigate-ERV-HRV-TRV-100-150-DS-opisanie-500.gif> (accessed June 2017.).

CHAPTER

8

Smart-Decarbonized Energy Grids and NZEB Upscaling

ABBREVIATIONS

AEC Architectural, Engineering, and Construction
ASHRAE American Society of Heating, Refrigerating, and Air-Conditioning Engineers
BIM Building Information Modeling
BMS Building Managements Systems
BPS Building Performance Simulation
BREAAM Building Research Establishment Environmental Assessment Method
CCS Carbon Capture and Storage
CDD cooling degree day
DEMS Decentralized Energy Management Systems
DGNB German Sustainable Building Council
DRP demand response programs
EFS energy flexible systems
EN European Union
EPBD Energy Performance of Buildings Directive
FSGIM Facility Smart Grid Information Model
GHG greenhouse gas
HDDs heating degree days
HVAC Heating, Ventilation, and Air Conditioning
HVACR Heating, Ventilation, Air Conditioning, and Refrigeration
IAQ indoor air quality
ICT Information and Communications Technology
IDP integrative design process
IEQ indoor environmental quality
IoT Internet of Things
ISO International Organization for Standardization
LEED Leadership in Energy and Environmental Design
M&V measurement and verification
MEP Mechanical, Electrical, and Plumbing
MERV Minimum efficiency reporting value
NENA National Electrical Manufactures Association

Net Zero Energy Buildings (NZEB)
DOI: https://doi.org/10.1016/B978-0-12-812461-1.00008-3

NZE	Net Zero Energy
NZEB	Net Zero Energy Buildings
OECD	Organization for Economic Co-operation and Development
RES	renewable energy systems
RTP	Real-Time-Pricing
SESP	smart energy service provider
SMR	steam methane reformer
TES	transactive energy systems
ToU	Time-of-Use
WELL	WELL rating system

1 INTRODUCTION

To get the best economic and environmental benefit of NZEB, a major redesign of energy grids is necessary. NZEB do not stand alone in the built environment apart from the energy infrastructure. They will need to be integrated in the wider energy infrastructure that allows interactions between the grid and buildings based on a demand response mechanism to reduce load at peak hours. Climate change carbon reduction targets put pressure on cities and urban agglomerations to cut emissions and invest in high-performance buildings. NZEB play a major role in this energy transition. The challenge of decarbonizing the energy demand in cities basically depends on three components:

1. Design of smart NZEB.
2. Flexible and smart operation of NZEB.
3. Connecting to smart and decarbonized energy grids.

As shown in Fig. 8.1A, there is a strong relationship between the above mentioned goals. As discussed in Chapter 2, Evolution of Definitions and Approaches, the balance between energy savings in NZEB and local renewable energy generation depends on the flexible interaction and operation of energy grids and the level of their decarburization. By energy grids we mean electric and thermal grids. As shown in Fig. 8.1B, grid interaction is crucial for upscaling NZEB and making the best of them.

In this chapter, we explore these three challenges and investigate the opportunities to turn current grids into two-way interactive infrastructures. This chapter is dedicated to power companies' experts, residential, commercial and industrial consumers, producers of automated control and monitoring products, and facility managers.

We discuss the importance of designing NZEB as smart buildings and reinventing energy grids as smart, connected, and reliable energy systems. We also highlight the importance of integrating storage to stabilize the interaction and make the best benefit of NZEB energy neutrality. We explore the potential of energy grid decarbonization and the

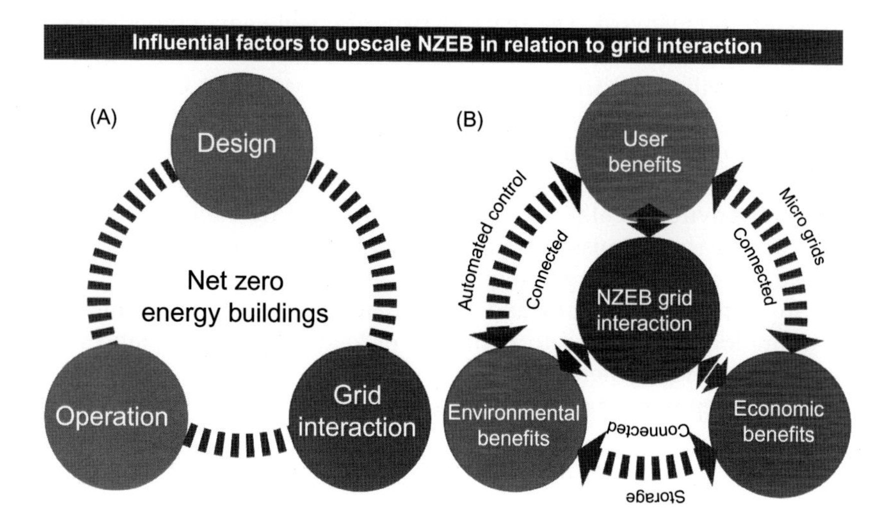

FIGURE 8.1 (A) Three influential factors to upscale NZEB and increase their proliferation, (B) the benefits of grid interactions and components to make grids smart.

implication of modernizing energy transmission and distribution systems that can meet future demand growth and assure the proliferation of NZEB while scaling them up.

2 PROBLEMS AND CHALLENGES

NZEB are grid connected and rely on the grid or networks to balance energy consumption. High-performance buildings that are well-insulated, airtight with optimized window openings design, rely on renewable energy systems (RES) such as solar panels, heat pumps, or the purchase of green electricity or green district heating or cooling (see Chapter 7: Energy Systems and Loads Operation). To ensure that the balance between generation and production is achieved at the lowest cost, the exploitation of the exchange flexibility has to be achieved on different scales including the building scale, district scale, and city scale. This means that bilateral reciprocal exchange and grid dependence of NZEB cannot be based on individual building entities. This makes grids central in the development of NZEB. Grids need to be smart, flexible, and decarbonized. There are three major benefits of smart and decarbonized grids.

On the economic level, managing demand through schedules and activities can reduce energy demand. Demand management coupled to pricing information based on the purchase contract or rate tariff will lead to minimizing cost. The management of on-site generation, storage, and power purchases based on grid monitors together with energy pricing signals will drive the decisions to import or export energy to the grid (Bushby, 2011). With the help of automated control systems in a NZEB, or in a smart grid, users and providers can achieve significant economic benefits.

On the environmental level, smart buildings (facilities) and smart grids can measure emissions and consumption to increase the use of variable renewable energy sources. Tracking and visualizing energy consumption and emissions and using real-time visualization and control technologies, can lead to reduction of load at peak times and manage on-site energy generation or storage (Bushby and Jones, 2016). This is a crucial step to decarbonizing an energy grid and influence occupant behavior in real time.

On the user and operational level, determining the demand target and available critical and curtailable loads can improve comfort while monitoring nearly optimal environmental and economic performance. Determining the capacity of on-site generation, energy storage, present demand, and user forecasts demand can influence the behavior of occupants and make it easier for service providers to improve operation while enabling better management of grid stress. Increasing the visibility of internal loads and grid loads allows users as well as operators to interact and improve operations on the building side as well as on the energy grid side (Bushby and Jones, 2016).

Next to the benefits of smart grids, there are five top consequences of NZEB that continue to push the building and energy industry forward:

- Active interaction and connection between the building and its occupants. Building users in NZEB are engaged through the ability to control the lighting and temperature wherever they are living or working. This active interaction is based on improving occupants' comfort and air-quality conditions—while saving energy at the same time.
- Control the sustainable technologies of the building including solar panels, heat pumps, boilers, electric batteries, and mechanical ventilation systems in a smart way. The smart control of systems is necessary to guarantee flexibility and reduce energy consumption and allow homes and businesses to manage their electricity use more effectively. This requires equipping NZEB with sensors and smart meters and real-time dashboards to provide feedback and act with the grid price signals, climate, and occupants' needs. This makes smart building control essential to automatically respond to demand and supply signals.

- This means that NZEB become an integrated part of the energy system on a national level. This energy system does not only deliver and supply energy, but also stores energy, exchanges it, transforms it, and shifts its use. The mismatch from hourly differences in energy production and consumption at building level require demand-side management (Lund et al., 2011). The demand side management and supply side management makes it essential to develop smart grids and smart micro grids. Therefore, an approach that only addresses the energy generation and consumption of a NZEB is incomplete.
- Storage is expected to operate at different scales across the energy system. NZEB do not necessary require energy storage systems. However, storage can ensure an increased percentage of self-supply and provide added value for the distribution grid. To run NZEB on a city scale, flexible, secure, and sustainable energy systems are required. With the expected fall of battery prices they will be used more and more in buildings to balance the energy consumption and generation mismatch.
- The Internet of Things (IoT) will help to connect devices and assure communication to realize an efficient NZEB. Monitoring, control and actuation systems will be tightly coupled to Building Management Systems (BMS) and grids. Bi-directional wireless sensors and actuation systems connected to central servers will collect and maintain databases of on-site solar generation, battery state of charge, and load power consumption data of a NZEB with the help of wireless networks. NZEB can optimally schedule loads between local generation and utility grid, thereby minimizing peak demand on the grid.

These five major consequences of the proliferation of NZEB require that governments set new standards and regulations that anticipate and consolidate the changes. NZEB must, therefore, include smart meters for occupants' greater control and record of energy use. Energy companies and suppliers should be allowed to set smart tariffs and encourage off-peak consumption. A whole set of regulations and decision chains need to be developed to operate distribution networks using interaction models, operational planning, real time control, and business models. This makes the challenge of achieving high-performance buildings, including NZEB, strongly associated with smart and flexible buildings and grid management.

3 SMART BUILDINGS

Smart NZEB have the ability to reduce the energy use of the buildings and to utilize RES, and combine them into packaged solutions.

They require continuous monitoring of the environmental performance of the building in order to provide feedback to users, or to interact with the grid signals. To achieve control-demand flexibility of NZEB, they require smart metering of devices and energy storage. In this section, we present the key elements of smart buildings and the importance of incorporating these elements into NZEB.

3.1 Control-Demand Flexibility

Smart NZEB bring several benefits to demand reduction and enable demand response, occupant interaction, and the integration of volatile renewable energy supply increase. By increasing on-site energy production and self-consumption, NZEB empower end-users and mitigate the energy stress of peal loads of energy systems. This includes self-consumption schemes that gives end-users incentives to use their energy in a more efficient and smarter way.

Smart NZEB depend mainly on smart metering and control of energy systems to achieve control-demand flexibility. Accurate measurement of the energy consumption using smart meters provides real-time feedback on energy consumption, comfort conditions, and building services responses. As a consequence, the purpose of a smart building is to make spaces more enjoyable, comfortable, usable, and productive for the users. The use of lighting controls and temperature controls improves the working conditions in offices while saving energy at the same time. Connecting users through smartphones and allowing them to virtually control their working environment to suit their working needs and preferences, is part of the current reality. The increase of indoor environmental quality (IEQ) expectations through recognized measurement rating systems and standards such as WELL, LEED, BREAAM, DGNB, EN, etc., is becoming the best practice to make building users lives more productive and enjoyable. NZEB should be equipped with smart and user-adapted metering and control systems based on universal communication protocols (De Groote et al., 2016). For example, HVAC systems being equipped with demand response technology within thermostats that connect to Wi-Fi. NZEB users can use their smartphone, tablet, or computer to control their HVAC systems. This allows energy companies to have better control over demand response events needed by users.

3.2 Smart Metering

Smart metering plays a crucial role in smart NZEB. Smart metering allows data acquisition and management of energy exchange between

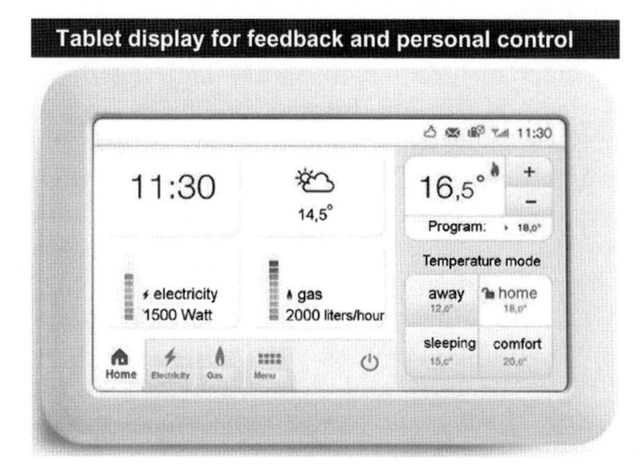

FIGURE 8.2 Illustrates the different aspects that need to be addressed regarding systems and loads operation of NZEB (Eneco Dashboard).

PV panel arrays and the utilities grid. Frequent and accurate digital measurements allow for automated reading of Time-of-Use (ToU) or Real-Time-Pricing (RTP) regime (Pitì et al., 2017). This contributes to reducing the mismatch between production and consumption (Allcott, 2009; Wang et al., 2015; Faruqui et al., 2009; Lijesen, 2007; Raj et al., 2015).

Another advantage of smart metering is its nature to provide direct feedback to building users making them aware of their energy consumption related behavior. The availability of information empowers end-users to control their consumption decisions and open the chance for a flexible use of devices and HVAC systems. Simple displays such as tablet and smartphone screens or web-based platforms fed by smart meters data can lead to better comfort, personal control and energy savings (see Fig. 8.2). According to the study by Hargreaves et al. (2010), D'Oca et al. (2014), and SCE (2012), real-time feedback is beneficial for building users when it is:

- Displayed on different devices (smartphones, tablets, PC devices, etc.).
- Visualized in an intuitive way through charts and readable aggregated results graphs.
- Available through a cheap process and easy-to-install settings.

Therefore, smart metering is an essential tool for NZEB. Smart meters allow for the balancing of the demand-generation-offer-curve. Also, they communicate with energy operators and distributions which in help saving costs and automating the I-devices and IoT.

3.3 Internet of Things

As part of smart metering and control and the demand response capabilities of smart NZEB, the Internet of Things (IoT) emerged. The IoT plays a major role in supporting users by raising their awareness regarding energy efficient applications and operation modes. The IoT is the inter-networking of physical smart devices in buildings including vehicles embedded with electronics, sensors, and actuators. As part of smart NZEB, the integration of the internet with building energy management systems is becoming vital. Real-time monitoring for reducing energy consumption and managing energy use is shifting in the AEC industry. The explosive growth in distributed energy resources and the broad proliferation of smart devices is transforming the way we design and operate NZEB. The emerging trend of IoT is leading to a more connected, responsive, distributed, smart, reliable, efficient, and sustainable built environment. As NZEB become connected to the internet, best practices for the IoT for installation and operation can be summarized below:

- Security of the connected devices should be the foundation of hardware and software installation and operation.
- Data normalization of products or devices with web services and exchange data with other devices.
- Data management and analytics infrastructure can ensure the deployments are effective, function properly, and connect the building software and hardware.

With the improvement of security and interoperability challenges, NZEB are expected to be equipped and managed with IoT technologies enabling cloud solutions and smart building management protocols.

3.4 Storage and Electric Vehicles

Energy storage is another important element of NZEB. A modest amount of energy storage can provide energy to a building when the sun is not shining during the day when a NZEB would be importing energy from the grid. Also, energy storage allows building owners to provide valuable services back to the grid, such as load shifting, demand response, spinning reserves, and frequency regulation. Depending on energy company or system operator incentives, these grid services can significantly improve paybacks for energy storage investments. For example, thermal storage can help cut overall building energy costs, by making ice at night to cool the building during day. Case study 5 in Chapter 11, NZEB Case Studies and Learned Lessons, shows a multiple storage system that combines short and long-term storage units.

For NZEB, it is important to discuss how the building can store energy. With a combination of wind and solar energy, it seems that there is a limited need for seasonal storage. Connecting wind, solar, and water energy, demand side management, and some batteries can bring us very far toward a secure energy system based on renewables.

Norway is planning to stop diesel cars by 2025, the Netherlands is planning to ban diesel cars by 2035, while France, Germany, and the United Kingdom have set the year at 2040. As a consequence, electric vehicle manufactures are announcing electrical energy storage batteries for buildings. Battery prices are currently expensive, but a significant price drop is expected in the coming years. Predictions are that cheap batteries will lead to an increase in electric vehicles and in the long-term, electric vehicles will dominate the market. This will require countries to increase their electric energy production between 30% and 40% by developing off-shore wind farms and inland solar farms to keep and respect the Paris Agreement's carbon emissions limits. If batteries become more compact, they can make buildings partially independent from the grid and lower the cost of on-site generated energy from PV arrays. In this way, batteries can turn NZEB into power stations and support the operation of micro grids (see Section 4).

Integrating batteries in NZEB are not meant to make them energy independent. However, battery storage can handle a building's uninterrupted renewable energy generation resource. In the long term, if energy storage systems become affordable they can provide long-term storage over days instead of hours (see Case Study 5 in Chapter 11: NZEB Case Studies and Learned Lessons). For example, a typical battery of 2 kWh per installed kWp PV in Central Europe can be used for a NZEB household. This can increase the percentage of self-consumed electricity to around 50% while reducing the amount of additional power imported from the grid (Peeters, 2017; Vaughan, 2017).

In the future, batteries will be at the heart of NZEB development because they can provide a balance to the variable nature of wind and solar generation. At the same time, batteries can decrease the energy demand supply mismatch. The impact of improving the building's storage capacity can improve the flexibility of national grids and local energy networks to cope with fluctuations. Using energy storage systems (ESS) can flatten out or shave the peak demand and avoid the cost for consolidating energy networks.

4 SMART GRIDS

A smart grid is an energy supply network that uses information technology to detect and react to local changes in building usage and energy generation stations. In this section, we explore the different concepts and challenges of smart grids and provide an overview of standards and best practices. Also, we explore micro grids as a new and important type of smart grid.

4.1 Concepts and Challenges

Prior to the discussion of smart grid challenges and future perspectives, we list key important definitions that characterize the new energy systems and networks in the built environment.

Smart grid: A smart grid is a grid which includes a variety of operational and energy measures to control the production and distribution of energy. The first official definition of Smart Grid was provided by the Energy Independence and Security Act of 2007 (EISA, 2007). A smart grid is characterized by (EISA, 2007; Kolokotsa, 2016):

(1) Increased use of digital information and controls technology to improve reliability, security, and efficiency of the electric grid.
(2) Dynamic optimization of grid operations and resources, with full cyber-security.
(3) Deployment and integration of distributed resources and generation, including renewable resources.
(4) Development and incorporation of demand response, demand-side resources, and energy-efficiency resources.
(5) Deployment of "smart" technologies (real-time, automated, interactive technologies that optimize the physical operation of appliances and consumer devices) for metering, communications concerning grid operations and status, and distribution automation.
(6) Integration of "smart" appliances and consumer devices.
(7) Deployment and integration of advanced electricity storage and peak-shaving technologies, including plug-in electric and hybrid electric vehicles, and thermal storage air conditioning.
(8) Provision to consumers of timely information and control options.
(9) Development of standards for communication and interoperability of appliances and equipment connected to the electric grid, including infrastructure serving the grid.
(10) Identification and lowering of unreasonable or unnecessary barriers to the adoption of smart grid technologies, practices, and services.

To ensure the balance between generation and production at the lowest cost, a smart grid should depend on a series of Decentralized Energy Management Systems (DEMS) or micro grids to assure better exploitation of energy systems flexibility. Based on Cornélusse et al. (2017), smart grids imply rethinking the whole decision chain used to operate and manage distribution networks. A smart grid comprises four elements:

- Operational planning
- Interaction models
- Investment in infrastructures
- Real-time control

Decentralized Energy Management System (DEMS): Distributed Energy Systems (DES) are independent energy generation, storage, and distribution systems. DES is based on monitoring and control solutions to assure two-way interactions between energy generation and production (see Fig. 8.3). Their main advantages are reducing transmission losses and lowering carbon emissions. Moreover, security of supply and grid stabilization on the national level is an additional benefit. DES covers energy in the forms of electricity, heating, and cooling. Initial installation costs are higher most of the time, but a special decentralized energy tariff creates more stable pricing (Karavas et al., 2015).

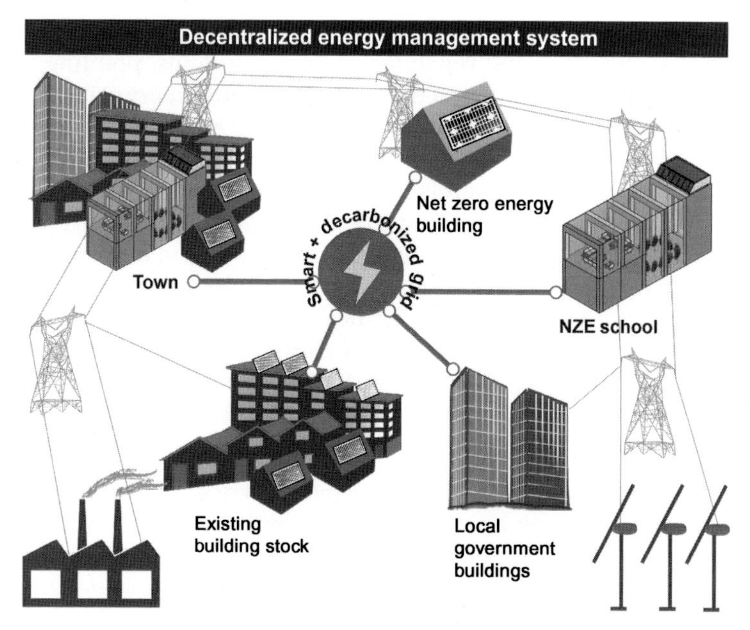

FIGURE 8.3 A schematic representation of Decentralized Energy Management System.

Smart energy service provider (*SESP*): Smart energy service providers are energy service companies that can provide multiple sources of energy, including renewables, batteries, and PVs. SESP sell bundled services enabled by smart meters and smart home technology. SESP provides better insight into energy management by the "smart" use of production resources and energy-consuming appliances. SESP is the enabler for demand response.

Demand response programs (*DRP*): Demand response programs are designed to decrease electricity consumption or shift it from on-peak to off-peak periods depending on consumers' preferences and lifestyles. Demand response is a reduction in demand designed to reduce peak demand or avoid system emergencies. "DRP are used by energy system operators as resource options for balancing supply and demand. Such programs can lower the cost of electricity in wholesale markets, and in turn, lead to lower retail rates. Methods of engaging customers in demand response efforts include offering time-based rates such as TOU pricing, critical peak pricing, variable peak pricing, RTP, and critical peak rebates. It also includes direct load control programs which provide the ability for power companies to cycle air conditioners and water heaters on and off during periods of peak demand in exchange for a financial incentive and lower electric bills (DOE, 2017)".

Energy flexible systems (*EFS*) or *transactive energy systems* (*TES*): The definition refers to the economic and control techniques used to manage the flow or exchange of energy within an existing electric power system (Atamturk and Zafar, 2014). It is a concept that is used in an effort to improve the efficiency and reliability of power systems, pointing toward a more intelligent and interactive future for the energy industry. In transactive energy, interoperability refers to the ability of involved systems to connect and exchange energy information while maintaining workflow and utility constraints (Atamturk and Zafar, 2014).

The consequence of increasing the market uptake of smart NZEB is to upgrade energy networks. Energy networks upgrade should enable smart grids that are not only applicable to the electrical power systems, but also to natural gas and district heating and cooling networks. The focus of OECD countries is mainly on electric networks toward flexible networks that adapts to the current development in energy production and consumption patterns and technologies. There are four major challenges related to smart grids:

- Enabling the flexibility of power systems to cope with intermittent and decentralized renewable generation and managing multifaceted interactions by developing new standards and best practices.
- Increase current network capacity to increase renewable generation and integration of smart grid systems technologies.

- Developing seasonal use pricing structure and regulations
- Exploration and investment into seasonal renewable energy storage technologies (see Fig. 8.4).

One of the assumptions of the NZEB concept is that it assumes the grid acts as a bank for local on-site energy production. Understanding that the grid is not a bank is key to recognize the impact of NZEB. NZEHB are high-performance buildings with on-site renewable generation. The RES is sized to balance their consumption annually using the grid as a credit system that stores energy for later use. In fact, the grid does not have the capacity to store all excess energy generated by NZEB. As a consequence, grids suffer from energy deficits during seasonal high use, leading to energy generation using fossil fuel sources. Therefore, NZEB under these conditions are still responsible for high carbon emissions generated by nonrenewable energy sources.

At the same time, there exists a large number of grid operators worldwide working in different environments, including SESP. For those operators, there is no "one size fits all" solution for grid management. There is a need for strategic knowledge for smart grid policies for grid operators, managers, and smart grids technology providers from one side and energy consumers on the other side. Therefore, we need to remember that the aim of smart grids is to increase the efficiency of

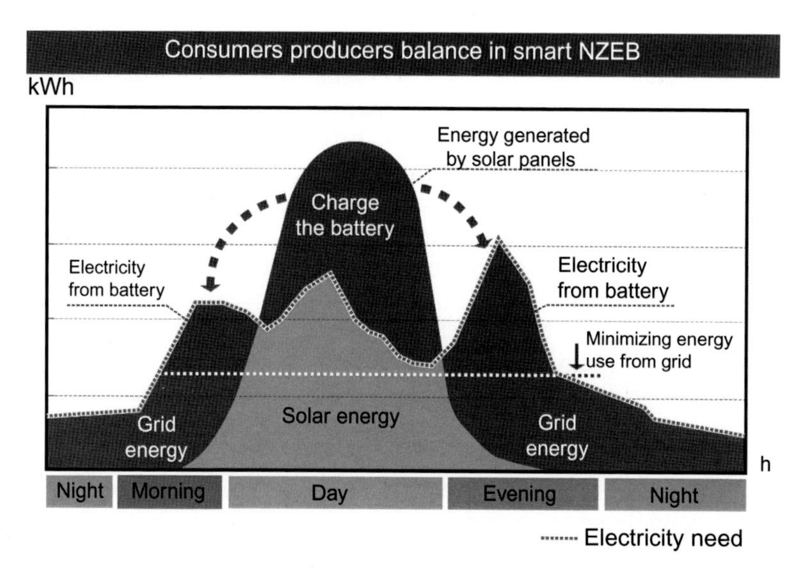

FIGURE 8.4 Example of an energy modeling web platform for the simulation of the consumers–producers balance in smart grids projects.

energy exchange and interoperability of the grid. New standards, protocols, and best practices can strengthen the existing electricity systems by considering cross energy carrier synergies for flexibility.

4.2 Smart Grid Standards and Best Practices

Numerous building level protocols exist in relation to smart grids. The best known is the ISO Standard 201-2016. A smart grid standard published by ASHRAE and the National Electrical Manufactures Association (NENA) has been approved as an ISO standard. There are several other international standards that aim to combine domain expertise to harmonize facilities and grid interactions. The list below identities the key standards that are applicable and used worldwide:

- FSGIM, Facility Smart Grid Infrastructure Model that describes, manages, and communicates electrical energy consumption and forecasts (NEMA, 2016).
- IEC, Common Information Model (CIM) for energy meter models (IEC, 2013a; NAESB, 2012).
- IEC, 61850 is a series of standards for energy generator models (IEC, 2003a,b, 2009, 2013b).
- OASIS WS-Calendar standard for representing schedules and time series of data (OASIS, 2015).
- OASIS Energy Market Information Exchange (EMIX) for modeling market interactions (OASIS, 2011a,b).
- Eurocontrol and American Federal Aviation Administration NextGen Network Enabled Weather (NNEW) and Weather Information Exchange Model (WXXM) for weather observations and forecasts modeling (Eurocontrol and NNEW, 2010).
- Energy Flexibility Platform and Interface (EF-Pi) aims to decouple Smart Grid services from customer appliances. The EFI effectively provides a common language for both sides, facilitating interoperability between all Smart Grid services and smart appliances (EF-Pi, 2017).
- TransEnergy P825—Meshing Smart Grid Interoperability Standards to Enable Transactive Energy Networks. This guide permits common transactive grid services to be exercised by connected Distributed Energy Resource assets behind the meter. The guide brings together a broad set of grid interoperability standards that will utilize the underlying IEEE1547 Interconnection conformity as an integration platform while leveraging multiple communications protocols.

NEMA, 2016, Facility Smart Grid Information Model (FSGIM) provides a common basis for electrical energy consumers to describe, manage, and communicate electrical energy consumption and forecasts (NEMA, 2016)—A standard which has four basic components that can be used to model real devices within a facility. The four components are:

- Load: Used to model devices that consume energy.
- Meter: Used to model devices that measure power, energy, or emissions.
- Generator: Used to model devices that produce or store energy.
- Energy Manager: Used to model devices that make decisions based on power, energy, emissions, price, weather, etc.

The standard has been approved by ISO/TC 205 Building Environment Design and has been published as ISO 17800 (2016). The standard provides one piece of a larger ecosystem of standards that support the transformation of the current electric grid into smart grid. The standard supports the two-way flow of both information and electricity as well as widespread use of distributed, renewable generation sources. The standard defines key information that must be shared between electricity providers and electricity consumers, along with internal operational and control information needed to control loads and generation sources in a building in cooperation with the smart grid. This standard is a seed standard to guide the evolution of control technology specific standards such as ANSI/ASHRAE Standard 135, BACnet—A Data Communication Protocol for Building Automation, and Control Networks.

Best practices in the domain of electricity transmission and distribution systems are emerging worldwide. They incorporate new sensing and control technology to monitor the grid, and use variable sources of energy generation —especially renewable energy sources. Best practices indicate the importance of using information technology for maximizing the feedback and interactions between NZEB and the grid. The most common information model designed to provide a standardized way to control all facilities in a smart grid environment is the ISO 201-2016 FSGIM (NEMA 2016). The model organizes the interaction of NZEB with the grid. The 201-2016 User's Manual provides examples of standard aggregations of demand, net demand, loads, curtailable loads, and forecasts to manage a NZEB in a smart grid. As shown in Fig. 8.5, the FSGIM represents the devices in a NZEB as combinations of four abstract components (Bushby and Jones, 2016). The goal of the FSGIM is to allow the aggregation of data from a wider variety of systems in a NZEB to facilitate the interaction with the electrical grid.

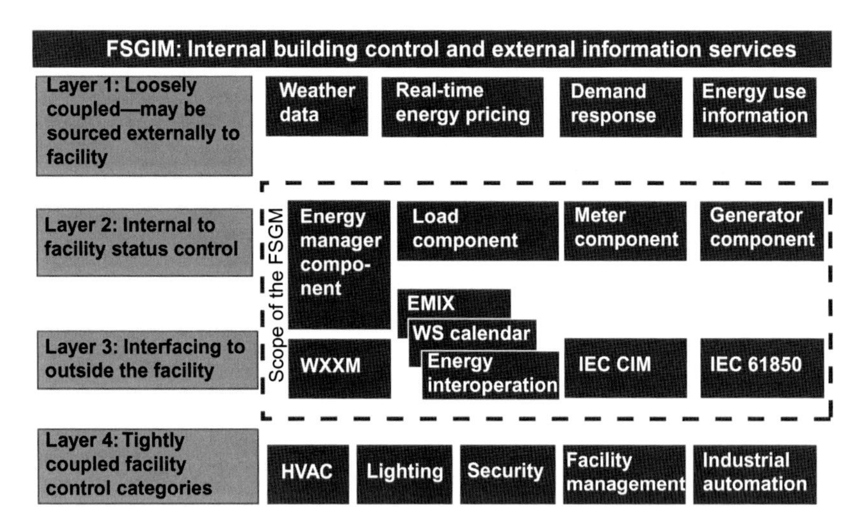

FIGURE 8.5 Representation of the FSGIM and the relationship between internal building controls and external information and services (Bushby and Jones, 2016).

4.3 Micro Grids

Batteries and PVs can turn a building into a power station. As shown in Fig. 8.6, Decentralized Energy Management Systems (DEMS) are independent energy generation, storage, and distribution systems often called micro grids (Karavas et al., 2015). A building that includes a battery or storage as well as RES such as PV, biomass, a heat pump, or diesel generator can also be called a micro grid. The definition of a micro grid is a group of interconnected loads and distributed energy sources within a building that act as a single controllable entity able to operate in both grid-connected mode or autonomous-mode (Barker, 2015).

The old definition of a micro grid was an electricity source, often a combined heat and power plant or reciprocating engine generator that provides full backup (Vince and Morsch, 2017). Today's definition is broader, incorporating renewable energy, smart technologies, and more flexible demand. Therefore, micro grids are a key driver and infrastructure to support NZEB proliferation and resilience. NZEB must be coupled to a micro grid to achieve efficient and low emissions electric systems. Public–private partnerships develop micro grids to power strategic facilities to get more resilient and affordable power together with environmental benefits. For example, LED lighting in streets, autonomous vehicles, and micro grids that use solar panels and battery storage are becoming a key element of the smart built environment. This includes energy storage technologies while helping to integrate more wind power.

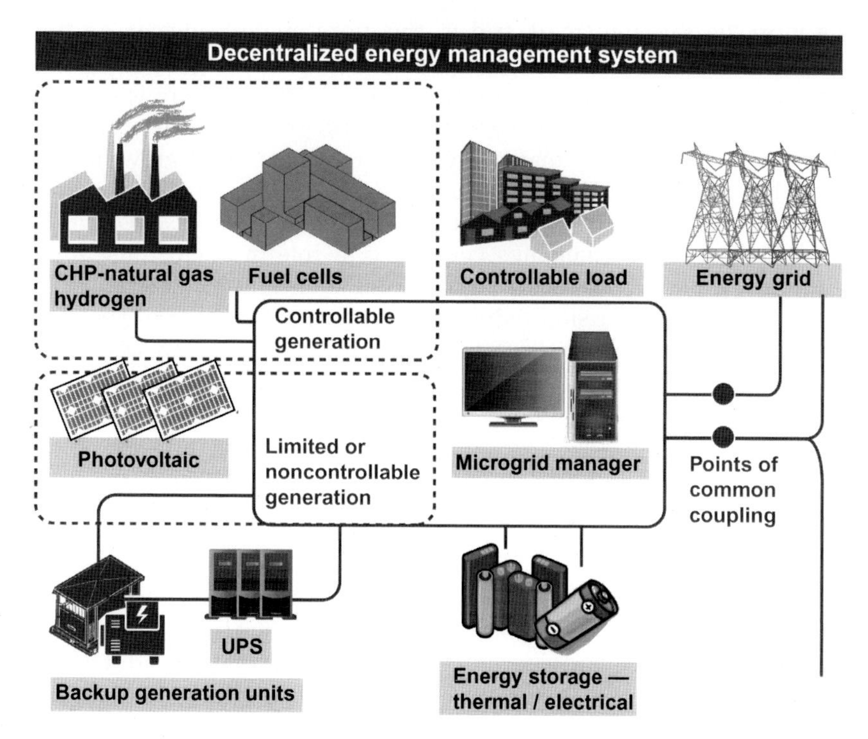

FIGURE 8.6 A schematic representation of Decentralized Energy Management System based on (Vince and Morsch, 2017)

Micro grids improve reliability and resilience of neighborhoods and districts and can increase the resilience, efficiency, better manage electricity supply and demand, and help integrate renewables, creating real chances to cut greenhouse gas emissions and reduce dependence on fossil fuels. The main advantage of micro grids is on the district scale because consumption and production peaks are managed locally using local storage capacities, demand-side management, and smart modulation of production. Micro grids provide a single access point on the district scale to the smart grid which can group many users and consolidate the structure and stability of smart grids. Micro grids can set up an intermediate solution to group multiple loads as well as several distributed energy sources, and operate them efficiently and cost-effectively in parallel with the broader utility grid.

Another advantage of a smart grid that relies mainly on micro grids is creating electric power systems which are resilient. The resilience allows communities to balance their consumption locally and face risks. Micro grids can deal with risks such as technological failures, terrorist or cyber-attacks, global shortages of supply, and even natural disasters.

The more flexible the smart grid is, the more micro grids can go fully off-grid (DOE, 2016). This needs advanced control strategies and storage capacities on the grid and micro grid level. Currently, there is a lack of legal definitions and frameworks for micro grids in most countries. However, it is expected that micro grid developers will grow and get access to reasonably-priced backup power and to wholesale power markets to sell excess electricity or services (Vince and Morsch, 2017).

Public–private partnerships could play a role in overcoming financial hurdles. Linear programming models (Vince and Morsch, 2017) can help cost savings, emission reduction, and forecast measures. The effective management of the power supply and demand can assure the smart grid's robustness and the independence of micro grids from the larger grid.

5 DECARBONIZED POWER GRIDS

To mitigate climate change, cities will need to increase their reliance on renewable energy resources. Government officials, energy companies, energy service providers, environment regulators, and research institutions stress the importance of promoting NZEB and climate-friendly policies to help promote renewable energy and low carbon grids. Climate change mitigation should be based on increasing the share of renewable energy, promote sustainable public transportation, and performance a transition from a gas to a hydrogen grid. In this section, we discuss the challenges of decarbonizing energy grids, the different scenarios for low carbon grids, and their implications and relation to NZEB.

5.1 Electric Grid and Renewables

Increasing renewable energy solutions to enhance energy security and build climate resilient cities is essential. As the shift toward adopting green technologies that can reduce GHG emission and increase renewables share, the carbon intensity of the electricity grid should continue to decrease. Grid electricity will be dominated by renewables in the near and mid-term future. The recent reduction in the carbon intensity of many cities has to do with the replacement of coal and liquid fossil fuels with gas. Meeting the climate target of the Paris Agreement will mainly be achieved by decarbonizing the electricity grid. The future vision of electricity grids is mainly coupled to gas and renewables. This inevitable change has a positive impact on NZEB. It allows reaching carbon neutrality faster than expected while promoting low carbon intensive green technologies.

At the same time, there is no technical challenge to get all of the required energy for local micro grids from renewables. All power could come from sources like wind, solar, and storage which are able to manage the reliability of the grid. Wind, solar, and battery prices plunged, while software to control storage and the grid has also advanced. Software can make grid adjustments and bring battery power online fast.

The spread of batteries and electric vehicle batteries that can be charged during the day can help balance daily and seasonal variation using dynamic demand management technology. If the electricity infrastructure for vehicles or NZEB gets upgraded, all power could come from wind, solar, hydro, and biomass. When thinking about the future of electric grids, we need to consider transportation.

Transport also needs to decarbonize. However, the existing capacity of most existing electricity grids is limited. This means that we will need to add 30%–40% extra grid capacity from renewables, or in the worst-case scenario from natural gases. As mentioned in Chapter 7, Energy Systems and Loads Operation, the electrification of NZEB through the use of heat pumps and PVs will be coupled to electric vehicles leading to the full reliance on renewable energy of electric grids. Autonomous micro grids will also be part of the solution landscape. For example, in Chapter 11, NZEB Cast Studies and Learned Lessons, we present an autonomous multi-family dwelling that is off-grid. Another advantage of electrifying transportation and heat for NZEB and reduce peak demand is that the electricity grid will get expanded and decarbonized.

5.2 Hydrogen Gas Grid

Hydrogen is created by a steam methane reformer (SMR) that transforms it into an odorless and low carbon intensive gas. The transition from gas grids to hydrogen grids raises the opportunity to capture up to 90% of the carbon dioxide created during the process. Hydrogen can be stored or distributed at a higher pressure because hydrogen's energy transmission capacity is approximately 20% lower than methane.

The idea of replacing natural gas with hydrogen as an energy carrier can be coupled to cooking, heating, and mobility. A hydrogen boiler, as shown in Fig. 8.7, can be an effective solution if the supply chain gets upgraded. Hydrogen must be distributed at a higher pressure so gas pipes (if used) need to be replaced or demand needs to decrease.

Safety issues require further development because hydrogen is colorless, odorless, and burns with an invisible flame. However, if these

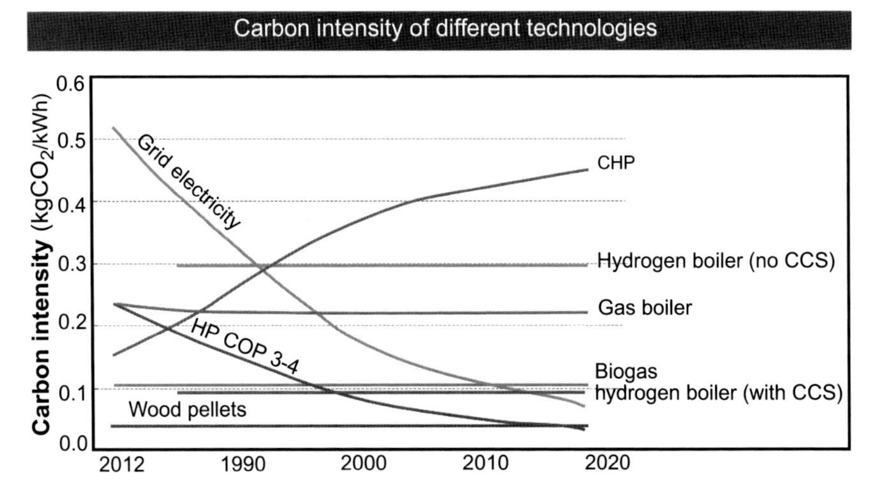

FIGURE 8.7 Comparison of carbon intensity of different heating technologies based on Davies and Pepper (2017).

safety issues are resolved, the benefit would be avoiding carbon monoxide poisoning from faulty boilers, fewer distribution losses compared to electricity, and lower carbon emission than gas and electricity grids. Also, storage should be taken into account to meet the demand fluctuations (Davies and Pepper 2017). According to the H21 Leeds Report (2016) the calculated carbon factor for converting the city of Leeds, United Kingdom to hydrogen is 0.086 $kgCO_2/kWh$. This can reach up to 58% carbon savings compared to natural gas.

The use of hydrogen in transport is already growing for buses and cars. Hydrogen with Carbon Capture and Storage (CCS) can lower vehicle emissions by 79% compared to the petrol alternative. Meanwhile, electric vehicles reduce carbon emission by 80%–95% depending on the grid decarburization. The switch from petrol to hydrogen transport is easier than converting the gas grid. However, a legal framework and incentive programs should stimulate the demand of clients and producers supply.

5.3 Gas Grid Future

By 2050, it is expected that all residential building will become off-grid. Already, more than 100 municipalities in the Netherlands declared the intention to be disconnected from the gas grid by 2030. A recent study published by Stedin (de Bruin, 2017) confirms the cost-effectiveness of avoiding a gas boiler for new NZEB. In their study, they proved that it is co-effective for building owners based on a 15-year calculation, which represents the life of a gas boiler. Instead of using gas boilers for heating, the use of heat pumps is more effective. The study investigated 5–6 housing typologies and included financial parameters such as mortgage rates, maintenance costs, and energy costs. The calculation was based on the 2016 data for energy costs, energy taxes, investment cost, and fiscal subsidies. A sensitivity analysis was done to estimate the effect of gas price increases and make sure that going all-electric is a feasible solution. The study did not include district heating systems. The study outcomes show that the all-electric solution is already a good alternative for new constructions. Based on our discussion in Chapter 7, Energy Systems and Loads Operation, the study conforms with the strategic interest to avoid gas grids, gas boilers, and network connections.

Already, most gas pipelines in many European cities were installed in the 1990s and are becoming outdated and will require replacement by 2015. The outdated infrastructure of gas pipes and installation of new pipelines will require huge investment. Therefore, the decision of local governments and municipalities on how to create a new smart technology infrastructure will guide the development of electricity or gas grids.

In short, new residential NZEB can be built without gas stoves. The residents are financially better off. However, this does not mean abandoning gas grid fully. Existing buildings and large-scale buildings might find it more cost-effective to rely on gas grids. We cannot generalize the study's findings which are based on the Dutch context. The study is another confirmation that going fully electric makes sense. It is up to countries, local authorities, and investors to decide their own energy mix and energy system infrastructure in a way that can decarbonize energy grids.

6 DISCUSSION

To mitigate climate change worldwide, there is a reliance on renewable energy to increase its share in the national energy mix, promote sustainable public transportation, and decarbonize energy grids. All evidence suggests that renewable energy will thrive and that energy grids will be an interoperable environment. There is a chance that a percentage of consumers will leave the central grid and opt for DEMS and micro grids. But, there is a technical challenge in many areas of the world to get all required energy from renewables. However, the main challenge remains in modernizing current energy grids and connecting them in a flexible way with NZEB to control storage and stabilize the grid. There are three main challenges that influence the growth of smart and decarbonized grids and consequently upscaling NZEB. Next, we will present these challenges as the core of this chapter's discussion section.

Forecasting tools and data storage systems need to be scaled up to accommodate for Big Data with the drastic increase of data quantities. The number of points of control on the supply side of energy companies is increasing exponentially. Managing millions of points of control in the energy generation system and linking them with communication and visualization platforms of million points of demand will be a serious challenge. Creating interoperable networks that allow communication between all points of supply and demand in which information can be exchanged in real time requires planning and investment as a critical part of the future, smart, and decarbonized energy grid.

The second challenge is to define energy flexible systems (EFS) or transactive energy systems (TES). Smart grids are exponentially more complex than the traditional control of generating sources because the demand side of the grid offers millions of points of control. This requires energy companies, grid operators, balancing authorities, government authorities, and standards organizations to define the basis of EFS and putting them into practice. Despite the review of standards and best practices we provided in Section 4.2, there is no global standard for EFS (Delony, 2017). There is a need to develop a common approach to all of the components that can be part of the grid infrastructure including consumer-centric models for NZEB. There is a need for consumer-centric models where occupants can consume, trade, generate, and store electricity. This should include interconnection and communication protocols that assure interoperability features for EFS. A standard that defines EFS can bring stakeholders together and frame the application of an energy flexible approach to real grids (Hammerstrom et al., 2016).

The third challenge is to implement and integrate the energy flexible standards and accelerate their adoption. The transformation of current energy grids into flexible and smart grids requires robust platforms that operate in real-time and exchange price signals to dispatch distributed resources and allow peer-to-peer functionalities. This includes allowing DEMS to provide services on a local level and encourage the penetration of renewable energy generation. Aggregating the loads in reality and responding to them effectively will require flexible demand and flexible supply control. Also, regulatory policies need to be investigated and updated to regulate the interaction of different stakeholders to encompass real-time net-metering, pricing, and control.

Finally, energy flexible and smart grids are the inevitable future. With recent advances in information technology it is possible to achieve a higher level of coordination, demand flexibility, and distributed energy resource management. It is important to investigate the different scenarios related to renewable thrive, DEMS, and consumer-centered NZEB. Planning the transition of energy grids will depend on many factors described in this chapter. It involves the intersection of technologies, economics, and social responsibility.

7 LESSON LEARNED # 8

In this chapter, we investigated the impact of upscaling NZEB. In reality, investigating NZEB widespread requires studying different scenarios to make sure the grid is robust and can meet the building energy demand. Evidence suggests that future energy grids will exchange information and interact with one another regarding supply and demand to support grid reliability, robustness, and low environmental impact. In future energy grids, supply and demand will respond to each other dynamically, benefiting customers, energy companies, grid operators, and the environment.

One of the key lessons learned about smart and low carbon grids is that the impact of interaction models at operation stage requires careful planning. A characterization of the energy network is necessary to apply interaction models that adapt to flexible network characteristics. Therefore, due consideration has to be given to the ICT infrastructure and information technology modeling regarding their implications of interaction models. Another lesson learned is that the right assessment has to be made to decide on the degree of control required for various flexible loads and generation. Based on the investment in smart metering and sensing technology, the degree of control and interaction requires to be planned and assessed. Smart grid infrastructure will require continuous monitoring of the grid data for improvement. Consequently, long-term forecasting tools and data storage systems need to be available to estimate the growth of load, generation, and investment in equipment. The prediction and clustering capacities need to be continuously improved. The forecasting tools and data storage systems need to be scaled up to accommodate for Big Data with the drastic increase of data quantities (Cornélusse and Ernst, 2017). Therefore, energy distribution network planning should be based on load and generation profiles over time.

Real-time control should be implemented to effectuate flexible responses and configurations. This should be based on smart metering and visualizations. Smart energy solutions are often complex, integrating various factors such as different energy sectors, infrastructure and components, data exchange as the different actors representing building users, service providers, and energy generation stations. Visualization and smart metering are crucial to communicate smart energy solutions in a simple way while illustrating the technical details for better real-time control. The future of the demand response interaction of consumers and energy company operators is growing. Demand response allows for flexibility of use and will continue to be smooth and easy by building users and grid operators. Finally, all regulatory policies need to be investigated and updated to regulate the interaction of different stakeholders to encompass real-time corrective control.

References

Allcott, H. Rethinking, 2009. Real Time Electricity Pricing. MIT Center for Energy and Environmental PolicyResearch, Cambridge, MA.

NEMA Standard 201-2016, Facility Smart Grid Information Model (FSGIM).

Atamturk, N., Zafar, M., 2014. Transactive Energy: A Surreal Vision or a Necessary and Feasible Solution to Grid Problems. California Public Utilities Commission Policy & Planning Division, Los Angeles.

Barker, M., 2015. Batteries and buildings turning the world on its head. Build. Afr. Mag. 16–18.

Bushby, S.T., 2011. Information model standard for integrating facilities with smart grid. ASHRAE J. 53 (11), B18.

Bushby, S.T., Jones, A., 2016. Facility information model standard. ASHRAE J. 58 (8), 24.

Cornélusse, B., Ernst, D., 2017. GREDOR: Outcomes and Recommendations, Report. Available from <http://hdl.handle.net/2268/205034> (accessed September 2017.).

Cornélusse, B., Ernst, D., Warichet, L., Legros, W., 2017. Efficient management of a connected microgrid in Belgium. In: Proceedings of the 24th International Conference on Electricity Distribution, Glasgow, 12–15 June 2017.

Davies, D., Pepper, A., 2017. Power of good—future of UK heat. CIBSE J. Cambridge Publishers Ltd, UK.

de Bruin, P., 2017. Kostenvergelijking van de alternatieven voor aardgas in nieuwbouwwoningen. DWA, Bodegraven, The Netherlands.

De Groote, M., Mariangiola, F., Volt, J., Rapf, O., 2016. Smart Buildings in a Decarbonised Energy System. Buildings Performance Institute Europe, BPIE, Brussels, Belgium.

Delony, J., 2017. A Transactive Energy Future: The Inevitable Rise of Economic-Based GridControl. Available from <http://www.renewableenergyworld.com/articles/print/volume-20/issue-5/features/solar-wind-storage-finance/a-transactive-energy-future-the-inevitable-rise-of-economic-based-grid-control.html> (accessed September 2017.).

DOE, December 2016. Maintaining Reliability in the Modern Power System. US Department of Energy, Washington, D.C.

DOE, 2017. Demand Response. Available from <https://energy.gov/oe/activities/technology-development/grid-modernization-and-smart-grid/demand-response> (accessed September 2017.).

D'Oca, S., Corgnati, S.P., Buso, T., 2014. Smart meters and energy savings in Italy: determining the effectiveness of persuasive communication in dwellings. Energy Res. Soc. Sci. 3, 131–142.

EISA, December 2007. Security act (EISA). In: 110th United States Congress. Energy Independence and Security Act of 2007.

EF-Pi, 2017. Energy Flexibility Platform & Interface. Available from <http://fpai-ci.sensor-lab.tno.nl/builds/fpai-documentation/development/html/> (accessed September 2017.).

Eurocontrol and NNEW, 2010. WXXM WeatherData Model Version 1.1.1. <http://www.wxxm.aero/public/subsite_homepage/homepage.html>.

Faruqui, A., Hledik, R., Tsoukalis, J., 2009. The power of dynamic pricing. Electr. J. 22, 42–56.

Hammerstrom, D.J., Corbin, C.D., Fernandez, N., Homer, J.S., Makhmalbaf, A., Pratt, R.G., et al., 2016. Valuation of Transactive Systems (No. PNNL-25323). Pacific Northwest National Lab.(PNNL), Richland, WA.

Hargreaves, T., Nye, M., Burgess, J., 2010. Making energy visible: a qualitative field study of how householders interact with feedback from smart energy monitors. Energy Policy 38, 6111–6119.

H21 Leeds Report, 2016. Northern GasNetworks, Leeds City Gate. Available from <https://www.northerngasnetworks.co.uk//wp-content/uploads/2017/04/H21-Report-Interactive-PDF-July-2016.compressed.pdf> (accessed September 2017.).

IEC, 2003a. 61850-7-3, Communication Networks and Systems in Substations—Part 7-3: Basic Communication Structure for Substations and Feeder Equipment—CommonData Classes.

IEC, 2003b. 61850-7-4, Communication Networks and Systems in Substations—Part 7-4: Basic Communication Structure for Substations and Feeder Equipment—Compatible Logical Node Classes and Data Classes.

IEC, 2009. 61850-7-420, Communication Networks and Systems for Power Utility Automation– Part 7-420: Basic Communication Structure—Distributed Energy Resources Logical Nodes.

IEC, 2013a. 61968-9, Application Integration at Electric Utilities—System Interfaces for Distribution Management—Part 9: Interfaces for Meter Reading and Control.

IEC, 2013b. 61850-90-7 edl.0, IEC 61850 Object Models for Photovoltaic, Storage, and Other DER Inverters.

ISO 7800, 2016. ISO 7800 Facility Smart Grid Information Model.

Karavas, C.S., Kyriakarakos, G., Arvanitis, K.G., Papadakis, G., 2015. A multi-agent decentralized energy management system based on distributed intelligence for the design and control of autonomous polygeneration microgrids. Energy Convers. Manag. 103, 166–179.

Kolokotsa, D., 2016. The role of smart grids in the building sector. Energy Build. 116, 703–708.

Lijesen, M.G., 2007. The real-time price elasticity of electricity. Energy Econ. 29, 249–258.

Lund, H., Marszal, A., Heiselberg, P., 2011. Zero energy buildings and mismatch compensation factors. Energy Build. 43 (7), 1646–1654.

NAESB, 2012. Business Practices and Information Models to Support Priority Action Plan 10—Standardization Energy Usage Information Standards, Revisions 1.1. North American Energy Standards Board.

OASIS, 2011a. Energy Market information Exchange 1.0. Committee Specification 01 <http//tinyurl.com/zjfw4k7>.

OASIS, 2011b. Energy Interoperation Version 1.0. Committee Specification 01 <http://docs.oasis-open.org/energyinterop/ei/v1.0/energyinterop-v1.0.html>.

OASIS, 2015. WS-Calendar Platform Independent Model (PIM) Version 1.0. OASIS Committee Specification 02 <http://docs.oasis-open.org/ws-calendar/ws-wscalendar-pim/vl.0/cs02/ws-calendar-pim-v1.0-cs02.html>.

Peeters, L., 2017. A Vision of Our Future Energy Systems the Story. Available from <http://horizon2020-story.eu> (accessed September 2017.).

Pitì, A., Verticale, G., Rottondi, C., Capone, A., Lo Schiavo, L., 2017. The role of smart meters in enabling real-time energy services for households: the Italian case. Energies 10 (2), 199.

Raj, C.A., Aravind, E., Sundaram, B.R., Vasudevan, S.K., 2015. Smart meter based on real time pricing. Proc. Technol. 21, 120–124.

SCE, Southern California Edison, 2012. Future Outlook for Residential Energy Management Research. SCE, Rosemead, CA.

Vaughan, A., 2017. Household batteries will be key to UK's new energy strategy. The Guardian, 24 July 2017.

Vince, D., Morsch, A., 2017. Microgrids: What Every City Should Know. Center for Climate and Energy Solutions, Arlington, VA.

Wang, Q., Zhang, C., Ding, Y., Xydis, G., Wang, J., Østergaard, J., 2015. Review of real-time electricity markets for integrating distributed energy resources and demand response. Appl. Energy 15, 695–706.

Further Reading

ANSI/ASHRAE 135-2012—BACnet, A Data Communication Protocol for Building Automation and Control Networks.

Kaneda, D., Jacobson, B., Rumsey, P., Engineers, R., 2010. Plug load reduction: the next big hurdle for net zero energy building design. In: ACEEE Summer Study on Energy Efficient Buildings, pp. 9–120.

CHAPTER

9

Construction Quality and Cost

ABBREVIATIONS

AEC Architectural, Engineering, and Construction
EPC Energy Performance Certificate
EPD environmental product declaration
EPBD Energy Performance of Buildings Directive
ETICS External Thermal Insulation Composite System
EU European Union
GHG greenhouse gas
HVAC Heating, Ventilation, and Air Conditioning
IAQ indoor air quality
ISO International Standardization Organization
NCR nonconformance report
nZEB nearly Zero Energy Buildings
NZEB Net Zero Energy Buildings
MEP mechanical, electrical, and plumping
PH Passive House
PQP Project Quality Plan
RES renewable energy systems
RFI request for inspection
SME small and medium enterprises

1 INTRODUCTION

NZEB require high construction quality through new construction technologies, high-tech components, specialized competencies, and high level expertise. To achieve NZEB, the use of energy-efficient technologies and materials is necessary. These technologies and materials must respond to the exigencies of the NZEB and satisfy the NZEB market demand. There are barriers regarding the know-how of professionals and the limited number of architects and engineers that are able to deal with new technologies and standards (Da Silva et al., 2015). The lack of knowledge or poor implementation quality during the construction

Net Zero Energy Buildings (NZEB)
DOI: https://doi.org/10.1016/B978-0-12-812461-1.00009-5

process can jeopardize NZEB performance and lead to potential loss of the benefit of energy-efficient and energy-generating building components. For example, the Passive-On (Ford et al., 2007) project published a guideline for designing and constructing Passive Houses (PHs) in Southern Europe and considered the construction quality a serious challenge. Through the QUALICHeCK project, Erhorn et al. (2015) identified four barriers that lead to poor quality on construction sites namely:

1. Ineffective specifications, regulations, and standards.
2. Lack of know-how and technical competence in the construction sector.
3. Critical economic planning conditions.
4. Lack of control and quality assurance.

As a consequence, the gap between design and operations remains a serious challenge that needs to be more solved during the construction phase. On paper, many NZEB might be code compliant without effective insulation installation or airtightness inspections. The construction sector should address these common barriers and avoid business-as-usual practices. NZEB are not about adding another layer to the existing construction and management activities on-site. NZEB need to be designed involving contractors, builders, material purchase agents, and suppliers to make sure that the construction will lead to a high-quality building in a cost-effective way. This requires advanced knowledge on NZEB and more interaction, coordination, and inspection during design and on-site. A coordination team from the contractor side should involve architects, equipment representatives, suppliers, HVAC contractor, renewables contractor, site managers, third-party commissioning authority, site supervisors, inspection authority, builders, and procurement representatives. All these experts should come together and receive training while overseeing how the building comes together. Simple detailing increases the likelihood of a good NZEB. For NZEB, the upfront cost and construction quality should not be a deterrent. By using proper passive and bioclimatic design features and selecting the most cost-effective construction system, the design team can transfer the cost from mechanical operations.

Therefore, in this chapter, we will identify the barriers and implications of NZEB construction. We will present best practices and guide the reader on how to achieve outstanding construction quality while keeping the budget under control. We will illustrate the way to create a formalized quality assurance strategy. Based on lessons learned from the Design-Build team for new NZEB, we will provide a better understanding and guiding overview for future projects delivery, inspection, quality assurance, and implementations involving tools and techniques that can be used in practice.

2 ACTUAL CHALLENGES

NZEB rely on a complex system that requires comprehensive design and construction quality. However, many building professionals cannot lead the NZEB implementation process from construction to operation to meet the expected market demand (QUALICHeCK, 2015). The lack of understanding and quality assurance make it difficult to implement NZEB in Europe. Even architects and engineers who opt to comply with the PH Standard or NZEB face serious challenges with implementation due to the lack of convocational training and capacity building for NZEB among builders. The NZEB integrated project delivery approach and the knowledge of the exigencies of designing a NZEB is not common among architects and engineers which will in many cases lead to inefficient solutions, nonoptimized buildings, and higher costs due to extra measures for integration of energy efficiency measures and renewable energy systems (RES) (Silva et al., 2015).

There are several multidisciplinary actors besides builders that should be involved in NZEB' implementation, adding to the complexity of the process. In many investigated cases, there are no national or regional strategies to empower consultation, builders' skills, and construction services for NZEB. Most contractors in Europe are far away from NZEB best practices and technical construction rigor. Unless rigorous quality assurance measures are addressed during the NZEB project delivery, a number of serious risks can become associated with these high-performance buildings (Siddall, 2015), such as:

1. Costing significantly more money than estimated.
2. Running into delays and crossing the project planning timeline.
3. Most importantly, failing to achieve the expected performance requirements and standard compliance.

Finally, the construction industry is not equipped with experience and products to deliver and supply the expected market demand. In relation to the first barrier, the national industry in European countries is not well positioned to cater for high-tech buildings requiring innovative products and systems. Overall, the top-down legislation framework and targets set by Europe do not meet national and local bottom-up infrastructure and environment. The Macleamy Curve (Fig. 3.2) graphically shows the advantages of the integrated design process over the traditional design process concerning cost and efficiency. It illustrates the fact that, the earlier decisions are taken, the better it is. Indeed, it is much easier, less costly, and more efficient to change the drawings of a building than the constructed building itself. By furnishing the main efforts during pre-design and early design stages, changes in later project delivery stages are not efficient. The common practices of modification of material choices, building construction details, or construction systems jeopardize the high performance of NZEB, and increase their cost.

2.1 Construction Quality

Estimating the building performance of NZEB during the design phase is crucial. Building performance simulation and energy modeling tools are essential to help designers and builders size and design building elements, components and systems properly. However, as mentioned in Chapter 3, NZEB Performance Thresholds, the energy performance gap remains the most challenging issue when it comes to NZEB. The difference between the expected energy performance thresholds for designed buildings and real monitored buildings is related mainly to: (1) occupant behavior, (2) HVAC systems and building services (sizing and installation), (3) weather uncertainties, and (4) construction implementation quality (on-site) (Fabi et al., 2012; De Wilde, 2014). In this chapter, we would like to focus on the construction quality of NZEB. Ultra-energy efficiency has serious implications and require extra insulation, low airtightness, delineated air/water/thermal control layers, smaller HVAC systems, more efficient ducts, and photovoltaic systems. As a consequence, more complex details and specifications are required and the as-design and as-built difference should be eliminated by reducing the margin of error. This requires better interaction between designers and builders and the readiness to use and integrate new materials and products intelligently in the buildings. Despite the focus on performance and the qualification of the design and construction teams, the New Building Institute conducted a study on NZEB and listed the challenges associated with their design, construction and operation as shown in Fig. 9.1 (Higgins et al., 2015). One of those challenges is the construction quality of NZEB and the rigor of the commissioning process.

As European and global energy standards move toward NZEB, designing and building a robust envelope is the first step to achieve high-performance buildings. During the design and construction of NZEB, three major strategies should be studied regarding the envelope construction:

1. Continuous thermal insulation that touches all six sides of the cavity or place where is installed.
2. High-performance windows and appropriate shading.
3. Continuous airtightness.
4. Fire Protection.

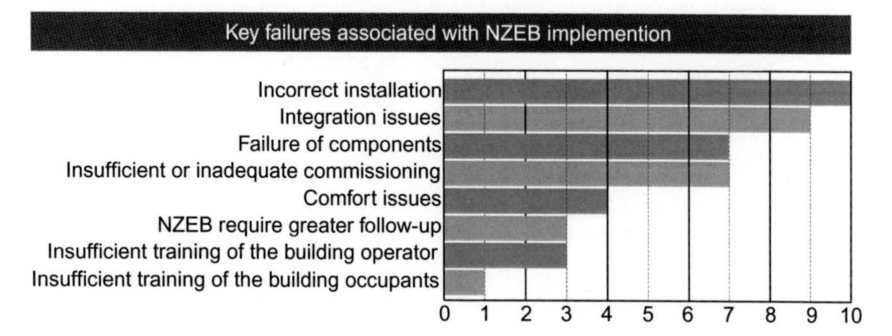

FIGURE 9.1 Challenges facing the design and construction teams follow-up on NZEB (Higgins et al., 2015).

Continuous insulation is meant to reduce heat losses and heat gains. High-performance windows should be properly sized, selected and placed to avoid heat gains when not needed and benefit from heat gains when needed. Continuous airtightness prevents air leakages and moisture problems. Thermal buildings should be eliminated to avoid weak points in the building envelope. Airtightness reduces the amount of air that infiltrates through the envelope. It is an important quality measure that can impact the mechanical ventilation, comfort, and the total energy consumption of the NZEB. More importantly, it can extend the life of the envelope and its durability and controls the transport of contaminants between the indoor air and outside air of the buildings (Sherman and Chan, 2006). For NZEB, airtightness is a determining factor because it can influence the heat recovery system leading to unbalanced ventilation of exhaust fumes in the air. After that, we must make sure that the envelope or thermal insulation will not accelerate fire, and more importantly will not release toxic substances during a fire.

All these requirements make it necessary to control construction quality and guarantee that the different stakeholders of the construction process can build and commission the building properly. This involves local fire authorities that are becoming very stringent with high-performance buildings. However, experiences from several case studies on NZEB indicate the sophistication of assuring quality of NZEB. For example, they require good planning of working activities and a differentiation of thermal insulation and acoustic insulation. Airtightness must be evaluated using the blower door test. As shown in Fig. 9.2, there are often serious quality problems in installing insulation properly, avoiding thermal bridges, and assuring envelope airtightness. The end report of the Flemish Government Agency for Innovation and Science and Technology indicates the increase of energy performance gap of nearly zero energy buildings (nZEB) (Staepels et al., 2013) (Fig. 9.2).

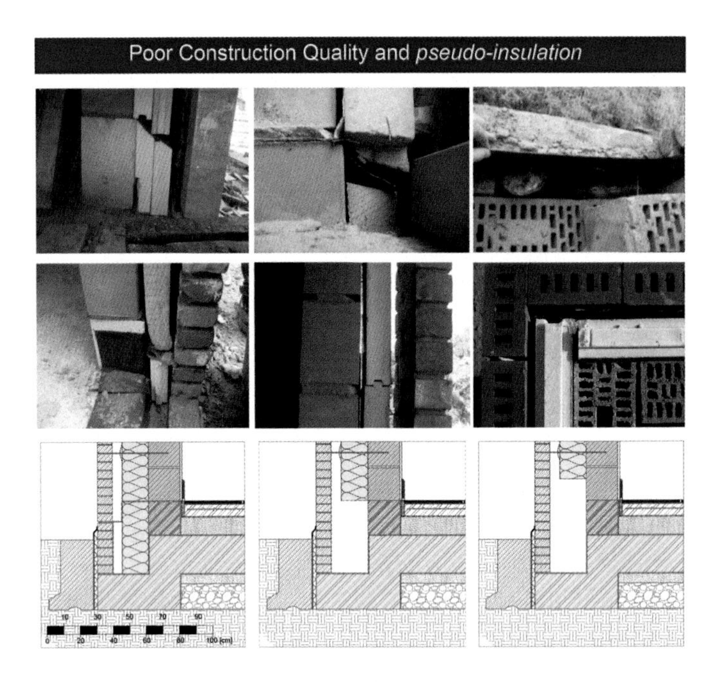

FIGURE 9.2 Examples of poor construction quality and *pseudo-insulation* (Staepels et al., 2013).

2.2 Modularity and Cost Reduction

NZEB are ultra-efficient buildings which also have a RES to offset imported energy. Ultra-efficient buildings are a good investment to achieve comfort and occupant well-being. The cost of a RES is effectively present-valuing the cost of energy consumption. However, estimating the cost of NZEB is very difficult. In fact, the cost estimation of NZEB involves societal, environmental, and economic cost analyses (Hamdy et al. 2017). NZEB reduce the environmental cost of energy production including air pollutants, emissions of toxics, thermal pollution, greenhouse gas (GHG) emissions, and toxic waste (including nuclear waste). Reducing these environmental effects reduces the costs associated with environmental mitigation and public health costs of illness and early death. At the same time, empowering occupants and improving their well-being reduces the social cost of absenteeism and increases the productivity and satisfaction in relation to all occupant-related metrics discussed in Chapter 5, Occupants Well-Being and Indoor Environmental Quality.

Failures in the estimation of cost and making the right assumptions and choices are the main problems associated with NZEB. The cost calculation is relative to the context and the life span of the cost estimation (initial cost, operational cost, maintenance cost, etc.), but NZEB do not necessary have extra costs in comparison with classical buildings

(Hamdy et al. 2017). For NZEB, the cost in itself is not the problem, because achieving 50%–80% reduction needed to achieve a NZEB can be done at little or with no cost increment (NREL, 2014). If the energy efficiency performance goals and energy performance thresholds are defined clearly at the beginning of the design process, NZEB can achieve super comfort, better IAQ, robust performance, and grid interaction. Lessons learned from practice indicate that the cost (increase or decrease) associated with most NZEB is mainly due to (Attia, 2018):

1. Selection building construction technology (modularity and prefabrication).
2. Selection of envelope design.
3. Selection of construction and structural system.
4. Selection of HVAC and RES.
5. Selection of operation of controls and systems.
6. Selection of project delivery type and risk distribution among stakeholders.

According to a study conducted by A2M Office, an architectural office based in Brussels, Belgium, the financial influence of the envelope improvement can reach ±10% of the project cost (Moreno-Vacca and Willem, 2017). This is followed by the influence of the construction and structural system choices that increase the building cost with 25%, while the choice of tender type can increase the building cost with up to 28%. However, Fig. 9.3 (left) indicates that the use of off-site modular construction and fast site assembly are the most influential factors to reduce the cost of NZEB. Fig. 9.3 (left) illustrates the result of a comparative study conducted by the authors on the influential factors that impact the project budget of NZEB. In fact, the prefabrication of building components and a smart choice for foundation and structural systems can decrease the cost of NZEB. Repeatable floorplates and modular precast wall panels with minimal finishes are cost effective. The simplification of envelope design and construction technologies also allow for cost tradeoff (Heymer et al., 2016).

Transferring cost from traditional envelope, foundation, and structural systems to modular high-performance envelope and smart-structural systems, can allow for extra insulation, triple glazing, taping, and membranes for airtightness. Integrated architecture and envelope with optimized glazing should be considered as primary efficiency measures (Fig. 9.3, right). As a consequence of architectural and construction cost optimization, the cost of mechanical and electrical systems is less (NREL, 2014). Moreover, the Design-Build project delivery process, described in Chapter 4, Integrative Project Delivery and Team Roles, has the most significant impact of cost reduction. According to Fig. 9.3 (left) the building contract that seeks to distribute the risk in an uneven, burdensome, or unwieldy fashion can lead to increasing or

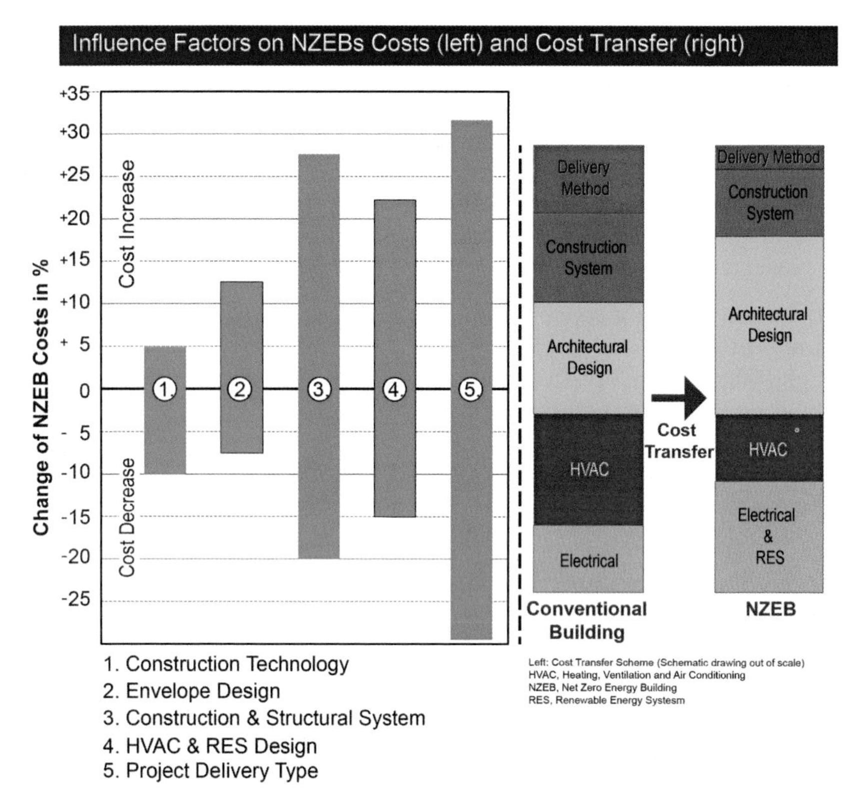

FIGURE 9.3 (Left) The influential factors on NZEB cost and (right) the cost transfer for NZEB (Attia, 2017; Moreno-Vacca and Willem, 2017; Heymer et al., 2016).

decreasing the project cost up to $\pm 30\%$ (Moreno-Vacca and Willem, 2017; Siddall, 2015). Sharing risk in an unbalanced way often creates uncertainties and leads to overestimated costs. The Design-Build process empowers builders early on in the process and reduces the risk which leads to transfer of cost to renewable systems. Instead of wasting time and effort to allocate blame following a Design-Bid-Build contract, a Design-Build contract can lead to more productive approaches that challenge the team to come up with a collective and collaborative working mindset (DOE 2012). Therefore, the cost of NZEB can be the same if more intelligence and effort is invested during the predesign phase.

3 CONSTRUCTION BEST PRACTICES

Construction best practices are related to the construction systems including structure, construction details, and implemented construction technology. In this section, we will explain each of these aspects and provide insights to guide and inform decision-making for NZEB.

3.1 Construction Systems

The selection of a construction system is a very influential decision that impacts the energy performance of cost of NZEB. As shown in Fig. 9.3 (right), construction system choice can increase the cost up to 25% or decrease it up to 10%. Contractors and builders should lead the process of construction and structural system selection together with structural engineers. Construction systems can be classified (as shown in Table 9.1) as lightweight, heavyweight, or hybrid. They can also be classified by type of structural system material.

According to Table 9.1, there are several criteria that can influence the selection of a construction system for NZEB. NZEB do not require a specific type of construction, however, the contractor should be aware of the local know-how of builders, materials availability, cost of materials, and the procurement route that fits the project. Together with the building physics expert, the design and builder teams should decide for the fit-to-context construction system in relation to function, occupancy, climate, and construction detailing. Details and junctions should be designed as early as possible in relation to the choice of structural system and construction technology. Modularity, prefabrication and on-site assembly are key issues that need to be addressed together with cost and performance. As mentioned earlier and according to Fig. 9.1, for a NZEB last minute changes and on-site decisions under pressure of cost, time, materials availability and site conditions, will compromise the construction quality and details. Thus, they must be avoided by

TABLE 9.1 Classification of Construction Systems in Association with their Weight, Performance, and Cost

	Lightweight	Heavyweight	Hybrid construction
Thermal mass	Low	High	Depend
Acoustic performance	Poor	Good	Poor
Environmental impact	Low	High	Depend
Materials	Timber or steel	Masonry and Concrete	Mixed
Cost	Low/medium	High/Medium	High
Soil stability	Low	High	Medium

following an integrated design process and adopting a Design-Build contract that guarantees wise choices during the early design phases of the project delivery trajectory.

Choosing the right construction system for NZEB is based on several factors including the soil characteristics and foundation technique in order to properly distribute the weight of the building on the ground. The foundations of NZEB can have a cost ranging from 5 to 25% of the total building cost. For example, a lightweight construction would not require extensive foundations. Thus, even if the lightweight construction system is more expensive than a heavyweight construction system, a light construction method would ultimately cost less than the traditional method foreseen with heavy foundations. This example already shows that several factors are involved in the selection process of construction systems and that these factors are inseparably linked.

Masonry construction: Masonry construction consists of a stack of building blocks, facade bricks, and concrete elements assembled with mortar or glue. The exterior walls are generally divided into three parts: A supporting part on the inside, an insulating part in the middle, and a protective part on the outside. This requires careful construction details of windows and openings. The structural stability and thermal break of the outer leaf and large openings must be inspected (see Table 9.2).

Timber construction: This construction system is based on load-bearing walls composed of small wooden beams and solid studs. On both sides of this structure, the builder fixes panel plates to give the assembly the required stability. Roof overhangs with projecting joists should be avoided (see Table 9.3).

Steel construction: This construction system consists of mounting the structure of the dwelling in columns and steel beams. The exterior face of this structure is completed with facade elements, and inside panels are installed. Potential thermal bridges with a connection of columns to

TABLE 9.2 Masonry Construction Characteristics

Stability	+ +	The structure is stable thanks to its own weight
Insulation	+	For the same insulation material, the wall structure will be thicker than in the case of a light construction method
Airtightness	+ +	Airtightness is ensured by the ceiling
Thermal mass	+ +	Concrete shows the best score; other materials such as terracotta or cellular concrete are less effective
Hygrothermal performance	+ +	Good hygrometric behavior

TABLE 9.3 Timber Construction Characteristics

Stability	+ +	Light foundations are sufficient and the frame is light. The structure is stabilized by covering the framework with panels
Insulation	+ +	Even in the case of thick insulation, the thickness of the wall remains limited. Thermal bridges are easily avoidable
Airtightness	+	Adhesive tapes should be placed correctly on the connections between the panels
Thermal mass	−	Low thermal mass due to lightweight frame, however could be filled with clay elements
Hygrothermal performance	+ +	Timber is perfectly suited to regulate the hygrothermal performance

TABLE 9.4 Steel Construction Characteristics

Stability	+ +	The structure is light, local foundations are sufficient. Large spans or large openings in the front are possible without requiring special measures
Insulation	+ +	Fits between steel columns or used facade materials
Airtightness	+	Achieving airtightness is challenging. The measures to be taken depend strongly on the facade cladding and interior finishing
Thermal mass	+ / −	Depending on the materials chosen and the soil structure
Hygrothermal performance	+ / −	Depending on the facade materials used

foundations should be addressed. Heavy external cladding or excessive cantilevered structures should be avoided to avoid thermal bridging (see Table 9.4).

Prefab concrete construction: This construction system involves mounting walls—both inside and outside—using prefabricated concrete walls constructed off-site. Due to the quality of the walls manufactured in a protected environment and because of the speed of this dry method of construction, the cost remains low. However, this system is not flexible and is limited to modular design. Excessive cantilevered structures such as balconies or heavy hung off external foundations can cause thermal bridges and should be avoided (see Table 9.5).

3.2 Construction Details

The envelope of a NZEB needs to achieve high thermal and airtight performance, while providing weather protection. There are a series of

TABLE 9.5 Prefab Concrete Construction Characteristics

Stability	+	The structure has a high degree of stability resulting in its heavyweight, which can be a disadvantage for foundations
Insulation	+	The insulation must be installed on the outside. ETICS can be a good combination
Airtightness	+ +	Few connections are not sealed which provides a high degree of sealing without having to add a membrane or ceiling
Thermal mass	+ +	Provides the best possible level of thermal mass
Hygrothermal performance	+ +	Presents a good score in terms of hygrothermal performance

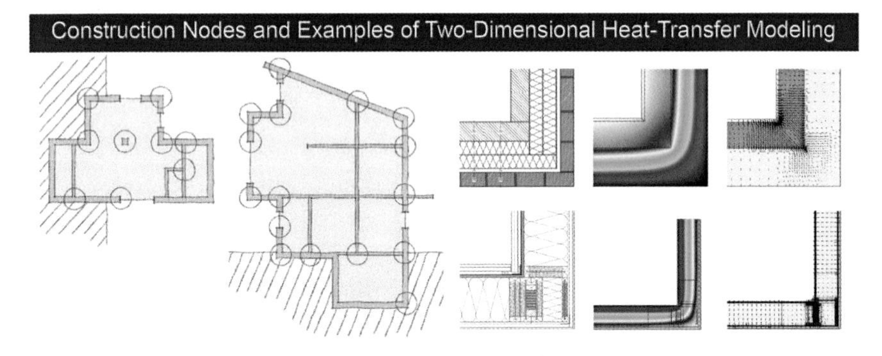

FIGURE 9.4 Construction nodes of NZEB and examples of two-dimensional building heat-transfer modeling.

measures that need to be addressed during the design of the envelope to guarantee its durability, ease of maintenance, and structural stability. Construction details should be simple, robust, and designed with adequate construction tolerance. The sequence of construction on-site should be taken into account when developing the construction details. Brick works, electrical conduits packing, window fixing, cladding, and plastering should be planned and the construction detail should assure the ease of work order and sequence for workers. The decisions on construction types, structural system, and envelope materials are often influenced by the building form, planning restrictions, hydrothermal properties, cost restraints, client preference, buildability, materials availability, construction speed, and aesthetics. In this section, we will focus on four major aspects that need to be addressed to provide high-quality construction details with a focus on joints and thermal bridges as shown in Fig. 9.4.

Building envelope: Un-insulated envelope loses heat primarily through conduction. Insulation works by slowing down the rate of conductive heat loss. Insulation therefore must be in contact with the exposed surfaces to lower the conductive properties of the whole assembly as much as possible. When the insulation is missing, heat loss will continue until both sides of the envelope surface are at thermal equilibrium. Therefore, insulation and envelope details should be designed to comply with a standard. In the case of NZEB, this could be the recommended *U*-value of the PH Standard or the ASHRAE 189 Standard. Thermal transmittance also known as *U*-value describes the rate of transfer of heat in watts through one square meter of an envelope for every one degree of temperature diffuse across the envelope (W/m^2 K). Surface thermal conductivity can be found in ISO-6946 too. Once the *U* values of each envelope element are determined, it is possible to calculate the one-dimensional transmission heat load and verify the design of the envelope construction details (Hopfe and McLeod, 2015). Finally, designers should guarantee continuous insulation, minimize the complexity of insulation details, and ensure simple, robust details.

Airtightness: Airtightness is a key quality in NZEB construction. It requires rigor and attention during design and construction. Construction quality should determine the choice of sealing products and processes to deliver the targeted airtightness. A formal air barrier strategy coupled to security air systems (SAS) should be addressed during design and construction. This strategy should include locating potential leaks using calibrated test equipment to measure the airflow through the envelope. A range of measuring equipment and testing organizations should be used during construction before the internal finishing activities to identify any lack and treat it (Prignon and Van Moeseke, 2017). When carrying out the building test, a steady state must be established of the air reentering the building through various cracks, gaps and openings (see Fig. 9.5). Typically for NZEB airtightness, the target for new construction should be:

$$< 0.6\ \text{ACH}^{-1} @ 50\ \text{Pa}$$

A formal air barrier strategy should address all these concerns by identifying potential leakages while technical details should be drawn carefully and made visible to inform and guide builders. Details should be inspected prior to finishing or plastering to ensure that there are no leaks or cracks. Pressurization and depressurization testing must be carried out and the airtightness testing must meet the target values.

Junctions and thermal bridges: Junctions and interfaces require specific attention in NZEB. Junctions and thermal bridges are associated with heat losses and condensation resulting into mold growth and a

FIGURE 9.5 Joints and nodes of linear thermal bridges (Architecture et climat).

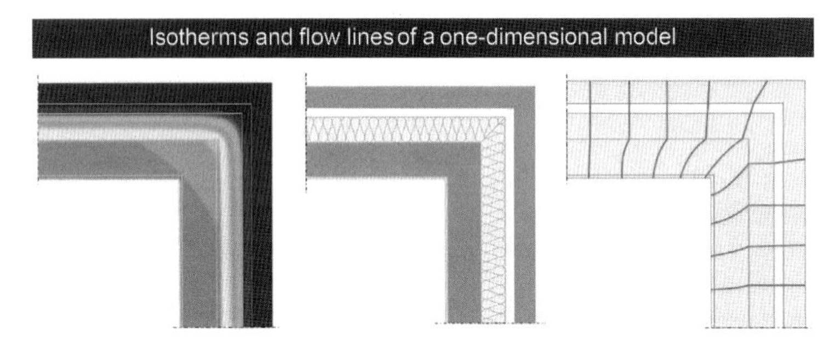

FIGURE 9.6 Isotherms and flow lines of a one-dimensional model and the calculation method based on the U values.

reduction of envelope efficiency. The increase of envelope heat loss can reach up to 15%. Therefore, thermal bridging should be fully eliminated. There are essentially three types of thermal bridge:

1. Repeating thermal bridges.
2. Linear thermal bridges—The Y-value (psi-value).
3. Nodal thermal bridges—The X-value (chi-value).

Linear and nodal thermal bridges are determined in accordance with ISO/EN 10211. Linear thermal bridges occur at the junctions between two building elements. Nodal thermal bridges arise when three building elements intersect (e.g., the corner between two walls and a floor) or where a single element (e.g., a steel beam) penetrates the insulation (Hopfe and McLeod, 2015). According to Fig. 9.6, the isotherms and the flow lines differ from a one-dimensional model and the calculation method based on the U values is no longer correct for the construction nodes induced by a variant geometry and by the presence of different thermal transmission constructing elements. When considering an outer wall with a homogeneous structure and a well-defined and homogeneous thermal transmission coefficient *U*, the isotherms (= lines of equal temperature) will always be parallel to the plane of the façade, and the heat flow lines perpendicular to this one. An outer wall with a homogeneous structure has isotherms parallel to the façade plane (in the middle) and flow lines perpendicular to the facade plane (right). Finally, thermal bridges should be calculated using a thermal bridges calculator and energy calculation software (Hopfe and McLeod, 2015).

Building services and renewable energy systems: The mechanical, electrical, and plumping (MEP) contractors are responsible for the installation and fabrication of all building services and RES on-site. The MEP contractors are responsible for detailing, fabrication coordination, and development of 2D and 3D construction drawings, which include building services and RES. Systematic quality assurance of the building services and automation technology systems during the construction phase ensures that the NZEB works correctly and as planned when received by the owner. Design and review assignments should be coupled to building automation monitoring, requests for inspection (RFI), and construction assignments. Systematic, proactive commissioning and phased inspection procedures can ensure that the design aims are met during the construction phase. The aim is that the heating and hot water systems, mechanical ventilation heat recovery, solar thermal, and photovoltaic are installed correctly and function as planned when commissioned by the owner. This requires revising duct sizing, pipework routing, splitting of the electrical ladder, equipment selection, among others. It is essential to focus on the functional tests and test runs of the HVAC systems and building automation systems as well as on the adjustment and fine-tuning of the systems. It is also important to reserve enough time for this work and the inspections during both the construction and commissioning phases.

3.3 Construction Technology

Lessons learned from different NZEB indicate the importance of construction technology to reduce cost, build fast, and achieve the expected performance requirements. One of the lessons learned from high-performance buildings is to maximize the use of modular and repeatable high efficiency design strategies. The NZEB design principles presented in Chapter 2, Evolution of Definitions and Approaches, should be integrated into repeatable design elements and building components. Expensive building elements and unique elements should be avoided. Punched windows or curved walls are not recommended and, as mentioned in Chapter 7, Energy Systems and Loads Operation, space efficiency and thermal zoning should be increased to the maximum.

The use of off-site modular construction and prefabricated building components is a best practice. Off-site prefabrication reduces the on-site construction time and makes the assembly faster and cheaper. The site coordination details and safety concerns are under control, and quality and construction precision increases significantly. A prefabricated construction system approach (i.e., one where the materials have been pre-selected and assembled into a holistic system) can help builders make their construction activities and decisions with confidence that the final outcome will meet the NZEB requirements. Moreover, and as discussed in Chapter 6, Materials and Environmental Impact Assessment, there will be more and more interest in low impact materials and sourced construction materials. Currently, there is a tendency in Central and Northern Europe to build prefabricated NZEB using timber. Depending on the project location and the status of current regulation, the contractor, structural engineer, and architect should keep this criterion into account and address it if necessary.

4 CAPACITY BUILDING, EDUCATION, AND TRAINING

The benefits of capacity building, education, and training for NZEB are personal and societal and lie in their ability to achieve robust facilities that can reduce carbon emissions associated with the built environment. Designers and builders require guidance during design and on-site during the building process. In Chapter 4, Integrative Project Delivery and Team Roles, we identified and discussed the NZEB team's role and design process. We provided an overview on the team's role responsivities during design, construction, and operation. In this section, we focus on the construction, operation, and maintenance phase. NZEB are advanced high-performance buildings with complicated systems and components. Therefore, it is essential to transfer knowledge for NZEB construction and operation teams. Handling innovative and high-tech solutions requires trained and experienced technicians, facility managers, and crafters. The fast developments in the construction sector in recent years make it difficult for many overstrained crafts experts. This includes the ability to read, understand, and apply specification of used technologies for NZEB. When it comes to capacity building, education, and training for NZEB, there are key criteria that must be met on a national level or regional level to guarantee the construction quality. These are discussed next.

4.1 Accredited Professionals

First of all, workers and crafters must have professional credentials or accreditation in their area of specialization. Only builders with credentials should be allowed on-site to ensure quality and safety so that NZEB components are placed in the correct order and function as designed. This can be technically achieved by using certified professional service providers, companies, and individuals. Emphasizing proven skills is necessary for NZEB. Lessons learned from practice indicate that things can easily go wrong with placing and installing insulation, achieving envelope airtightness, installing HVAC, and RES. In order to bridge the energy gap and achieve the robust energy performance of NZBEs, accredited professionals should have the skills that allow them to do the job properly. For example, 3 cm of insulation properly installed is much better than 15 cm of insulation installed incorrectly. With the increase of insulation thickness for high-performance buildings in the past few years, we forgot to train accredited insulation installers and created a new trend of what is called "pseudo-insulation." By pseudo-insulation we mean the wrong placement of large volumes of insulation resulting to poor thermal performance of the envelope. Becoming accredited, therefore, means that workers and builders have the skills set and a valuable base of knowledge to work on NZEB projects.

4.2 Vocational Education and Training

The second most important measure for construction quality is to improve the state of vocational education and training on energy efficiency and renewable energy for NZEB. Creating new curricula and NZEB-related training that meet and exceed the national building energy efficiency requirements is essential. This is one of the ways to bring high-performance building standards into practice, especially among young and foreign building professionals. Training of dedicated, on-site quality assurance responsibility is very important.

Worldwide, foreign workers contribute to the building construction sector. Most foreign workers are semi-skilled or unskilled which influences the construction quality. It is crucial to empower these workers through vocational training and accreditation while increasing the quality of the courses and the scope of acquired knowledge, skills, and competencies. The learning curve should be lowered to avoid extra cost, unpredicted changes, remediation, and extra time. Training courses should cover diverse fields and topics, including envelope hydrothermal performance, envelope airtightness, solar systems installation, biomass boilers installation, heat pumps, air conditions, ventilation and heating systems, building systems controls, and management. Building the capacity of construction workers ensures the performance of NZEB.

4.3 Knowledge Based and Institutional Infrastructure

The third measure is to create a knowledge base and institutional infrastructure—handbooks, virtual wikis, and catalogues that include advanced practices to attain good construction quality and provide expert technical assistance. Designers, contractors, and builders should get high-quality training and education on the latest, cutting-edge energy efficiency strategies and renewable energy sources. SMEs cannot sustain very high capital investments, have difficulties to access accreditation, have low technological expertise, and do not strive toward internationalization and technological innovation. Also, architects and building designers often do not have a full understanding of innovative technologies usable in buildings (Buonomano et al., 2016). It is critical that there is a workforce in place that has the skills to install, operate, and maintain the new lighting and building automation systems as well as the renewable systems in NZEB. There are a lot of multidisciplinary actors that should be involved in NZEB implementation and not only engineers, adding to the complexity of the process. There should be national or regional strategies to empower consultation, builders' skills and construction services for new and renovated NZEB. Local governments should offer financial incentives for construction learning centers. Contractors should be trained on latest NZEB best practices and technical construction accuracy.

5 CERTIFICATION AND QUALITY ASSURANCE

In order to manage the risk associated with NZEB and assure the quality of construction, a clear strategy and procedure must be fulfilled to achieve outstanding quality of works. Compliance, certification and project management should be part of the construction concerns. For a NZEB to be completed successfully, we suggest the following procedures:

5.1 Quality Strategy and Quality Team

A quality strategy should be in place as early as possible during the design process. This includes assigning responsibilities and identifying the responsible quality assurance team member. A dedicated on-site quality assurance supervisor is required. The role of the responsible team member is to supervise—in greater detail—the NZEB design principles, construction details, and rules of thumbs for quality construction. This responsible team member must be, at least, a full-time supervisor who is available during all key construction phases and follows the sequence of works. However, this supervisor should be part of a larger quality assurance team that identifies risks, allocate costs, and communicates with the client regarding the best methods to manage the risks. A continuous quality assurance plan should be part of the integrated design and construction effort.

5.2 Building and Components Certification

NZEB should be achieved through careful design, robust detailing of the building envelope, services, and RESs. An effective quality control strategy of the construction process should be included too. However, those goals cannot be achieved without certification as a formal assessment and quality assurance method. The certification of NZEB should be achieved on two levels. The first certification level should be achieved on the building level following a rating system or standard. There are several examples for certification schemes for NZEB, which were discussed in Chapter 2, Evolution of Definitions and Approaches (Section 4.1), including the PH Standard, Active House, LEED, BREEAM, DGNB, among others. A performance-based certification requires that overall planning is carried out with a calculated energy assessment instead of individually specified construction or technical details. Certification enables the design team and builders to elaborate the details and supporting documentation on components and services (USGBC 2013). This ensures that the planned building will actually perform as designed. The main factor in successfully achieving certification is planning, as only careful and considered planning will ensure you complete all the required factors to achieve the NZEB. Certification should be only given once the building has been constructed and all relevant information and evidence has been documented.

The second certification level should be achieved on the component level. It is often difficult to assess the energy efficiency and durability of a component or building system. Certified components can be two to three times more efficient than other commonly used products. This high level of efficiency is critical to achieving NZEB. Tested and certified products are of excellent quality regarding energy efficiency. Their use facilitates the designer's task and contributes significantly to ensuring high performance. The components certification should be coordinated with relevant ISO, EN, ASTM, or national standards (Siddall, 2015).

5.3 Project Management

Key stages design reviews should be part of the construction project management and quality assurance. The design team should recognize the boundary between conventional project management and quality project management. Quality project management should include structured reviews that interrogate construction quality. The reviews should inspect construction details and examine the sequencing, buildability, and construction program. The purpose of reviews is to identify and detect potential risks associated with the construction detail requirements mentioned in Section 3.2. A well-integrated design-build team can identify the key quality milestones and procedures within the project management plan (DOE, 2014).

The use of proper management checklists is another tool for construction quality management. A construction quality management checklist is like a road map. It is condensed and highly useful. Quality assurance checklists should be referenced during the design stages with information collected as the project develops. A road map cannot show every quality measure—however, it aids and saves brainpower. Checklists are critical quality control tools. As contractors begin their phases of work, they can review their checklists for important items to remember. In addition to heightened awareness checkpoints, checklists should verify compliance to quality control policies and procedures. General managers, owners, and general contractors can use checklist data to generate inspection report data to monitor work crew performance and make improvements before poor performance impacts future work opportunities.

5.4 Cost Control Reviews

The greatest capital expenditure occurs during construction. The highest financial risk is often during the construction process. Therefore, including a continuous value engineering process is a part of the quality assurance plan and cost reduction plan. The need to record and track changes related to value engineering stimulates the management of change. Setting a balance between cost models and energy models during the early design phases as well as the construction phases is crucial (Heymer et al., 2016). Cost models and energy models have to be developed from the beginning with the right team and the right experience. NZEB quality briefing and cost reviews are recommended to take place at the end of the predesign phase (preparation and brief) and it is advised that audits take place at the end of the concept design phase, developed design stage, and the technical design phase, right before construction starts. As recommended in Chapter 4,

Integrative Project Delivery and Team Roles, a Design-Build contract can facilitate cost management and should be based on fixed prices with required energy goals. During the project management process, the cost control review should be repeated during key milestone stages.

5.5 Site Inspection and Procurement

Detailed site inspection and reporting should be used to provide feedback to the NZEB project stakeholders. Site managers should discuss these reports to establish a quality and feedback loop. For NZEB, a close relationship between the design team members is essential based on periodic meetings. Site reports should identify errors, record lessons learned and report on measures to prevent the same errors occurring in the future. As shown in Fig. 9.7, the site inspection should be performance-based and coupled to the procurement process. Every time, at any project stage, the quality team should confirm that energy requirements are met. During construction, the design team, contractor, project managers, site managers, and operations staff should verify performance. Refined quality assurance and site inspections can only happen with experienced professionals. Professionals should have been

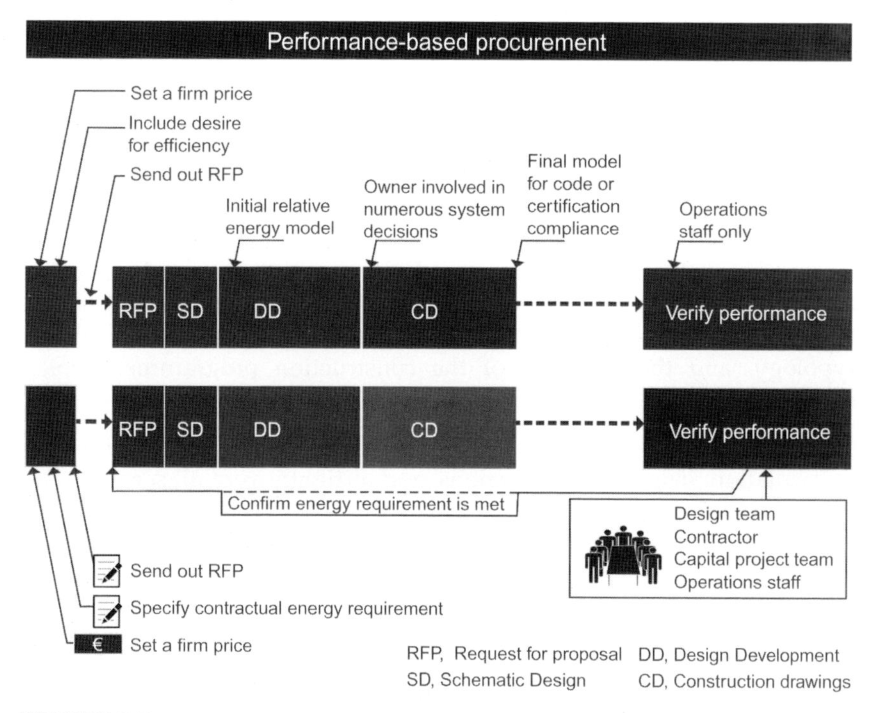

FIGURE 9.7 Performance-based procurement of NZEB (Heymer et al., 2016).

given specialized training and should have a clear definition of their scope of responsibility for managing the performance-based procurement process.

The following recommendations serve as a guide for contractors:

- Nominate a responsible quality assurance supervisor.
- Provide training to the quality assurance team and make sure they have the time and resources to perform their tasks properly.
- Develop a formal strategy to address insulation, airtightness, thermal bridges and junctions, HVAC installations, and RES installations.
- Inspect details prior to closing any openings to avoid cracks and discontinuities. Special attention should be given to the phenomena of pseudo-insulation.
- Avoid improvisation and apply specifications based on the drawing details.
- Protect insulation material from weather or mechanical damage.
- Corners require specific attention.
- Pressurization and depressurization testing must be carried out by a third-party body and testing results must meet performance requirements.

5.6 Commissioning

Accurate seasonal commissioning and testing of air leakages, building services, and RESs must be in place. Commissioning allows verifying the performance of materials, products, and workmanship relevant to ISO, EN, ASTM, or other national standards. Manufactures and suppliers should be part of the installation and commissioning process to understand the implication of the quality assurance requirements and become part of the quality assurance process. Intermittent site inspections should also be undertaken at agreed milestone dates. The frequency of these site inspections should be based on the building type, the construction typology, and the duration of the construction programme. Typical considerations will address (Siddall, 2015):

1. *Envelope*: Below-ground and above-ground envelope, installation of insulation, insulation at junctions, and airtightness of all service penetrations.
2. *Services*: VHR and ductwork installation, and VHR commissioning.
3. Commissioning: Pressure test, VHR commissioning, commissioning of other services, and RESs commissioning.

Finally, there are several factors for effective quality assurance. A committed and experience team is the first step to assure quality. According to the Staepels et al. (2013) report, site managers or on-site quality supervisors are the weakest links in the quality assurance chain. Therefore, it is essential to empower them through training and formal involvement. They should have the authority and ability to perform an effective control and penalties mechanism in cases of noncompliance (CESBA 2014). They should be placed on top of a quality assurance strategy and team with clear procedures to decide on compliance and quality related actions (Erhorn et al., 2015).

6 DISCUSSION

NZEB are recognized as leading high quality and ultra-energy efficiency buildings. Unlike conventional buildings, NZEB are designed, constructed, and certified to perform as predicted. NZEB require high construction quality through new construction technologies, high-tech components, specialized competencies, and high levels of expertise. They follow rigorous and stringent quality assurance standards to bridge the energy performance gap and achieve comfort and well-being of users. With the increased and rapid implementation of NZEB worldwide, there is a need to guide the increased volume, scale, and complexity of NZEB projects. Also, we need to minimize the gap between design and as-built performance during the construction completion stage, based on as-built drawings. The guidance should be based on structured quality assurance strategy to maintain the required standards of design and construction. In the near future, the success of NZEB projects and their proliferation will depend on three key factors that we discuss next.

To date, NZEB depend on a motivated client and dedicated and motivated design and construction teams. They require a high level of experience and knowledge applied in a close working relationship between all parties. Harmonization and coordination sound like a simple task, however, it requires awareness and a systematic strategy together with an experienced team to make it a reality. A client's organization without experience might not appreciate the change NZEB impose on the design team or contractor. We think that Design-Build contracts are the best way to allow experienced teams work together, learn from the past and deliver NZBEs in a cost-effective way while maintaining construction quality.

Capacity building and professional accreditation is another dimension that will shape the future of the AEC industry and its ability to deliver massive volumes of NZEB. Capacity building requires extensive education and training, the type that reduces risk and ensures quality. A comparison between the AEC industry and the automobile and aviation industry reveals the importance of continuous quality improvements, with education and training as common concerns. In fact, the AEC industry is facing rapid changes and sectors and companies groupings that will change the project delivery methods of new buildings. There is a need to address the necessity of developing training schemes for professionals involved in the NZEB building process, transferring successful practices and knowledge from front runners to target countries that are less advanced in this area (IEE, 2008). At the same time, there is a growing demand worldwide to build nZEB, NZEB, and Plus Energy Buildings. The only way to lead this change is to increase the knowledge capacity of the construction sector through extensive education and training. These changes point toward the adoption of digitalization, automation, and prefabrication within the industry to provide fully integrated building services and building solutions. We need a transition from a product-based soloed sector to an integrated service-based building industry.

High-quality construction of NZEB that is cost effective and heavily reliant on quality can be achieved only through highly trained and educated professionals who can inspect, test, and commission buildings. The best practice is to favor a construction ongoing test to take corrective action as early as possible. This is the only way to foster the sector's transition toward standardization and assure quality. However, we are still far away from that. Today, in many Northern Countries, there are no regulatory mechanisms that require, for example, an insulation installer to be accredited or to install insulation to a certain standard. As an industry, we need to realize that high-performance buildings require much more diligence than conventional buildings. NZEB require proper insulation installation, blower door resting, and proper systems commissioning. Without capacity building, training and professional accreditation, we simply cannot supply well-constructed buildings, no matter how good they look on paper or on energy modeling software. Clients and construction organizations must recognize the extent to which inexperienced design and construction teams will require education and training based on an infrastructure of knowledge and training centers. Experienced and trained workers and professionals can robustly deliver NZEB and inform their cost effectiveness and quality.

The third key future prospective of NZEB and the AEC industry is related to the potential progress of the industry toward digitalization and smart prefabrication. Today, building owners, construction companies, and real estate agencies are focused more than ever on cutting costs. Digitalization and Building Information Modeling (BIM) technology are advancing in ways to allow more accurate spatial coordination and empower teams to model the building more effectively through all stages of the project delivery. They allow the seamless exchange of information and allow for better decision making. When coupled to off-site prefabrication and assembly using automated fabrication systems, they improve the construction details quality, construction accuracy, and efficiency. Consequently BIM and automated prefabrication can reduce the project cost in relation to standardization and modularity. 3D virtual models can eliminate chances of clashes and ensure that team members address the performance and spatial integration issues before construction. Disputes and delays at site can be prevented and an easy mean of communication can align the team. Combining 3D BIM models and off-site prefabrication is a very promising progress that can change the AEC industry and streamline the building process of NZEB to save time, effort, and money.

Until those future prospectives are evolved, some urgent measures and recommendations need to be applied on a national level for countries that seek to accelerate the implementation of NZEB in the near future:

- Legislation must require permits and certification for NZEB. This includes ensuring the quality of construction works through quality checks, compliance procedures, and proper commissioning. A project like QUALICHECK is a very good start to achieve the reliability of Energy Performance Certificate (EPC) declarations and the quality of the works (QUALICHeCK, 2015).
- Ensuring better enforcement and refurbishment in the frame of the revised EPBD to create regulatory conditions to ensure better IEQ in Europe.
- Ensuring regular inspections and continuous commissioning of technical building systems of NZEB to maintain the envisaged IEQ parameters, EE, and RES production.
- Better prepare building professionals and provide convocational training while simplifying the design and construction process of NZEB. This includes educating experts and one shop service providers and builders.
- We suggest creating cooperatives with a focus on renovation service for members to provide a strong direction toward deep renovation and to bring capital and investments for middle and large NZEB renovation projects.

7 LESSON LEARNED # 9

Lessons learned from practice show the importance to integrate NZEB principles as early as possible and include them in the details. Exemplary standards and certification of on-site quality assurance are required in order for NZEB to perform as expected. Diligence and excellent standards of craftsmanship is the only way to achieve this. Design-Build contracts can empower the team which leads to avoiding site compromises or later changes under pressure such as time, cost, or materials availability during construction. We already highlighted this in Chapter 2, Evolution of Definitions and Approaches, however, in the context of construction and cost reduction it is also highly significant as discussed in Section 2.2. Also, the selection of a construction system that is appropriate for the project regarding its type, form, function, procurement, and contractor is influential. Designers should avoid mixing different structural systems and investigate the foundation options. Avoid hybrid systems and thermal bridges by using unnecessary cantilevered structures or penetrations. Detailing is another key to the success of NZEB. Details and junctions require special attention by designers and builders, while building energy modeling software should be used to validate all construction nodes. The focus is to achieve a highly insulated, airtight, and fire-resistant envelope. Foam insulation makes firefighting more hazardous and difficult. Therefore, contractors together with the design team and the fire authorities must check the fire resistance of the envelope, as well as products that can get be released in the air during a fire. Qualified and certified professionals and workers should be engaged as early as possible. Training off-site staff plays a role in the success of project completion. This includes site inspectors and on-site dedicated quality assurance supervisors. The learning curve of all those professionals needs to be addressed. The aim is to lower the learning curve and guarantee an effective completion with minimized cost, rework, and remediation. Finally, to minimize the gap between design and as-built performance, certification, quality inspection, commissioning, and testing are essential. All these aspects are the constituents of a successful delivery process that maintains construction quality and reduces cost.

References

Attia, S., Eleftheriou, P., Xeni, F., Morlot, R., Ménézo, C., Kostopoulos, V., et al., 2017. Overview and future challenges of nearly Zero Energy Buildings (nZEB) design in Southern Europe. Energy Build 155, 439–458. Available from: https://doi.org/10.1016/j.enbuild.2017.09.043. ISSN: 0378-7788.

Attia, S., 2018. Regenerative and Positive Impact Architecture: Learning from Case Studies. Springer, London.

Buonomano, A., De Luca, G., Montanaro, U., Palombo, A., 2016. Innovative technologies for NZEBs: an energy and economic analysis tool and a case study of a non-residential building for the Mediterranean climate. Energy Build. 121, 318–343.

CESBA, 2014. Initiative Policy Paper Towards a Common Sustainable Building Assessment in Europe. Available from <http://www.fedarene.org/wp-content/uploads/2014/09/CESBA-Policy_paper-final.pdf> (accessed 20.05.17.).

Da Silva, S.M., Almeida, M.G., Bragança, L., Carvalho, M., 2015. nZEB Training Needs in the Southern EU Countries—SouthZEB project, Latin-American and European Encounter on Sustainable Building and Communities—Connecting People and Ideas, vol. 3, pp. 2469–2478. ISBN: 978-989-96543-8-9.

De Wilde, P., 2014. The gap between predicted and measured energy performance of buildings: a framework for investigation. Autom. Constr. 41, 40–49.

DOE, 2012. The Design-Build Process for the Research Support Facility, p. 60.Retrieved from <http://www.nrel.gov/docs/fy12osti/51387.pdf>.

DOE, 2014. Quality Management Systems for DOE Zero Energy Ready Homes, Video: <https://youtu.be/SQW0TZ1awTc> (accessed July 2017.).

Erhorn, H., Erhorn-Kluttig, H., Doster, S., 2015. Towards Improved quality of the works—Documented Examples of Existing Situations Regarding Quality of Works, Brussels.

Fabi, V., Andersen, R.V., Corgnati, S., Olesen, B.W., 2012. Occupants' window opening behaviour: a literature review of factors influencing occupant behaviour and models. Build. Environ. 58, 188–198.

Ford, B., SchianoPhan, R., Zhong, D., 2007. The Passivhaus Standard in European Warm Climates: Design Guidelines for Comfortable Low Energy Homes. School of the Built Environment, University of Nottingham, Nottingham.

Hamdy, M., Siren, K., Attia, S., 2017. Impact of financial assumptions on the cost optimality towards nearly zero energy buildings – a case study. Energy Build 153, 421–438. Available from: https://doi.org/10.1016/j.enbuild.2017.08.018. ISSN: 0378-7788.

Heymer, B., Pless, Sh, Hackel, S., 2016. Zero net energy building cost and feasibility. Webinar. Available from: http://www.seventhwave.org/sites/default/files/zero-energy-webinar-slides-052616.pdf (accessed 20.01.18).

Higgins, C., Miller, A., Lyles, M., 2015. Zero Net Energy Building Controls: Characteristics, Energy Impacts and Lessons. NBI, New Buildings Institute, Continental Automated Buildings Association, Portland, OR.

Hopfe, C.J., McLeod, R.S. (Eds.), 2015. The Passivhaus Designer's Manual: A Technical Guide to Low and Zero Energy Buildings. Routledge, Abingdon.

IEE, 2008. QUALICERT Common Quality Certification & Accreditation for Installers of Small-Scale Renewable Energy Systems, Project IEE/08/479/SI2.528546.

Moreno-Vacca, S., Willem, J., 2017. Hyper-Efficient Building Workshop: Strategies for Complying with Local Law 31 Day 2, 30/08/2017. AIA NY, NYC, USA.

NREL, 2014. Cost Control Strategies for Zero Energy Buildings, High-Performance Design and Construction on a Budget. NREL, DOE, USA.

Prignon, M., Van Moeseke, G., 2017. Factors influencing airtightness and airtightness predictive models: a literature review. Energy and Buildings, 146, 87–97.

QUALICHeCK, 2015. Overview of Existing Surveys on Energy Performance Related Quality and Compliance. Available from <http://qualicheck-platform.eu/2015/06/report-status-on-the-ground/> (accessed September 2017.).

Sherman, M.H., Chan, W.R., 2006. Building air tightness: research and practice. Building Ventilation: the State of the Art. pp. 137–162.

Siddall, M., 2015. Passivhaus Quality Assurance: Large and Complex Buildings, Passivhaus Trust Technical Panel.

Silva, S.M., Almeida, M.G., Bragança, L., Carvalho, M., 2015. nZEB Training Needs in the Southern EU Countries—SouthZEB project, Latin-American and European Encounter on Sustainable Building and Communities—Connecting People and Ideas, 3, pp. 2469–2478. ISBN: 978-989-96543-8-9.

Staepels, L., Verbeeck, G., Bauwens, G., Deconinck, A.H., Roels, S., Van Gelder, L., 2013. BEP2020: betrouwbare energieprestaties van woningen. Naar een robuuste en gebruikersonafhankelijke performantie.

USGBC, 2013. LEEDv4: Reference Guide for Building Design and Construction (v4). US Green Building Council, Washington, D.C.

Further Reading

Attia, S., Eleftheriou, P., Xeni,F., Morlot, R., Ménézo, C., Kostopoulos, V., Betsi, M., Kalaitzoglou, I., Pagliano, L., Almeida, M., Ferreirag, M., Baracuh, T., Badescuh, V., Crutescui, R., Hidalgo-Betanzosj, J.-M., 2016. Overview of Challenges of Residential nearly Zero Energy Buildings (nZEB) in Southern Europe. Sustainable Buildings Design Lab, Technical Report, Liege, Belgium, ISBN: 9782930909059.

CHAPTER

10

Occupant Behavior and Performance Assurance

ABBREVIATIONS

AEC Architectural, Engineering, and Construction
EPBD Energy Performance of Buildings Directive
BIM Building Information Modeling
BMS Building Managements Systems
EPC Energy Performance Certificate
EU European Union
FM Facility management
HVAC Heating, Ventilation, and Air Conditioning
IAQ indoor air quality
IEQ indoor environmental quality
IoT Internet of Things
ICT information and communications technology
NREL National Renewable Energy Lab
NZEB Net Zero Energy Buildings
M&V measurement and verification
POE postoccupancy evaluation

1 INTRODUCTION

Occupants' impact on building energy use continues to increase as building components and systems become more efficient. As a consequence, the uncertainty of Net Zero Energy Buildings (NZEB) extends beyond plug loads and lighting, and often include heating and cooling. For NZEB, HVAC technology and controls have developed, but it is important to put people first. It is important to always look beyond the technology —to work with IEQ requirements that are system independent and to strive at optimizing the energy and health in buildings to achieve well-being, productivity and satisfaction simultaneously. Maor

Net Zero Energy Buildings (NZEB)
DOI: https://doi.org/10.1016/B978-0-12-812461-1.00010-1

(2016) found that an average of four HVAC technologies were installed per building, based on reviewing 90 high-performance building case study projects. This reflects the layered approach to meeting occupant needs that, in turn, results in complex operation. This means that NZEB require fine tuning, which takes time. The process of fine tuning is called soft-landing and will be described in detail in this chapter. Moreover, we present concepts, applications, and methodologies to better understand human building interaction in NZEB.

The impact and complexity of occupant behavior requires continuous follow up of building performance during operation. The magnitude of occupant behavior and their influence on energy savings makes it challenging to maintain the energy neutrality. When occupants have control of building features it becomes difficult to predict their behavior (Day and O'Brien, 2017). Therefore, performance assurance measures should go through a particular process. Postoccupancy evaluation (POE), monitoring, and soft-landing best practices will be presented in relation to controls and systems fine tuning. Design team members and building operators will better understand the importance of continuous tracking and measurement of NZEB performance in relation to occupant behavior. This includes performance data visualization while providing centralized access to NZEB performance indicators. Engagement and feedback tools also allow building owners to ensure the building is on track with health, comfort, and energy performance requirements.

In this chapter, we elaborate on the influence of different occupant behavior modes and interaction with controls and systems. Educating users and allowing them to understand the impact of their actions through dashboards and apps is part of the success of NZEB. The engagement of occupants as active participants helps to operate NZEB projects efficiently and maintain IEQ requirements. Finally, we discuss future challenges related to occupants' behavior and smart controls NZEB. Based on the analysis of several case studies, we discuss the growing importance of building energy management and smart grids in relation to occupants' adaptive behaviors under the influence of incentives.

2 IMPACT AND COMPLEXITY OF OCCUPANT BEHAVIOR

The AEC industry is aware of the recurring mismatch between predicted and in-use energy consumption of NZEB. The energy performance gap is not only due to inappropriate design assumption, it is also due to operation problems. For building owners and building operators working at a more detailed level to improve performance optimization as-built performance, engaging NZEB occupants in the performance optimization, represents a challenge. Occupant engagement represents a sophisticated task that can significantly minimize the gap between design, as-built, and occupied performance.

At the same time, the future generation of high-performance buildings should focus on comfort and health prior to thinking about energy saving. To avoid Sick Building Syndrome (SBS), we have to be aware of the new and flexible working and living modes. User-centered or user-defined workspaces that are designed based on the activities or tasks performed by occupants is becoming mainstream. The proliferation of smart buildings that provide technologies that are designed to interact with users will influence the spatial use of NZEB. The raising awareness of the 4 Cs design strategy will change the way office buildings are designed and operated. The 4 Cs design strategy provides users with more choices between closed spaces for concentration and contemplation and more open space for collaboration and communication (Ross, 2016; Myerson and Ross, 2006). As shown in Fig. 10.1, this design strategy provides diversity between the idea of being alone or being together on the right hand side. This approach has been proved to be effective for group building users in a heterogeneous workplace while saving energy. The 4 Cs design strategy can help create NZEB that are not just energy efficient, but are also healthy, comfortable, and meet relevant IEQ requirements. The tendency to use working and living spaces more efficiently is growing with a focus on user-centered building design.

Occupant behavior in NZEB is extremely important and is often overlooked. The use of technological solutions and innovations in NZEB are insufficient, because NZEB are dynamic systems, and occupants behave in complex ways (Hong et al., 2017). Understanding occupant behavior is a key starting point to bridge the energy performance gap identified in Chapter 3: NZEB Performance Indicators and Thresholds. The breakdown of energy use elements of NZEB confirms the general trend of appliances and ventilation dominance. (Garde et al., 2017). As shown in

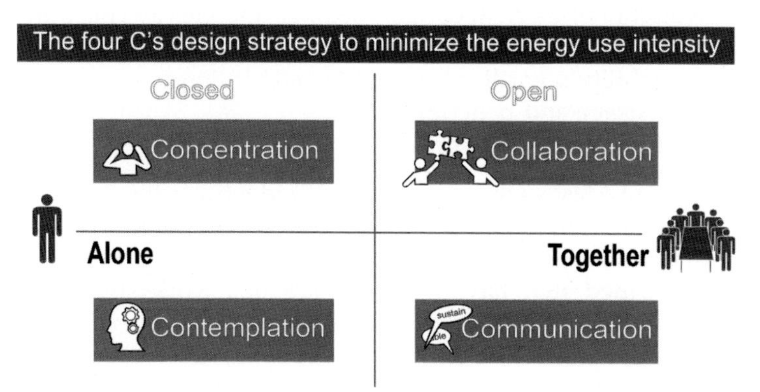

FIGURE 10.1 Future working modes in relation to energy saving and occupant well-being and productivity (Ross, 2016).

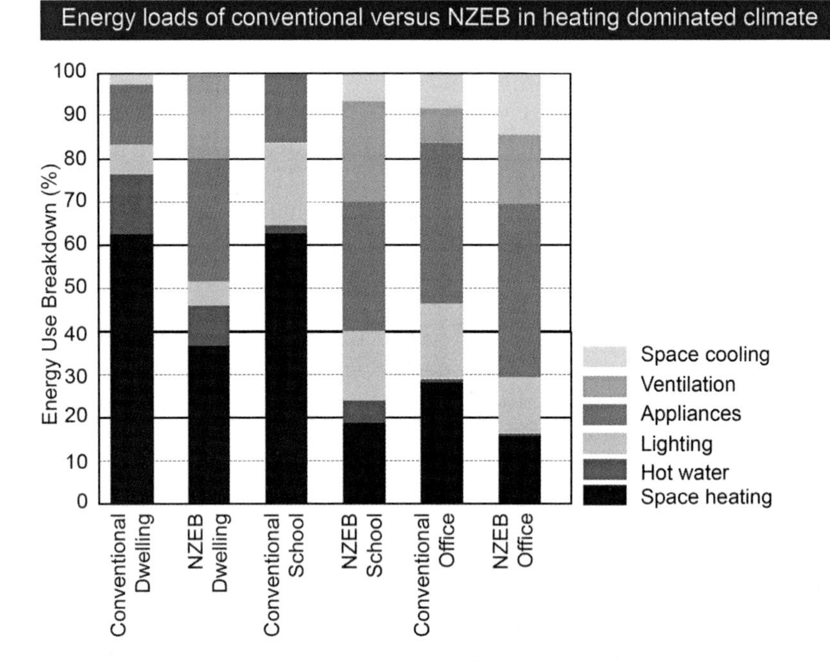

FIGURE 10.2 Energy use breakdown of NZEB in cooling and heating dominated buildings versus conventional buildings.

Fig. 10.2, appliances including plug loads are the most dominant load in NZEB, followed by servers' electricity and space heating/cooling. Based on several comparisons between calculated loads and actual loads, researchers found that the influence of occupant behavior is directly related to appliances use and space heating or cooling system use.

Therefore, occupant behavior is pivotal to determine:

- Large variably of thermal comfort settings and energy consumption.
- Occupant's interactions with energy control systems.
- Large variability regarding air quality and air flow.
- Appliance usage and penetration.
- Models of occupancy and presence.

The paper of Masoso and Grobler (2010) identified the "dark side" of the influence of occupant behavior and pointed out the impact of energy consumption during nonworking hours and working hours. Several studies confirmed that having an unconscious work style or unconscious living style is related to excessive energy use and a rebound effect. Leaving lights and equipment on, heating all building zones, or using central controls can result in a counter effect. So, energy-related occupant behavior influences building energy performance.

Studies have shown that energy consumption at the household level increased in low-energy housing due to the rebound effect. The rebound effect is the reduction in expected gains from new energy conservation measured in newly constructed or renovated high-performance buildings because of behavioral responses. Occupant behavior usually tends to offset the beneficial effects and can increase the consumption up to 30% above the expected or calculated performance threshold (Haas et al., 1998).

Depending on the magnitude of the occupant behavior influence, the rebound effect can be classified under five categories (Saunders, 2008):

1. *Super conservation*: The actual energy savings are higher than expected savings.
2. *Zero rebound*: The actual energy savings are equal to expected savings.
3. *Partial rebound*: The actual energy savings are less than expected savings.
4. *Full rebound*: The actual energy savings are equal to the increase in usage.
5. *Backfire*: The actual energy savings are negative, because usage increased beyond potential savings.

Studies have shown that energy consumption in households vary largely based on user behavior. This is a universal phenomenon that is not bound to geography and economy. A study by Andersen et al. (2009) compared calculated energy use versus measured energy use in low-energy buildings in Denmark and demonstrated a factor of three variations of energy consumption for residential buildings. A similar study confirms the same trend in variation up to a factor of two in the German residential sector. The study findings confirm that user behavior in NZEB is influential. High-performance buildings and NZEB have the potential to decrease energy consumption and heating or cooling loads significantly while achieving good comfort quality and user well-being. However, this can only be achieved if the building is operated as designed (Hong et al., 2017). The probability of achieving the predicted performance or calculated performance of NZEB depend on the occupant knowledge and awareness to operate passive design systems, such as windows and shading devices, as well as occupant expectations and perception with regard to comfort, control, and satisfaction with the indoor environment (Boerstra et al., 2013; Day and Gunderson, 2015).

Notably, occupancy density is another indicator that influences the EUI in NZEB. As discussed in Section 4.6 of Chapter 3, NZEB Performance Thresholds, varying occupancy density can have unintended consequences on occupant behavior and primary energy intensity or square meter. With higher occupant densities, the delivered air

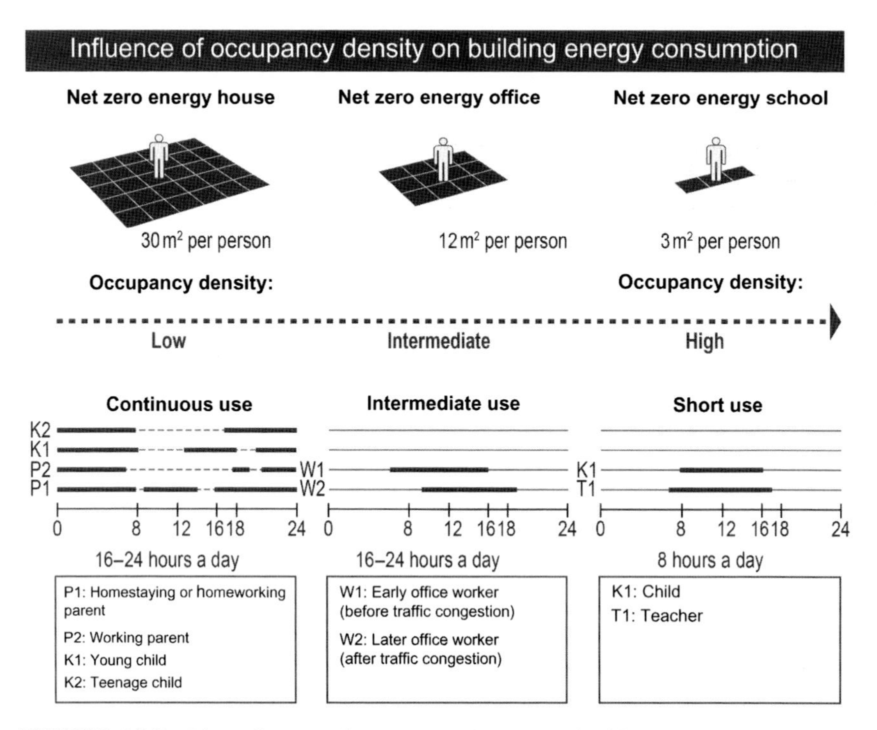

FIGURE 10.3 The influence of occupancy density on building energy consumption (inspired by Agion, 2015).

increases and the cooling or heating loads increase in association with latent and sensible loads increase resulting in overall EUI increase. As shown in Fig. 10.3, the number of people in a room influence occupancy behavior and their sense of control besides the energy use needs associated with their presence (Boerstra et al., 2013).

As a consequence of all the factors mentioned above and in order to achieve the performance goals, we need to monitor NZEB performance and communicate the performance to users.

3 PERFORMANCE ASSURANCE MEASURES

Energy use calculation assumes that everyone behaves similarly. After handing the project to the building owners and with the start of occupation is the moment to investigate the accuracy of prediction and calculation in comparison to what actually happens in NZEB. POE, monitoring, performance certificates, performance visualization, and feedback can ensure performance control. They are part of proactive building management and NZEB design reaped results. In the following sections, we will present the key performance assurance measures that should be taken into account for NZEB.

3.1 Postoccupancy Evaluation (POE)

POE is a quantitative and qualitative process that provides feedback of a building after it has been occupied for some time (Bordass and Leaman, 2005). The purpose of POE is to ensure that the new building fits with its functional usage and with individual user requirements. POEs often combine measurements of environmental variables (temperature, noise level, lighting level, etc.) and qualitative method such as surveys, questionnaires, and interviews following a systematic and empirical approach. The nature and goals of POE depend on the stakeholder that it considers:

- *The owner*: Way to assess the design quality and potential gains (value for money invested) enabled by a better building.
- *The building operator*: Lowering energy consumption and maintenance costs.
- *The building user*: Improve well-being, health and therefore productivity.

Depending on the stakeholder interest and project complexity, several aspects can be addressed for POEs. Also, POE has the role of integrating different stakeholders. POE imposes integration between the pre- and posthandover phases in the building life cycle. It integrates various stakeholders in the building process, particularly the designer, owner, operator, and occupant. POE combines subjective and objective techniques to evaluate the building performance, user experience, and environmental measurements. The vital importance of POE is related to the ability to investigate the simulated performance conceived for NZEB versus the real functioning when users interact with the building. This allows bringing building design and performance assumptions closer to actual practices and performances. The benefits are also for the owner of the building. Indeed, the advantage of POE is the improvement of building performance and maintaining longevity. In general, HVAC systems that are monitored have an impact on energy consumption and on operating cost. Moreover, there are obvious variable benefits related to occupant productivity and satisfaction.

POE is often performed as part of the building commissioning plan during the first 6 months of occupation. Then, after at least 1 year of occupation, another postcommissioning POE round should be performed (Meir et al., 2009). Continuously commissioned buildings opt for POE every five years. However, in recent years, a new direction for certification and benchmarking of buildings focused on building performance based on real-time (or ongoing) data and benchmarking. The new trend of POE includes building performance monitoring and feedback platforms. It keeps tabs on how an entire building (or even just your offices) are doing in terms of energy use, water use, waste reduction, transportation impacts of users, and also their view of the human experience and well-being while inside the building. POE often regroups several aspects, such as those in the list below (Leaman and Bordass, 2001):

- Building function: Producing the best possible building use within the existing economic, statutory, technical and other constraints.
- Building performance: Indoor environmental quality (IEQ), air quality, or thermal performance.
- User productivity: Subjective aspects related to user productivity.
- User satisfaction: Subjective aspects related to user satisfaction or sensation.

In NZEB, the focus on energy performance can lead to underestimating the importance of occupant health, comfort, and productivity. However, POE describes several parameters associated with productivity that need to be taken into account to ensure occupant well-being. Below, we list the most important parameters of POE in relation to occupant well-being and productivity:

- *Cognitive functions*: In high-performing, green certified buildings, studies indicate higher cognitive test scores. Fewer symptoms are reported in high-performing, green certified buildings (MacNaughton et al., 2017).
- *Indoor air quality (IAQ)*: According to Brager (2013), the benefits of improving occupant comfort and well-being in buildings in relation to IAQ is significant. Productivity improvements of 8% − 11% are not uncommon.
- *Thermal comfort*: Modest degrees of personal control can return single digit improvements in productivity.
- *Daylighting*: Daylighting is crucial for occupant satisfaction—understanding of the importance the circadian rhythm in relation to daylighting is growing all the time.
- *Biophilia*: Studies identify productivity gains, particularly the connection to nature.
- *Noise acoustic*: Poor acoustics in offices can decrease productivity by up to 66% (Banbury and Berry, 1998; Treasure, 2009)
- *Interior layout*: Fit-out issues that can have an effect on well-being and productivity include workstation density and configuration of workspace, breakout space, and social space.

Therefore, POE is a powerful tool to assess the performance of NZEB and to make sure that users are empowered by ensuring suboptimal conditions that increase productivity and at the same time allows users personalized control and assess to the energy performance of NZEB.

3.2 Monitoring

Monitoring is critical to ensuring high-performance buildings. It supports energy management and identifies opportunities or threats for energy savings by tracking building energy use and occupant behavior. Tracking energy consumption over time helps building operators and users illustrate variations in usage patterns. Real-time monitoring and energy consumption or generation visualization allows staff to track energy savings and occupants to adopt their behavior.

Building operators gain detailed feedback, which enables them to check and validate operational parameters, depending on the needs and change of occupancy groups (USGBC, 2013). Disparity between how buildings are designed to operate and actual performance are common and are named energy performance gaps. Monitoring can help to identify inadequate commissioning, inaccurate assumptions about occupant behavior or the daily generation and performance of building systems and controls. The collection and analysis of performance data is a complementary work to monitoring and allows benchmarking or comparisons of building performance across similar building typologies and articulate lessons learned to improve NZEB' performance. There are several available measurement and verification (M&V) protocols. Three M&V protocols can be considered for implementation in NZEB monitoring:

(i) The International Performance Measurement and Verification Protocol (IPMVP).
(ii) The ASHRAE Guideline 14 on Measurement of Energy and Demand Savings.
(iii) The M&V protocol specifically developed for NZEB within the IEA SHC/ECB Joint Project Task 40/Annex 52—Towards Net Zero Energy solar Buildings (IEA, 2017).

A key component of a M&V protocol is the M&V Plan, which has to be shaped according to the requirements of the specific project. Based on the existing monitoring protocols indications (EVO, 2002, 2014; ASHRAE, 2002a,b), the main information to be reported in the plan should include the selection of the IPMVP Option that will be used to determine savings and the measurement boundary of the savings determination. Any possible interactive effect beyond the measurement boundary, together with their possible effects, has to be described. Indications of the baseline conditions within the measurement boundary is necessary, i.e., (1) period, (2) energy consumption and demand data, (3) independent variables and static factors (occupancy and operating conditions) coinciding with the energy data conditions, (4) adjustments, (5) characteristics of building envelope and equipment, and (6) measurement equipment information. The baseline documentation typically requires well-documented short-term metering activities. Also, a description of the exact data analysis procedures, algorithms, and assumptions to be used in each savings report has to be stated. The facility operation managers have to indicate the metering parameters, metering points, and period if metering is not continuous, and assignment of responsibilities for reporting and recording during the reporting period (Pisello and Piselli, 2016) (see Tables 10.1 and 10.2).

A wide variety of sensors is available for the measurement of energy flows, which are selected depending on the specific case study characteristics, budget, and expected results. The metering technologies usually used for the monitoring of energy flows within a building depend on metering granularity. Submetering is recommended for NZEB to separate metering for at least end use and total building energy

TABLE 10.1 Sensors for Energy Flows Measurements (IEA, 2013; Pisello and Piselli, 2016)

Type of Meter	Measurement Technique
Electricity	Electronic meters
	Electrochemical induction meters
Gas	Displacement flow meters: diaphragm meters
	Thermal mass flow meters
Solid flow	Conveyor based methods
	Free fall solid measurement
	Detectors of the level of solids in tanks
Liquid flow	Electromagnetic flow meters
	Ultrasonic flow meters
	Vortex-shedding flow meters
	Differential pressure meters
Heating and cooling	Liquid flow meters
	Temperature sensors

TABLE 10.2 Different Monitoring Levels with Associated IEQ Parameters (IEA, 2013; Pisello and Piselli, 2016)

Level 1—Basic Monitoring	Level 2—Advanced Basic Monitoring	Level 3—Detailed Monitoring	Level 4—Advanced Detailed Monitoring
• Indoor air temperature • Outdoor air temperature • Global irradiation	• Indoor humidity • Operative temperature	• Indoor air velocity • CO_2 concentration • Outdoor humidity	• Volatile organic compounds • Spatial daylight autonomy • Annual sunlight exposure • Mean radiant temperature • Global & diffuse solar radiation • Wind speed and direction

consumption, water, and gas. The design team should provide owner and operation teams with end-use budgets that are determined through the energy goal substantiation process (see Chapter 4: Integrative Project Delivery and Team Roles) (Scheib et al., 2014).

In NZEB, monitoring of energy performance can lead to significant electric savings. Through the calibration of the real performance to the end-use budget identified during the design and simulation stage, better control of end-uses can be achieved. Based on the analysis of different case studies identified in Chapter 11, NZEB Case Studies and Learned Lessons, we highlight the key saving areas:

- *Reducing small power use*: Monitoring can lead to lowering the demand of idle PC stations.
- *Reducing lighting use*: Monitoring can lead to improving system control, installing better LED lights and daylight-linked dimming, removing central management and reverting to personal control, and optimize the external lighting operation.
- *Reducing HVAC use*: Monitoring can lead to improving chillers, AHU fan inverters, and pump control.
- *Reducing lighting use*: Monitoring can lead to improving system control, installing better LED lights and daylight-linked dimming, removing central management and reverting to personal control and optimize the external lighting operation.

Measurement of energy-related occupant behavior must be collected using objective and subjective measuring methods. They can be classified under four categories: (1) Measurements, (2) Questionnaires,

(3) Observations, and (4) Stochastic Models. The measurements of occupant behavior can happen by detecting presence and movement, interaction with the building envelope, and the use of control systems. On the other hand, subjective measurements happen through surveys. Surveys are designed to collate self-reported behaviors, and occupancy patterns can be predicted based on the interpretation of the survey. Both techniques are commentary and classified under the POE method.

3.3 Energy Performance Certificates (EPCs)

Legislation governing the energy performance of buildings in Europe arises from the European Energy Performance Regulations. Different countries have adopted different methods to express EPC rating. Most EU Member States use a ratio method, where the building energy (and or carbon emissions) rating is assessed relative to the baseline energy calculated to be used in a national Reference Building. This is defined to be of the same size, shape, layout, and function as the building being rated, but with construction and plant specifications defined by a national calculation methodology. In the EU, a new EPC layout and scheme harmonized on the national territory is mandated for every member state. EPC is compulsory for new or majorly renovated buildings, when buildings are sold or rented out to a new tenant, and for buildings occupied by Public Authorities. EPC boosts awareness of energy saving potential among end users. At the same time, EPC supplies key data on the energy performance of the building stock that are extremely relevant for policy-making on energy efficiency (Hugony, 2017). Similarly, in the United States the Energy Star rating plays the same role.

On a national level, an EPC database is accessible and published as open data. This is a unique case and best practice in Europe, since it also allows local authorities, traders, and market actors to use the data and thereby improves the insight on the EP of the building stock. The global energy performance indicator, expressed in kWh/m^2 per year of nonrenewable and total primary energy, is indicated as EP_{gl} and is calculated as follows (IT, 2015):

$$EP_{gl}\ (kWh/m^2) = EP_H + EP_C + EP_W + EP_V + EP_L^* + EP_T^* \qquad (10.1)$$

where EP_H is the energy performance indicator for heating; EP_C the energy performance indicator for cooling; EP_W the energy performance indicator for domestic hot water; EP_V the energy performance indicator for ventilation; EP_L the energy performance indicator for lighting; EP_T the energy performance indicator for transport services (lifts, escalators); and * valid for nonresidential buildings.

$EP_{H,nd}$, $EP_{C,nd}$, and $EP_{W,nd}$ are the energy performance indicators of the envelope capacity to hold back specific energy needs—heating, cooling, and domestic hot water. These indicators, related to the energy efficiency of the technical building systems, represent the specific building energy performances cited in Eq. (10.1).

A new building (or majorly renovated building) satisfies the minimum requirements if $EP_{H,nd}$, $EP_{C,nd}$, and EP_{gl} are lower than those calculated for the reference building.

To define the energy class of a building it is necessary to calculate the following parameters:

1. $EP_{gl,\ nren,\ rif,\ std(2019/2021)}$: Global energy performance indicator (nonrenewable primary energy) of the reference building with a standard thermal plant and building elements, and with the minimum energy requirements typical of a NZEB.

   The EPC indicator shows the energy efficiency rating on a linear scale of 0–150 divided into bands A–G with A being the most efficient. This indicator will constitute the threshold value between the B class and A1 class for new construction, where the entire classification goes from G grade to A class.
2. $EP_{gl,\ nren}$: Global energy performance indicator (nonrenewable primary energy) of the real building.

Regions in every member state are the responsible authorities for the independent control of EPC quality. They should assure a minimum of 2% annual EPC control in their territories. Fig. 10.4 shows an example for the Italian EPC layout that was designed to provide improved information to the final consumer. Technical information is accompanied by an explicit indication of performance of the envelope, both in winter and in summer, performance of single energy services, energy sources used, a comparison to similar units/buildings (new and existing), recommendations to improve the energy performance, and the class (Hugony, 2017).

EPCs should provide for better decision-making on energy performance of NZEB and the building stock at national level. In some countries, the EPC rating is accompanied by a recommendation report based on a national list of improvement measures such as that used in the United Kingdom. Moreover, EPC needs to be renewed every 5 years in existing buildings. Despite the significant roles of EPCs to assure the performance of NZEB, they remain limited to the building energy performance and carbon emissions without providing information on occupants. There is a perception among building users that EPC grades represent real performance. Therefore, it is important to make NZEB users aware that the EPC grade does not represent real consumption. The monthly and annual feedback on real energy consumption through performance visualization and feedback is more effective.

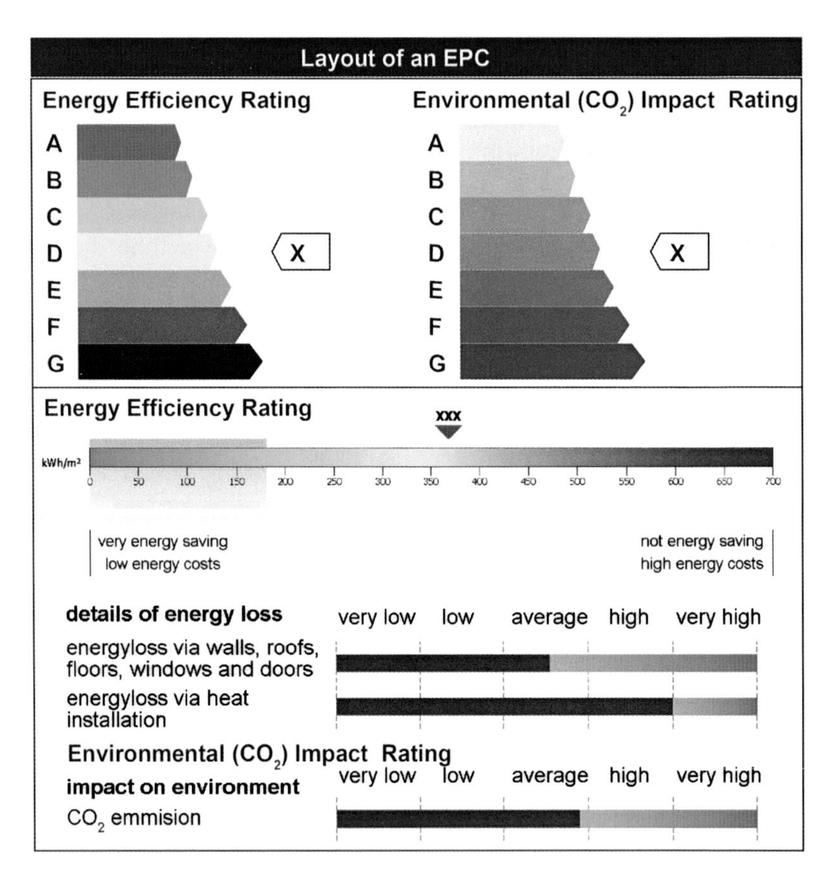

FIGURE 10.4 Shows the layout of an EPC.

3.4 Performance Visualization and Feedback Loop

Performance visualization for feedback is vital for assuring the performance of NZEB and occupant well-being. The importance of relating monitored information and feedback to building users and operators every day is the only way to close the loop and maintain the predicted building performance thresholds. Identifying and benchmarking building performance data to be visualized is a complicated and time-consuming task for building managers (El Gehani, 2013). Key performance indicators for evaluating the building performance and operation are extensive. Many energy parameters of a virtual building can be visualized—i.e., lights, plug loads, heating, cooling, and CO_2 emissions. They depend on spreadsheet software to export, manipulate, and plot data (Lehrer and Vasudev, 2010). According to Fig. 10.5, performance data postprocessing and visualization is based on three main criteria.

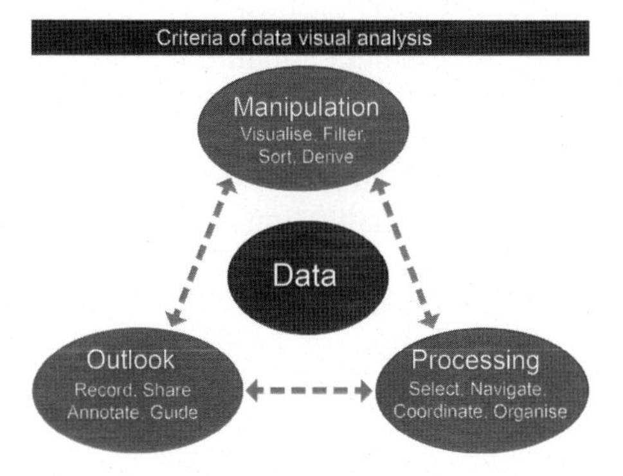

FIGURE 10.5 The criteria of data visual analysis. Source: *Adopted from El Gehani, H., 2013. Data Visualisation for Building Performance Analysis.*

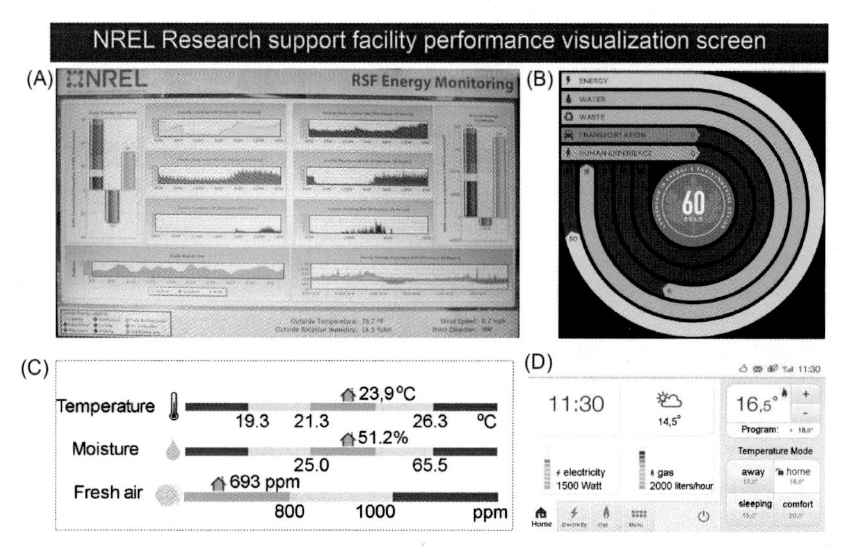

FIGURE 10.6 (A) NREL Research Support Facility performance visualization screen (up left), (B) LEED Dynamic Plaque (up right), (C) another visualization of indoor environment suitable for residential housing (down left), and (D) interactive energy data dashboard by Eneco energy company (down right)

These three criteria reflect the complexity performance visualization. Considerate measures should be taken to ensure data is more accessible to this segment of users. This requires the implementation of easy-to-read graphical displays that interpret complicated building energy information via simple equivalent energy costs, water consumption quantities, and light bulb usage hours (El Gehani, 2013). The NREL Research Support Facility is an example of a NZEB with a performance visualization screen in the building's lobby (see Fig. 10.6A). However, few building occupancy-based performance visualizations exist for building performance assessment (Ioannidis et al., 2016). Building Management Systems (BMS) ignore nontechnical user requirements. They offer, at most, a visual representation with no control options. Conversely, there are several financial and technical obstacles which stand between advanced visualization means and nonexpert users (El Gehani, 2013).

In 2013, the US Green Building Council published the "LEED dynamic plaque," as the future of LEED performance tracking and recertification (See Fig. 10.6A). The focus of the plaque is monitoring how a building performs over time, providing operators and occupants with a tool to control performance and engage with that. Reporting through in-house LCD screens and mobile apps allows for real-time data transparency and disclosure of building performance.

The visualization of monitored performance data significantly assures the performance of NZEB. The visualization facilitates understanding of occupant behavior and their relationship to building performance indicators in real-time. Recently, simple feedback controls have been available on the market to allow users' feedback. For example, Fig. 10.6B shows a common feedback tool for satisfaction concerning services and comfort conditions in buildings. The LEED plaque is an example that provides a form of complex feedback using smartphones and apps to review multivariate data. Arguably, this is a large area of research that tries to combine data visualization with real-time feedback and control (Ioannidis et al., 2016). There is a need for more mainstream technology and interfaces or methods that allow presenting energy performance data in universal dashboard or control panels. A number of energy service companies have power monitoring hardware with user-friendly data analytics and visualization software platforms that enable visualization of how a building uses energy (see Fig. 10.6C and D).

4 PERFORMANCE ASSURANCE PROCESS

In order to assure the performance of NZEB, it is important to plan the process that will allow tracking, controlling, and validating performance. The management of the process is important because, for NZEB, occupants are expected to play a proactive role regarding their behavior and, more importantly, the design team and building operators should understand the occupant needs and the system and control reaction for calibration. In the following, we will highlight the importance of soft-landings, building handover, and maintenance as part of NZEB's performance assurance.

4.1 Soft-Landings

Soft-landings aim to extend the scope of service so that feedback and follow up can become part of the delivery of a project. This approach is a continuous process of the building project. It aims to make designers and builders more involved in the implementation of real, concrete, and precise goals pursued from a performance point of view, security, and health as the indoor quality of the building. A soft-landings team, which involves designers and builders, is based on-site for residents during the early operation period in order to deal with emerging issues more effectively (BSRIA, 2017). Through soft-landings, buildings should get monitored using energy performance indicators for the first three years of occupation—identifying opportunities for fine tuning of the building and future projects.

Soft-landings started in 2003, with the aim that the AEC industry would use occupants as a tool or mean to test the building's performance and at the same time seek their satisfaction and well-being. Soft-landing is a process that depends mainly on POEs to soften the rough ride between the building commissioning step until 3 years after occupation. The aim to create a smooth transition from design to operation has been gaining momentum in recent years (Cheshire, 2017). It is a professional part of the process changes for projects that is focused on aftercare and feedback in the first months and years of occupation. It is an important milestone to connect the briefing requirements or end-use budgets that are determined through the energy goal substantiation process, the handover, and aftercare. It requires preparation during conception until construction, and users must get involved in the process (Way and Bordass, 2005).

Planning this important process of soft-landing and including it in the project's timeline and contract allows the design team to investigate and validate the dynamic simulation assumption and models to better evaluate the optimal performance. It helps to compare and explain the energy use and required occupancy adaptations. By matching predictions and expectations, a well-designed NZEB can offer its occupants well-being and increase productivity, and so align with the main goal of the company. The soft landing approach is the best way to fulfill all the building process goals (BSRIA, 2017). However, it should not be underestimated. The soft-landing method needs time and energy to be well settled, and high cost investments. The soft landing approach is probably more expensive and takes longer than questionnaires, but significantly improves the quality of high-performance buildings.

During the 3 years of postoccupancy monitoring and fine tuning, building performance can be tweaked and occupancy schedules can be calibrated to refine ventilation rates, lighting contorts, bring down energy consumption and carbon emission, and optimize comfort and IEQ. Based on the lessons learned from some case studies presented in Chapter 11, NZEB Case Studies and Learned Lessons, here are some key insights (Cheshire, 2017):

- Feedback from POEs showed that controls and submeters were often incorrectly installed or not calibrated, creating issues with comfort and poor data respectively.
- It is difficult to get buildings working optimally as they are prototypes and can only be tested when occupants are in the building.

Feedback on the operational performance of NZEB is essential to enable building professionals to refine and approve designs.

Soft-landings should be used to push NZEB into the mainstream while empowering occupants, ensuring their well-being, and saving energy. Soft-landings inform occupants and raise user awareness while providing invaluable insight into postoccupancy experience of the building design, construction, and operation partners. NZEB should be evolving with learning lessons from the past, so that all NZEB assure an outstanding and robust minimum EUI while maintaining occupant productivity and satisfaction.

4.2 Handover and Continuous Commissioning

Each NZEB is a prototype, meaning that the handover process needs tailor made actions and measures. The different loads distribution for heating, ventilation, and cooling in relation to systems, equipment and components, and occupant behavior should be checked. The project handover should allow occupiers to concentrate on the actual move without having the added complication of learning new controls and techniques on the day. An educational guide should form a part of the general documentation which should be handed over on the day occupants move into the building. A familiarization session with as many occupants as possible allows them to meet technical experts and learn about the expected building controls and adaptive behavioral measures. An educational guide that is based on a checklist can be used to explain each of the main functions. Occupants should be involved in demonstrations to operate and control building functions. Questions and interaction between the experts and users can clarify operational issues and assure the availability for further points of contact. Repeating familiarization sessions at seasonal change between winter/summer provides is an opportunity to renew or refresh occupant experience and behavioral adaptation measures for controls (Passivehaus Trust, 2017).

Moreover, monitoring-based commissioning procedures should develop the points to be measured and evaluated to assess the performance of energy- and water-consuming systems. Continuous commissioning should be planned to include the procedures and measurement points in 5 years based on a commissioning plan that can be coupled to the EPC renewal. The continuous commissioning plan should address the measurement requirements (meters, points, metering systems, data access) and the limits of acceptable values for tracked points and metered values. A user manual should be developed updating modifications or new settings, and give the reasons for modifications from the original design.

4.3 Maintenance and Facilities Managers

Facility managers and maintenance teams repeatedly report that they feel abandoned with new NZEB regarding control technologies and guidance on operational optimization (Eley et al., 2017). Operators are not impressed by hundreds of data points/sensors that provide them with high resolution minute-interval data. They want user-friendly displays and simple key performance indicators that help them to improve the building performance and achieve occupant satisfaction. They require an understanding of the sensitivities and contribution of components to the overall performance.

At the same time, they are responsible to ensure the comfort and energy qualities of NZEB during operation. The access to the heat recovery units and HVAC systems for changing filters and other maintenance tasks should be planned and well prepared. Intermediate cleaning of filters requires effort, therefore it is recommended to change filters frequently depending on the recommended use period. The provision of stickers or laminated graphic instructions affixed to plant, installation tubes, and ducts should be realized (Passivehaus Trust, 2017). For large-scale NZEB projects, technical documents should be included as appendices within the overall building manual and log book. The maintenance team should have access to the building manual, log book, and occupancy operational schedules. The maintenance team should be informed on the expected performance and ensure that fault response teams are well informed.

Smart controls and internet connected appliances such as the IoT offer maintenance teams and service providers means to gain operational efficiency and meet the high expectations real-time monitoring of installations. This allows for predictive maintenance and lower repair costs. Reconciliation between the design team's BIM model and the facility management (FM) team combining technical know-how and digital skills, could quickly take place and facilitate the maintenance team's task. Smart predictive maintenance of NZEB is a newly emerging practice that can improve HVAC systems throughput and maintenance. In the near future, cloud computing and artificial intelligence will play a significant role in NZEB maintenance.

5 BEST PRACTICES TO INFLUENCE OCCUPANT BEHAVIOR

Occupant behavior is the last predictable aspect of NZEB. However, if we have the opportunity to inform people and find the balance between automatic and manual control, we can respond to user needs and save energy. This requires real-time monitoring and feedback. In this section, we will present best practices related to occupant experiences, occupant engagement, and control. We also highlight the importance of establishing a relationship between occupant adaptive measures in relation to building controls and operation modes.

5.1 Occupant Experiences

Assuring comfort and occupant well-being in NZEB is a challenge. From one side, the integration of different conjunctions of technology and building solutions is complex. From another side, the definition of comfort and occupant satisfaction is an elastic and highly varying socio-cultural concept. Based on the POE and analysis of different studies, we identify major reasons for dissatisfaction of occupants that frequently appear in NZEB (Attia, 2018):

1. *Overheating*: In NZEB, the occupant's main focus is to be comfortable rather than optimizing energy use. Most complaints about indoor temperature tend to appear during summer. Findings from POE confirm the correlation between NZEB and overheating (Attia et al., 2017). NZEB with low ratings for summer indoor temperatures often have a problem with the estimation of passive cooling demand. These findings confirm the need for strict sizing and design to meet the space cooling demand. In actively cooled buildings, there is often a reserve for cooling power. Passive cooling usually does not. With passive techniques, it is important to take future climate scenarios into account.
2. *IAQ and draft*: Most complaints about air quality and ventilation are related to draft, humidity, or the low rate of air renewal. In NZEB, high ventilation rates should be avoided to provide proper humidity,

while respecting the maximum carbon dioxide concentration thresholds. Therefore, we recommend measuring the exchange rate of the indoor air after the installation of the ventilation system. The installation of humidifiers or dehumidifiers should be investigated during the design stages.

3. *Noise from building services*: The noise from ventilation systems is a recurring issue that affects user appreciation in NZEB. We recommend that the sound pressure level Lp for building service units should be equal to, or below, 35 dB(A) and that HVAC units are placed in a separate sound-insulated room for building services (Mlecnik et al., 2012).
4. *Use and occupant control*: Perceived control capability can alleviate occupants' discomfort complaints (Luo et al., 2016). Occupants should be offered opportunities to interact with their personal control. Users would override solar shading control to enjoy the view or improve privacy. As a consequence, this can increase the risk of overheating as well as larger heating energy use. However, occupants with an effective level of control and training change their behavior. They were satisfied with their indoor environment and tolerated slightly more discomfort (Day and Gunderson, 2015). The role of building designers is to make it possible for users to control their personal environment, privacy, and view to fulfill comfort needs.

Understanding the reasons behind occupant dissatisfaction in NZEB is challenging because it depends on many different factors (Stazi et al., 2017). It requires evaluating the thermal environment quality of the buildings as perceived by users. POE allows us to understand how building occupants interact with various building systems and components. More importantly, they allow us to understand the causes of occupant dissatisfaction and pinpoint areas of concern that may affect comfort and productivity. To inform future decisions regarding occupant behavior related issues, we identified the following key influential parameters:

- Access to nature influences occupant tolerance for IEQ and their expectations.
- Satisfactory level of control over the work or living environment.
- Adjustment and coping behavior regarding clothing, metabolic level, energy intake, etc.
- Working culture and the influence of colleagues through peer pressure in relation to shared workspace.
- Organization factors such as dress code or working day schedules.
- Awareness and level of engagement.

5.2 Occupant Engagement and Control

Building control should be able to fulfill both the objective needs for indoor environment and energy use in the space in cooperation with mechanical systems. Before expecting any interaction or engagement, designers should make sure to fulfill the subjective needs of the individual and make users feel they are in control by having the possibility to override and interact. The perception of control is more important than real control, because if people feel that they are in control, they tend to be more tolerant to slightly irregular temperature. Therefore, some level of occupant engagement is required to achieve sustained performance outcomes for the operation of high-performance buildings.

Occupants are asked to become more aware and educated about how their building is meant to operate. The effective influence of passive design usually depends on occupant interaction to open and close windows, or raise and lower blinds. To help enable desired behavior and engagement, building designers should provide a feedback loop. Energy dashboards, smartphones, or PC software can keep the user engaged to act on time and to become an aware building occupant. Customization of building information and making it accessible to users is a reality in many NZEB.

Technology-based approaches are not the only options for better occupant engagement and control (Zhao et al., 2017). A phenomenon called *social diffusion* can also facilitates behavioral change by analyzing and responding to human habits. It depends on interaction and dialogue at a community level through social groups and organizations. When occupants are asked to attain a certain performance goal related to, and based on, this a conversation is generated. Through the power of dialogue and peer discussion, behavior is changed. Social diffusion among building occupants might be as simple as a Facebook group, suggestion box, an email list-serve, or web-based comment box.

Building owners, operators, and facilities managers should organize the context in which users are expected to make their behavioral decisions. Simple actions such as placing signs or leaflets for selection or performance specific behaviors can be an easy option to influence users. Interactive energy data dashboards that provide emojis or colored bars influence occupants to take actions to improve their score to lower energy consumption. Also, transparency is very important, because when users understand the control or operation systems, they become more flexible and tolerant. If building operators communicate with users in an interesting way and use compelling language or visualizations to provide simple and relevant information, people will respond.

Building operators should not expect that technology alone will lead to building-aware occupants. Without communication, storytelling, and meaningful feedback, occupant engagement will remain passive. The following recommendations should be addressed by building owners and facilities managers for a better awareness and engagement of occupants in NZEB:

- *Communication and operation*: A qualified person should explain and demonstrate the handling of the system as soon as a tenant moves in.
- *Feedback loop and error detection*: There should be a prompt or a display indicating the failure of components like ventilators. Tenants do not care as much about the devices' proper functioning as owners do.
- *Occupancy engagement strategy*: The development of an occupant engagement strategy is important to identify occupant behavior adaptations and measures. A test-run or pilot program could be useful during the soft-landing phases.
- *Performance reference*: In case or error detection or system failure, building operators should have good reference data and good performance reference. This allows easier maintenance.

5.3 Building Controls and Operation Modes

The design of NZEB should include controls and systems that are appropriate for the skills and understanding of the occupants. For the occupant, an understanding of the contorls and techniques available to users ensures a better perception of control and a better IEQ with ultra-low energy demands. Small size NZEB (less than 4000 m^2) typically lack an on-site full-time operator (Eley et al., 2017). There are two main approaches to empower occupants and provide optimal control of NZEB.

The first approach is minimalistic—almost seeking manual control of HVAC systems allowing users to engage and regulate their environment. NZEB are high-tech based facilities that depend significantly on their operation. According to Fig. 10.7, we advise the use of manual controls in NZEB for buildings without FM staff or skills. In this case, the use of familiar comfort controls that can be easily operated following a shallow learning curve is recommended. Controls need to be clear to the occupant and have defaults (Eley et al., 2017).

For NZEB that can be operated by building professionals or facility managers, smart and automated controls are recommended. In time, NZEB will get more and more connected with the internet of things (IoT). Design teams must make sure that there is full-time equivalent FM for those types of buildings. Finally, both approaches are good, but they depend strongly on users and the FM capabilities and availability. The role of the control strategy for NZEB is to offer the right balance between automatic, semi-automatic, and/or manual control while responding to user needs in a transparent and understandable way.

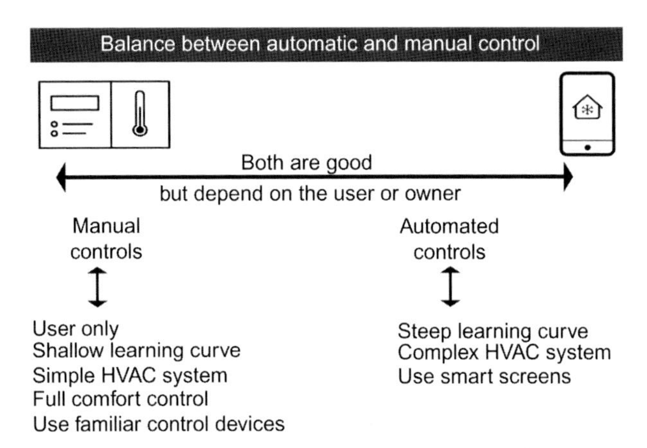

FIGURE 10.7 Design teams should find a balance between automatic, semi-automatic, and/or manual control while responding to user needs in NZEB.

6 DISCUSSION

The current state and future challenges in predicting occupant behavior and assuring the performance of NZEB include POE, performance monitoring, EPC, and performance visualization to close the feedback loop. Planning the performance assurance process and understanding occupant behavior is as important as performance validation. NZEB require a close follow up, control, and validation of the IEQ and energy performance. In this section, we will discuss the major future challenges that need to be addressed (or further addressed) to make sure we design for people first, and how to programme and monitor their use or occupant behavior interaction for optimal performance assurance in NZEB.

6.1 Intensive Examination of Technology

Facilities managers and building operators are considered the technical experts that can bring in-depth improvement to NZEB performance. NZEB depend on systems and control technologies that were never designed to work together or to last forever. The proliferation of NZEB technologies make them sensitive to failures and therefore, require intensive examination. Sometimes we get attracted to technology or anything coming from the ICT or automation worlds and integrate them in our buildings as if they will last forever. We might ask ourselves what is the best BMS and BMS technology for NZEB, but we should not forget that it is not about software or technology.

For the operation of middle and large-scale NZEB projects, it is all about building operators to assure performance and occupant's well-being. It is about experienced facility managers who are thoughtful and savvy. They are responsible to ensure that products and technologies are tested and used properly. They are responsible to respond to systems failures and errors based on their education and skill set. They will be running the digital control systems and maintaining HVAC installations. A committed, well-educated, full-time, building operation professional or team should be engaged from the commissioning phase throughout the soft-landing process. It is expected that the FM team that can handle and analyze data and at the same time act on occupant dissatisfaction and system failures and replace filters or installations. This will require the development of a robust and easy-to-use toolkit for tracing systems and appliances, and evaluating IEQ. Large-scale metering and sensor networks requires extensive data processing to be usable. Without qualified and educated building operators and maintenance services, energy management cannot be effective.

6.2 Flexibility of Use and Building Operation

New living and working modes will change our patterns of building use where tenants buy or rent a squared meter of service and no longer a squared meter of building (Section 3 of Chapter 5, Occupants Well-Being and Indoor Environmental Quality). Ideas like hot desking, telepresence, virtual workplace or flexible working are changing the way people use buildings. Flexibility is becoming essential to adapt to telecommuting and part time work needs. Flexible working can increase satisfaction, motivation, and productivity of staff and will shift the attention from energy savings to IEQ and occupant experience. Workspaces will be used based on employee's schedule: sitting desk, standing desk, work booth, meeting room, balcony seat, or "concentration room."

Moreover, the operation of NZEB requires a certain flexibly on the energy management and comfort management level to satisfy occupant needs and meeting these needs to stabilize the surrounding energy grid. Using smart apps and IoT, and based on personal preferences for light and temperature, smart controls will regulate the indoor environment accordingly. This will require collecting big data on how occupants interact in buildings, and then communicate these using central dashboards. Tracking personal preferences and appliances such as coffee machines, printers, and parking space will allow FM teams to optimize operation, cut the costs of heating, cooling, lighting, and cleaning. We should be prepared that NZEB will rely more and more on smart control that will need to be robust and easy to use.

6.3 Understanding Occupant Experience

In a NZEB, we should expect an extended POE. In the near future, we will need to achieve comfort and meet the flexible needs for grids. The building operation should take into account occupant behavior and include an autonomous operation within the BMS. However, for users' acceptance and ease of use of building control, the autonomous operation of NZEB and smart management needs to simple. Users must have control over their indoor environment. This is a serious challenge, because after construction the owner should continue to follow up the building to evaluate the IEQ. It is extremely important to understand the satisfaction levels of occupants with various aspects of IEQ. Firstly, the operation team will need to evaluate the thermal environment quality of the building as perceived by users. This is the moment to revise the energy end-use budget and performance thresholds identified during the design phase. This first step involves understanding the interaction of occupants with the various building systems and controls. Occupants interact with light, color, sound, temperature and humidity, toxins and contaminants, plants, and nature. Secondly, the causes of occupant dissatisfaction and the reasons that may affect comfort and productivity should be identified. The third step is to ultimately improve the IEQ and user satisfaction. The only way to deeply understand how users interact with the building is performance extended POE involving continuous monitoring. Only after gathering and analyzing this data can we revise the goals and evaluate the building performance.

Based on the analysis of several NZEB, we found many NZEB projects that had severe overheating or overcooling problems after they were occupied. This could be related to design, construction, or operation problems. It could be also related to climate change and the failure to predict the climate. The majority of architects and engineers are unaware that the risk of not achieving comfort in NZEB is higher than conventional buildings. There is also likely to be low awareness that overheating has a negative impact on health even in warm and hot climates. Therefore, there should be continuous and systematic work to monitor internal temperatures in NZEB. Comfort and health can only be assured if we understand the occupants' experience and act accordingly.

6.4 Occupant Adaptive Behavior

Due to energy pricing and grid interaction with NZEB, it is expected that occupants will adapt through behavioral adjustment. Occupant behavior and autonomous operation modes will result in a flexible demand-supply balance of energy. Weather forecast information and pricing signals from the grid will influence occupant behavior.

Occupants are expected to come up with adaptive occupant behavior influenced by signals from the grid. People are adaptive if we give them control over clothing, heating, or cooling system control. With the proliferation of IoT and the stronger connection between smart buildings and smart grid, we assume that occupants will receive nudges or recommendations to change their behavior. The stochastic nature of occupant behavior, the number of people occupying a space, and the duration occupied will become more challenging. The newly expected motivational patterns of behavior will make it more difficult to find common understandable building occupant patterns and will require energy management systems that can record data at precise time intervals to identify the operational issues of NZEB.

7 LESSON LEARNED #10

Ultra-low energy consumption alone does not make a good NZEB. Continuous IEQ assessment should be undertaken to assure occupant satisfaction. Therefore, POE and monitoring can help maintain the performance target for energy, comfort, and IAQ. Any change of occupancy density or systems operation can be detected and real performance can be calibrated to end-use budget. Maintaining comfort and healthy indoor conditions in NZEB could be a challenge for some spaces because of the nature of their design and operation. POE can identify occupant perceptions or enjoyment of their spaces through qualitative surveys and field measurements. The advantage of POE is that it can assure the NZEB performance and, at the same time, identify local issues with ventilation, RH, air temperature, and indoor noise levels. It can identify any lack of personal controls in specific work areas or other occurring problems for occupants. POE can provide evidence of end user satisfaction with all general building aspects and indoor conditions, as well as with most of their specific living or working area conditions.

In the operation of NZEB, the feedback loop should be closed to allow users to interact with the systems. If NZEB users receive feedback through an energy display, this will encourage them to conserve energy and stay within the net-zero target. Controls can cause problems in practical use rather than expected because users might have other priorities and are not always able to understand and use the technology. On the other hand, real-time information should be communicated to occupants through dashboards or apps, and allow them to report discomfort.

Technologies like envelope insulation, building airtightness, ventilation heat recovery, etc., are robust technologies with a well-defined performance. However, it is better to assume that NZEB will likely have higher energy use and lower IEQ in practice until monitoring data proves the opposite. Lessons learned from practice indicate that:

- NZEB performance is very sensitive to control and different levels of user interaction.
- NZEB systems operation and control are often difficult for users to understand.
- Consider the interplay with other technical systems in the building.
- For NZEB, building operators should continuously analyze, interpret, and improve performance.

User perception and interaction with the system should be taken into account in the design and operation of NZEB. The interplay with mechanical systems is delicate and requires improvisation to make sure IEQ is assured. Metering data allows the project team to focus on the largest energy consumers. Occupant behavior can be better understood after end-use metering is done. Also, domestication of the technology to local expectations can be critical for success. Therefore, we should be adapting control strategies to user practices. We need further knowledge on user practices and user interaction in NZEB. Finally, the role of the control strategy of adaptive façades is to offer the right balance between automatic, semi-automatic, and/or manual control while responding to user needs in a transparent and understandable way.

References

AGION, 2015. Pilootproject Passiefscholen: Bilan 2015, Agentschap voor Infrastructuur in het Onderwijs (AGION) Koning Albert II-laan 35 bus 75, 1030 Brussel.

Andersen, R.V., Toftum, J., Andersen, K.K., Olesen, B.W., 2009. Survey of occupant behaviour and control of indoor environment in Danish dwellings. Energy Build. 41 (1), 11–16.

ASHRAE, 2002a. Guideline 14-2014 (Supersedes ASHRAE Guideline 14-2002), Measurement of Energy, Demand, and Water Savings. Available from <http://www.techstreet.com/products/1888937>.

ASHRAE, 2002b. Guideline 14-2002, Measurement of Energy and Demand Savings. Available from <https://gaia.lbl.gov/people/ryin/public/Ashrae_guideline14-2002_Measurement%20of%20Energy%20and%20Demand%20Saving%20.pdf>.

Attia, S., 2018. Regenerative and Positive Impact Architecture: Learning from Case Studies. Springer International Publishing, London, UK, ISBN: 978-3-319-66717-1.

Attia, S., Eleftheriou, P., Xeni, F., Morlot, R., Ménézo, C., Kostopoulos, V., et al., 2017. Overview of challenges of residential nearly Zero Energy Buildings (nZEB) in Southern Europe. Energy Build. Available from: https://doi.org/10.1016/j.enbuild.2017.09.043.

Banbury, S., Berry, D.C., 1998. Disruption of office-related tasks by speech and office noise. Br. J. Psychol. 89 (3), 499–517.

Boerstra, A.C., Loomans, M.G., Hensen, J.L., 2013. Personal control over temperature in winter in Dutch office buildings. HVAC&R Res. 19 (8), 1033–1050.

Bordass, B., Leaman, A., 2005. Making feedback and post-occupancy evaluation routine 1: A portfolio of feedback techniques. Build. Res. Inf. 33.4, 347–352.

Brager, G.S., 2013. Benefits of improving occupant comfort and well-being in buildings. In: Proceedings of the 4th International Holcim Forum for Sustainable Construction: The Economy of Sustainable Construction, pp. 181–194.

BSRIA, 2017. Available from <https://www.bsria.co.uk/services/design/soft-landings/> (accessed July 2017.).

Cheshire, D. (2017) Is Soft Landings finally taking off? Available from <https://www.linkedin.com/pulse/soft-landings-finally-taking-off-dave-cheshire> (accessed September 2017.).

Day, J.K., Gunderson, D.E., 2015. Understanding high performance buildings: The link between occupant knowledge of passive design systems, corresponding behaviors, occupant comfort and environmental satisfaction. Build. Environ. 84, 114–124.

Day, J.K., O'Brien, W., 2017. Oh behave! Survey stories and lessons learned from building occupants in high-performance buildings. Energy Res. Soc. Sci.

El Gehani, H., 2013. Data Visualisation for Building Performance Analysis.

Eley, C., Gupta, S., Torcellini, P., Mchugh, J., Liu, B., Higgins, C., et al., 2017. A Conversation on Zero Net Energy Buildings (No. PNNL-SA-128289). Pacific Northwest National Laboratory (PNNL), Richland, WA.

EVO, 2002. Efficiency Valuation Organization (EVO), International Performance Measurement & Verification Protocol (IPMVP)—Concepts and Options for Determining Energy and Water Savings, vol. I. Available from <http://www.nrel.gov/docs/fy02osti/31505.pdf>.

EVO, 2014. International Performance Measurement & Verification Protocol—Core Concepts, EVO 10000—1:2014.Available from <http://evo-world.org/en/>.

Garde, F., Aelenei, D., Aelenei, L., Scognamiglio, A., Ayoub, J., 2017. Net ZEB case study buildings, measures and solution sets. In: Solution Sets for Net-Zero Energy Buildings: Feedback From 30 Buildings Worldwide, pp. 39–102.

Haas, R., Auer, H., Biermayr, P., 1998. The impact of consumer behavior on residential energy demand for space heating. Energy Build. 27 (2), 195–205.

Hong, T., Yan, D., D'Oca, S., Chen, C.F., 2017. Ten questions concerning occupant behavior in buildings: the big picture. Build. Environ. 114, 518–530.

Hugony, F., 2017. The EU energy performance of buildings directive: the Italian implementation. In: International Symposium on Energy Efficiency in Buildings, February 13–14, 2017.

Ioannidis, D., Tropios, P., Krinidis, S., Stavropoulos, G., Tzovaras, D., Likothanasis, S., 2016. Occupancy driven building performance assessment. J. Innov. Digit. Ecosyst. 3 (2), 57–69.

IEA, 2013. SHC/ECBCS Task 40/Annex 52—Towards Net Zero Energy solarBuildings. M&V Protocol for Net ZEB. A Technical Report of STA, 2013. Available from <http://www.nachhaltigwirtschaften.at/iea_pdf/endbericht_201417_iea_shc_task40_eb c_annex_52_anhang03.pdf>.

IEA, 2017. SHC/ECBCS Joint Project Task 40/Annex 52—Towards Net Zero Energy solar Buildings. Available from <http://task40.iea-shc.org/>.

IT, 2015. Italian Decrees 26 June 2015 <www.sviluppoeconomico.gov.it/index.php/it/energia/efficienza-energetica/edifici> (accessed July 2017.).

Leaman, A., Bordass, B., 2001. Assessing building performance in use 4: the probe occupant surveys and their implications. Build. Res. Inf. 29 (2), 129–143.

Lehrer, D., Vasudev, J., 2010. Visualizing Information to Improve Building Performance: A Study of Expert Users.

Luo, M., Cao, B., Ji, W., Ouyang, Q., Lin, B., Zhu, Y., 2016. The underlying linkage between personal control and thermal comfort: Psychological or physical effects? Energy Build. 111, 56–63.

MacNaughton, P., Satish, U., Laurent, J.G.C., Flanigan, S., Vallarino, J., Coull, B., et al., 2017. The impact of working in a green certified building on cognitive function and health. Build. Environ. 114, 178–186.

Maor, I., 2016. Evaluation of factors impacting EUI from high performing building case studies, High Perform. Build. Mag. Fall, 30–31.

Masoso, O.T., Grobler, L.J., 2010. The dark side of occupants' behaviour on building energy use. Energy Build. 42 (2), 173–177.

Meir, I.A., Garb, Y., Jiao, D., Cicelsky, A., 2009. Post-occupancy evaluation: an inevitable step toward sustainability. Adv. Build. Energy Res. 3 (1), 189–219.

Mlecnik, E., Schütze, T., Jansen, S.J.T., De Vries, G., Visscher, H.J., Van Hal, A., 2012. End-user experiences in nearly zero-energy houses. Energy Build. 49, 471–478.

Myerson, J., Ross, P., 2006. Radical Office Design. Abbeville Press, New York.

Passivehaus Trust, 2017. Handover. Available from <http://howtopassivhaus.org.uk/handover> (accessed July 2017.).

Pisello, A.L., Piselli, C., 2016. Effective monitoring protocols to be implemented in the outdoor areas of each settlement and an overall report about technical details and motivation of selected procedures. In: Achieving near Zero and Positive Energy Settlements in Europe using Advanced Energy Technology H2020, 67840.

Ross, P., 2016. EPFL smart living lab. Expert Interviews Video November 5, 2016.

Saunders, H.D., 2008. Fuel conserving (and using) production functions. Energy Econ. 30 (5), 2184–2235.

Scheib, J., Pless, S., Torcellini, P., 2014. An energy performance based design-build process: strategies for procuring high-performancebuildings on typical construction budgets, Proceedings of the ACEEE 2014 Summer Study on Energy Efficiency in Buildings, vol. 4. pp. 306–321. <http://aceee.org/files/proceedings/2014/data/papers/4-643.pdf>.

Stazi, F., Naspi, F., D'Orazio, M., 2017. A literature review on driving factors and contextual events influencing occupants' behaviours in buildings. Building and Environment vol. 118, 40–46.

Treasure, J., 2009. Julian Treasure: The 4 Ways Sound Affects Us. TED.

USGBC, 2013. LEEDv4: Reference Guide for Building Design and Construction (v4). US Green Building Council, Washington, D.C.

Way, M., Bordass, B., 2005. Making feedback and post-occupancy evaluation routine 2: Soft landings–involving design and building teams in improving performance. Build. Res. Inf. 33 (4), 353–360.

Zhao, D., McCoy, A.P., Du, J., Agee, P., Lu, Y., 2017. Interaction effects of building technology and resident behavior on energy consumption in residential buildings. Energy Build. 134, 223–233.

CHAPTER

11

NZEB Case Studies and Learned Lessons

ABBREVIATIONS

AEC Architectural, Engineering, and Construction
AHU air handling unit
BMS Building Management Systems
BIPV Building Integrated Photovoltaic
EPBD Energy Performance of Buildings Directive
ETICS External Thermal Insulated Composite System
COP Coefficient of Performance
DHW domestic hot water
EPC Energy Performance Certificate
EU European Union
EUI energy use intensity
FM facility management
HVAC Heating, Ventilation, and Air Conditioning
IAQ indoor air quality
IDP integrative design process
IEQ indoor environmental quality
LPD lighting power density
MVHR mechanical ventilation heat recovery
M&V measurement and verification
NREL National Renewable Energy Lab
NZEB Net Zero Energy Buildings
PH Passive House
PHPP Passive House Planning Package
PMV Predicted Mean Vote
PPD Percentage People Dissatisfied
POE postoccupancy evaluation
RFP request for proposal
RSF Research Support Facility
SAS security airlock system
SHGC solar heat gain coefficient

Net Zero Energy Buildings (NZEB)
DOI: https://doi.org/10.1016/B978-0-12-812461-1.00011-3

TABS thermal activated building systems
VAV variable air volume
WWR window-to-wall ratio
***U* value** thermal conductivity

1 INTRODUCTION

Designing Net Zero Energy Buildings (NZEB) requires planning and evaluation. The aim of the integrative project delivery process is not only to bring stakeholders together during early design stages, but also to involve energy consultants, commission agents, contractors, and building operators to assess the performance. Therefore, the measurement and verification (M&V) responsibility should be defined as early as possible in any NZEB project. Without M&V and real-time monitoring we cannot ensure performance and achieve transparency.

In this context, we selected five case studies that represent different typologies with high resolution monitoring results. We tried to normalize the different key performance indicators (KPI) to make the five case studies comparable and extract useful information that can be generalized for other NZEB projects. We also decided to build on the existing studies that document the performance of NZEB. There several studies that investigated the performance of NZEB. One of the earliest comprehensive studies was done by Voss and Musall (2013) as part of the International Energy Agency Task 40 on NZBs. This was followed by the work of Athienitis and O'Brien in 2015. Also the work of Ayoub et al. (2017) and Attia (2018) provide useful case studies as part of the IEA Task 40 (2008). Other studies that collected and compiled case studies include the work of Kurnitski (2013), Heaps (2015), Reeder (2016), and Maor (2016). The previous studies provide valuable insights on NZEB, however they remain focused on definitions and analysis of case studies. In this chapter, together with Chapter 12, Roadmap for NZEB Implementation, we exceed the observation and analysis of NZEB and provide procedural and performance-based management advise.

The five case studies in this chapter reflect a layered approach to meeting occupant needs while using less energy. The implication of HVAC technologies, plug loads, and the share of renewable energy systems (RES) in relation to occupant behavior are presented here. We consider the case studies valuable to come up with a road map for NZEB implementation. Even though the road map is presented in the next chapter (Chapter 12: Roadmap for NZEB Implementation), we consider this chapter as one of the most important chapters that provides evidence-based insight on monitoring and POE. In this chapter, we present five case studies on NZEB to demonstrate lessons learned from practice. The aim of this chapter is to provide informed guidance to the architecture, engineering, and construction communities. The project's design and operation are documented highlighting the importance of technology selection, performance communication, and occupant behavior adaptation measures.

2 CASE 1: SERAING MUNICIPAL BUILDING

2.1 Project Description

The municipality building in the town of Seraing, in East Belgium, was designed by the Greisch Office in association with the Neo Construct/IDES Engineering Energy Consulting firm. Based on an architectural competition, the winning design team proposed a design that would consume 85% less energy and reduce CO_2 production by 500 tons per year compared to typical office buildings. The project was inaugurated in 2014 and received the Belgian Passive House (PH) certification with a total occupied area of 4500 m^2 hosting 200 employees working in office areas and administrative services (see Fig. 11.1A and B).

The project integrates passive design strategies combining natural ventilation with thermal mass and an automated solar protection system. The building includes advanced technologies and building services combined with passive design measures such as the management of

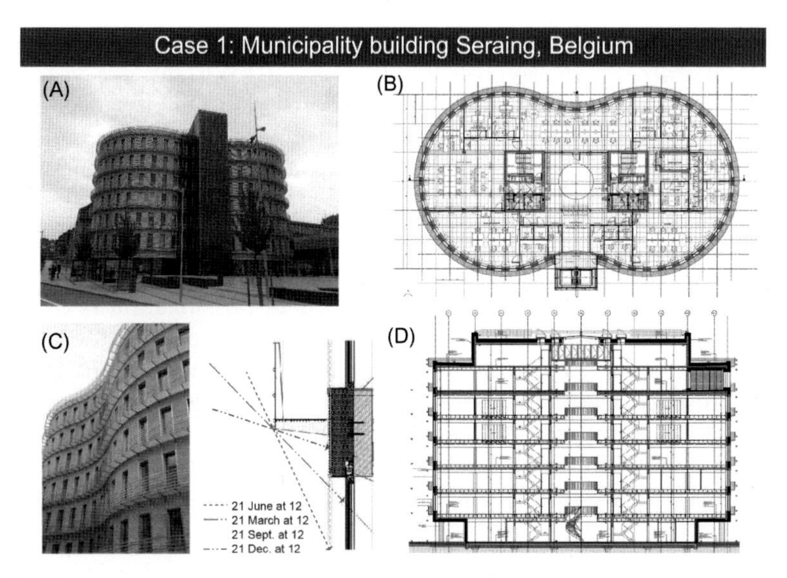

FIGURE 11.1 (A) Seraing municipality building, (B) plan of the typical floor, (C) exterior solar protection, and (D) cross section showing the two building wings (Neo Construct & IDES Engineering, 2009).

solar gains, high-insulation, passive cooling through natural ventilation, and evaporative cooling (Busch, 2010). The Belgian PH certification criteria for commercial buildings are:

- Net energy requirement for heating <15 kWh/m^{2}per year
- Net energy requirement for cooling <15 kWh/m^{2}per year
- Airtightness <0.6 vol/h
- Primary energy consumption <(90 − 2.5 × building compactness ratio) (kWh/m^2 per year)
- Overheating: the number of hours > 25°C may not exceed 5% of the working hours

Therefore, the design team looked at applying the following goals. The first goal included limiting the internal heat gain, optimizing the openings design to avoid overheating (orientation and sun protection) and combine natural and artificial lighting. The second goal was to introduce automated controls systems. The third goal was to implement ultra-efficient HVAC systems. Finally, increasing the renewable energy share through photovoltaic panels, BIPV, solar thermal panels, and a heat pump was the fourth goal.

2.2 Performance Characteristics

The total consumption for the administrative city of Seraing is equal to 97 kWh/m^{2}per year serving 200 occupants. We can already notice that the values are higher than those put forward during the design. The total primary energy consumption is 110 kWh/m^2 per year. The energy consumption breakdown is 14% for heating energy needs, 57% for cooling energy needs, 24% for appliance, mechanical ventilation, and plug load energy needs and 15% artificial lighting energy needs.

Climate

The project location (Lat: 50.61, Long: 5.51, Alt: 157 m) is classified as temperate climate. The Köppen-Geiger climate classification is Cfb. The average annual temperature is 9.8°C in Seraing. In a year, the average rainfall is 827 mm. According to ASHRAE Classification the site location is cold humid (Climatic Zone 5A) with 3000 <Heating Degree Days 18°C ≤4000. Temperatures are relatively mild during the whole year with the average low at 1°C in winter and 23°C in summer, while wind tends to be slightly stronger in winter.

Envelope

The design team decided to select a compact building form equal to 4.1. The window-to-wall ration is 30%. Automated shading blinds are installed from the outside. The management of solar access is ensured through an automated operation system with 50% opacity (see Fig. 11.1C).

Airtightness was quantified by a Blower Door Test before the building was occupied. This test provided information on the quantity of air that enters the building outside the systems of ventilation. In order to carry out the measurement, a difference in pressure was created artificially between the inside and outside, taking care to seal all the orifices such as vents. The leakage rate for an imposed pressure difference corresponds to the leakage. The Blower Door Test is standardized (EN, 13829) and has resulted into an infiltration rate under a pressure differential of 50 Pa equal to 0.28 vol/h. Proper sealing avoids short-circuiting the ventilation system by unmanaged infiltration (Fig. 11.1D). The ventilation in this building is managed by two air handling units. Poor airtightness would induce air infiltrations, but also disrupt the balance of air flow sized for building services.

The envelope is thermally resistant and its performance is based on thermal insulation coupled to airtightness resulting in highly thermal resistant walls. A number of optimization runs were made during the design stage and construction stage to design the envelope and coordinate the sequence of order on-site to achieve ultimate performance. As shown in Fig. 11.2A, thermal modeling calculations allowed improving the

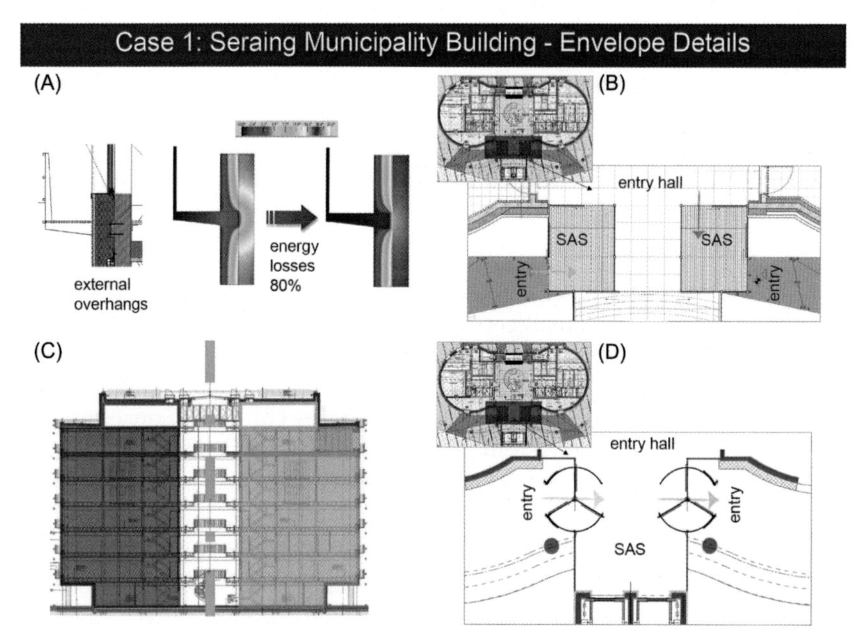

FIGURE 11.2 (A) Design detail for a thermal bridge (Neo and Ides, 2009; Douin, 2017), (B) the security airlock system with sliding door, (C) the east and west thermal zone, and (D) the suggested Security Airlock System with revolving doors (Douin, 2017).

detailed envelope design and localizing heat losses to avoid structural deterioration with the building fabric. In Belgium, the risk of mold growth is high, therefore, thermal modeling of thermal bridges is extremely important to ensure thermal comfort of end users, minimize risk, and save client costs. As part of optimizing the envelope design, a security airlock system (SAS) was placed in the building entry with two sliding doors (Fig. 11.2B).

The window frames and glazing are made of aluminum with thermal break and equipped with triple glazing units (Uw = 0.93 W/m^2 K) and solar factor (SHGC) $g = 0.59$. The frames are fitted with a medium reflective awning to limit the solar contributions and avoid glare. The automated rise and fall of the blinds are controlled according to the horizontal sunshine for which the solar brightness set point has been set at 59 W/m^2 (corresponding to 40,000 lux at 555 nm).

HVAC and RES Systems

According to Fig. 11.2C, the building is divided into two thermal zones. The set temperatures for heating and cooling are 21°C and 22°C, respectively. The Building Management System (BMS) manages the East zone and the West zone separately. There are different operating rules for the two plants. Each zone is served by a central air handing unit (AHU). Both AHUs are used to ensure thermal comfort and air ventilation. Each AHU is equipped with:

- A mechanical ventilation heat recovery (MVHR) system for exhaust air.
- An AHU with a heating coil supplied by a central heat pump.
- A free cooling system backed up with an adiabatic cooling unit.
- Mechanical cooling—a compressor integrated in each AHU in case of high heat.

With all the above systems, the building was designed to meet all of its heating needs by using only air preheating of fresh air (Neo Construct, 2009). Thus, the MVHR is coupled to a heat pump to ensure additional heating if it is too cold. During summer, the adiabatic cooling system supported by night cooling assures comfort. Consequently, the fresh air ventilation system is optimized by being able to play two roles at the same time, which is delivering air and regulating temperature.

During summer days, the internal concrete walls store heat. In the evenings, when offices are unoccupied, windows open automatically by the BMS regime to create air movements and dissipate heat based on night ventilation air. The atrium is centralized to act as a chimney profiting from the difference in pressure between the bottom and the top of the atrium. When windows are opened, the upper floors rely on cross ventilation while individual offices rely on-site ventilation. If the passive cooling is not sufficient, the adiabatic cooling system operates to meet cooling needs (Lesage, 2015). The BMS monitors the indoor environment with sensors and coordinates all systems (windows, doors, shutters, fresh air ventilation, etc.). An on-site weather station provides feedback to the BMS. RES comprises a BPIV on the elevator shaft, solar thermal panels, and PV panels arranged on the roof.

Domestic hot water (DHW) requirements (L/m^2 day) are met by a gas boiler exclusively dedicated to the production of DHW (0.75 L/m^2 daily for the restaurant and its kitchens) (Neo Construct, 2009).

Plug Loads and Lighting

The metabolic activity of occupants is defined according to the function performed by the occupants and varies between 120 and 140 W/m^2. The contribution of appliances and plug loads (computers, printers, etc.) were modeled with a radiant fraction of 0.5 and a power density of 3.5–5 W/m^2. For lighting, 500 lux are provided in all offices. The illumination is lower in circulation spaces, storage rooms, etc. The luminaires are surface mounted and the lighting power density (LPD) was imposed at 8 W/m^2 (Neo Construct, 2009; Douin, 2017).

Controls

The operation of the motorized blinds depends on external brightness. The blinds are lowered by sector (NE-E-SE-S-SO-O-NO) in the same way on each floor. They close when the outdoor brightness is greater than 40,000 lux. If the brightness falls below this threshold for more than 12 minutes, then the blinds are raised. In certain specific situations, the blinds rise to the open position during strong wind (speed above 100 km/h) and in cases of rain with a risk of frost—when the outside temperature is below 5°C.

Occupants can take control of the blinds operation. Manual operation overrides automated operation and, in this case, the awning no longer responds to the signals emitted by the automatic mode except during specific weather situations. Between the state changes (blinds deployed or folded), a 15-minute delay is programmed to avoid unwanted movements (Neo Construct, 2009; Douin, 2017).

Feedback and Occupant Satisfaction

A postoccupancy evaluation (POE) was performed following a twofold approach. The first focused on air temperature measurements over a year to calculate the percentage of time when the temperature is above 25°C as required by the PH Standard. The second was the measurement of comfort parameters during the winter season from January 2016 to April 2016 to calculate PMV/PPD and draft risk for air movements.

2.3 Lessons Learned

The design and sizing of the SAS has a significant impact on heat loss. In the case of the Seraing Municipality Building, the SAS could not operate as an effective air barrier due to the high visitor traffic. The existing sliding door and air curtain could not effectively reduce heat losses. The 220 employees and the 100 visitors that walk through the SAS daily resulted in the sliding doors almost being open during operation hours (Rebours, 2016). The POE revealed that the reception personnel placed a glass curtain in front of the reception desk and installed electric heating units. This was a reaction to the continuous cold air current that crossed the SAS. Occupants were allowed to equip their workstations with personal heaters using glass screes to block the air current (see Fig. 11.2B). Therefore, we recommend future NZEB project to include revolving doors and secondary swing doors that open only during emergency and for users with disabilities (see Fig. 11.2D). We recommend restricting occupants equipping their workstations by performing soft-landing that can give the operation team a chance to experience the real comfort condition after moving in, and adapt the building operation to meet occupant needs (Douin, 2017).

The second lesson learned is related to the operation of external blinds. In fact, the POE revealed that the blinds are lowered by sector (NE-E-SE-S-SO-O-NO) in the same way on each floor. They close when the outdoor light intensity is greater than 40,000 lux. If the light intensity falls below this threshold for more than 12 minutes, then the blinds go up automatically in relation to weather conditions. However, in reality this operation rule is not convenient for several occupants. The raised blinds allowed glare and increased the solar penetration in the working spaces which caused discomfort and overheating during summer and autumn. On the other hand, the lowered blinds decreased natural light and distribution within the working spaces making artificial lighting necessary. Due to the 12-minute delay, the exterior blinds system does not completely prevent the risk of glare, resulting in occupants using adaptive means to ensure their comfort.

In order to solve the overheating problems associated with excessive solar radiation, while limiting the risk of unwanted glare from the occupants, it may be appropriate to review the programming rules of the automated blind operation. The risk of glare can be minimized by refining a glare probability and relying on the readings of an on-site weather station. More importantly, we recommend performing soft-landing and customizing blinds operation to occupants needs per floor and per orientation, and not for the whole façade

The third lesson learned from this project is related to the effectiveness of night cooling. As planned, at 10 p.m., the windows should open to allow fresh outside air to enter. However, in reality the opening time varied between 2 and 4 hours in general, and the air temperature did not decrease below 20°C. The periods of night cooling did not decrease the indoor temperature as expected to cool the thermal mass of the internal concrete walls (Douin, 2017). Moreover, once the windows were closed, the indoor temperature profile rises quickly, sometimes above 25°C. In conclusion, the night cooling did not achieve the expected effect. We recommend verifying and validating more closely the effect of the night cooling potential of the site as well as the microclimate. Also, the programming of window openings and command rules should be more flexible allowing longer periods of ventilation. This would reduce the operating temperature at the beginning of the day and would make it possible to take more advantage of the thermal storage capacity of the building.

The fourth lesson learned is mainly related to the HVAC system. The complexity of NZEB requires delivering fresh air while at the same achieving comfort. In the case of the Seraing Municipality Building, only two AHUs were installed to deliver air to the right and left building wings. However, the POE results showed that comfort was not attained according to the PH requirements (Rebours, 2016) due to the disparity of temperature between the different building's thermal zones. The BMS received varying heating and ventilation signals from the local sensor and could only respond to the grouped zones signals. This means that, at least 25% of the occupants will be always be dissatisfied, as results of a situation of discomfort which can be amplified since the AHUs always impose a maximum air flow for all floors (Douin, 2017). The building thermal zones are heterogeneous (solar contributions, internal contributions, thermal stratification, etc.) while the HVAC system used a centralized VAV. We recommend, therefore, simplifying the HVAC by separating space heating and cooling from ventilation and providing at least two AHUs per floor.

3 CASE 2: RSF, COLORADO

3.1 Project Description

The Research Support Facility (RSF) is a state-of-the art office building to host researchers of the National Renewable Energy (NREL) Lab.

The RSF in Golden, Colorado was designed and constructed between 2006 and 2010 after calls for proposals and a process of selection. The vision of the selected project operates within the energy efficiency paradigm aiming to build an energy neutral office building, or a NZEB. The design brief emphasized an integrative design approach to design, build, and operate the most energy-efficient building in the world. The proposal had a design-build acquisition strategy that connects the building to the electricity grid for energy balance through a power purchase agreement (PPA). The Design & Build Team was comprised of Haselden Construction, RNL Architects, and Stantec as Sustainability Consultants and MEP engineering. The design process involved an integrative approach looking to:

- Avoid needs for energy by integrating passive heating, cooling and ventilation.
- Improve energy efficiency.
- Incorporate renewable energy and green power.

The project is a NZEB and obtained the LEED Platinum Certificate (V.2) and Energy Star Plus certification. The design brief also required maximum use of natural ventilation and 90% of floor space fully daylit.

3.2 Performance Characteristics

The building is 20,400 m^2 hosting 800 persons. The building energy use intensity had to perform less than 80 kWh/m^2 per year and additional 20 kWh/m^2 per year was allowed for a large data center that serves the entire NREL Campus. The RSF facility had to perform 50% better that ASHRAE 90.1-2007 energy performance requirements. The building monitored energy consumption and production has been monitored since its construction. The average annual consumption is 109 kWh/m^2 per year including the data center, serving 1325 occupants.

Climate

The building is located in latitude 39.74 and longitude −105.17 and is 151 m above sea level (see Fig. 11.3A). The site receives 660 mm of rain per year with an average snowfall of 1371 mm. The number of days with any measurable precipitation is 73. On average, there are 242 sunny days per year in Golden, Colorado. The July high is around 30°C and January low is −8°C, while humidity during the hot months, is a 58 out of 100.

FIGURE 11.3 (A) the South Wing is optimized for a window-to-wall ratio of 25% and the east glazing surface is minimized, (B) two wings optimized to allow natural lighting and ventilation, RSF, NREL, Golden, Colorado, United States, (C) the open space office was optimized for maximum natural lighting and natural cross ventilation, and (D) PV array of mono-crystalline panels of 17% efficiency.

Envelope

The building has an H-form with long thin wing to optimize daylight penetration and natural ventilation in the workspaces (see Fig. 11.3B). The main wings are 18 m wide so that occupants are always closer than 9 m to a window and dramatically reduces the energy use of the building (see Fig. 11.3C). The roof shape is slanted at 10 degrees to the south to increase the solar electricity production of the photovoltaic system (see Fig. 11.3D).

The concrete panels on the exterior of the building are a sandwich panel that includes a layer of rigid insulation between two layers of precast concrete. The sandwich panel is held together by hard plastic pins that minimize thermal transfers through the insulation layer. The 15cm of concrete on the interior side of the panels serves as a thermal mass.

The south facing windows of the RSF are divided into a daylight harvesting upper panel and a traditional lower panel. Windows are triple-glazed with individual overhangs. A reflective louver behind the upper

Case 2: RSF learned lessons

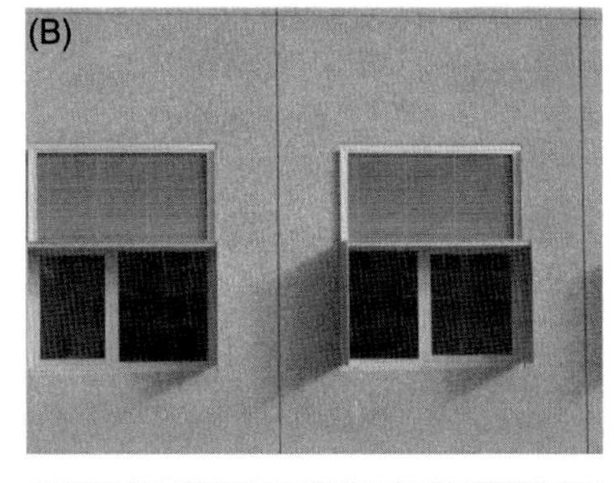

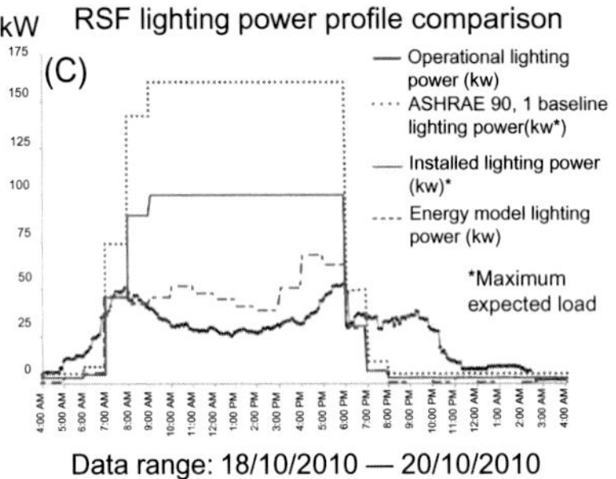

FIGURE 11.4 (A) the South Wing is optimized for a window-to-wall ratio of 25% and the east glazing surface is minimized, (B) optimized window design to distinguish view window from clerestory window and include exterior light shelves that provide more shade and less glare than interior light shelves for year-round light distribution, (C) monitoring and soft-landing results identified the jump in lighting loads at night (Torcellini, 2010), and (D) additional parking space was created and covered with PV panels to compensate for the discrepancy in the annual energy balance.

pane bounces light deep into the interior. Sunshades on the exterior are designed to limit glare. The north facing windows act like a clerestory to take advantage of the naturally diffused north light (see Fig. 11.4A and B).

HVAC and RES Systems

Thermal comfort is addressed using an integrated system of thermal mass, radiant slabs, night purging, and natural ventilation. Heating has a whole systems approach to energy conservation. A radiant heating and cooling system is installed in the roof slab. Natural ventilation is achieved during the day through manual window control and at night through automated control for night cooling and thermal mass activation. Mechanical ventilation is demand based and air is displaced through an underfloor air distribution (UFAD) system. A heat recovery system is installed on the outside air intake and exhaust from restrooms and electrical rooms.

Both office wings are heated and cooled by a radiant ceiling slab, using water heated by NREL's Renewable Fuels Heating Plant (central woodchip boiler) and cooled by high-efficiency chillers (DOE, 2012). A dedicated outdoor air system (DOAS) and natural ventilation provide ventilation for the RSF. Ventilation air is distributed by an UFAD system with swirl diffusers. During mild weather, operable windows provide natural ventilation.

A third party owned PPA provided a full rooftop array of 1.7 MW of mono-crystalline panels of 17% efficiency (see Fig. 11.3D). The current power is purchased from a fossil mix—60% coal, 22% from natural gas, and 18% from renewable energy resources (EIA, 2014).

Plug Loads and Lighting

The whole building energy use is 283 continuous watts per occupant. Laptops of 60 W with 35-W thin screens are used in workspaces. The artificial lighting system is based on motion and daylight intensity sensors. Sensor controlled LED task lights of 15 W are used for workstations lighting. Voice-over-internet protocol phones that consume 2 W each were used instead of standard phones and standard 300-W computers were replaced by 30-W laptops.

The RSF features low-energy elevators rather than the standard hydraulic elevators commonly seen in low-rise office buildings. The elevators have energy-efficient fluorescent lighting, and both the light and the fan turn off when the elevator car is unoccupied. Also, the building has a new highly energy efficient data center designed to minimize its energy footprint without compromising needed service quality

Controls

Lighting is an integrated system of daylighting, daylight control systems, occupancy controls, and high-efficiency lighting. The vision glass is manually operable and gets automatically controlled depending on indoor and outdoor environment. Vacancy sensors were installed in private offices (daylit), and night sweeps turn off almost all lights. The building "sleeps" at night with nearly all equipment and lights off thanks to effective lighting controls, programmable power strips, and other building controls (DOE, 2012).

Feedback and Occupant Satisfaction

One of the requirements in the original RFP was to provide an energy use display. These displays clearly show the electrical distribution within the building and are useful for the building's operations staff and the building's occupants to see how the building is performing. After the building was complete, a separate meter was added to measure the data center's chilled water (DOE, 2012).

3.3 Lessons Learned

The Research Support Facilities Building (RSF) at the National Renewable Energy Laboratory (NREL) in Golden, Colorado achieved a 67% reduction in energy use (excluding the solar PV offset) at zero extra cost for the efficiency measures, as the design team was contractually obliged to deliver a low-energy building at no extra cost (Torcellini et al., 2010). Torcellini and Pless (Pless and Torcellini, 2012) present many opportunities for cost savings such that low-energy buildings can often be delivered at no extra cost. Other examples of low-energy buildings (50%–60% savings relative to standards at the time) that cost less than conventional buildings are given in McDonell (2003) and IFE (2005). There have been several lessons learned in the case study of the RSF.

The first lesson is the extreme importance of the energy performance-based procurement of NZEB projects. In the case of the RSF, energy performance requirements were prioritized in the preplanning phase with specific Energy Use Intensity targets (EUI). The energy target was included in the contract and the design and contractor teams were chosen based on their ability to meet the target. After the project was complete, the project teams had to measure the energy performance to verify that targets identified in the contract were successfully met. The performance-based contract and the Design & Build project delivery forced the team to use energy modeling extensively. Energy modeling required to substantiate goals that were coupled to energy end-use metering requirements. Understanding the importance of controlling unregulated plug and process loads resulted in successful soft-landing that was partially based on educating occupants and coaching them to change their behavior and adapt to the net zero energy performance target.

The second lesson learned is that ongoing measurement and verification are essential in realizing the full benefits of a net zero energy design. Planning for retro-commissioning is beneficial to keep the performance up to net zero energy. For example, monitoring revealed that vacancy sensors that were installed in private offices (daylit) were successful, and night sweeps were turning off almost all lights. Only 2–3 kW of lights remain on night, with about 700 W used for controls. Some employees have requested two task lights to help light work surfaces when needed. Engineers have determined that all lights should be controlled based on daylight, including egress lights and stairwells. Another example is the jump in lighting loads at night, as shown in Fig. 11.4C, which occurred during the first year of monitoring. By calibrating the real performance and energy consumption profiles versus the modeled energy performance profiles, a mismatch was detected and the building operator could identify the janitorial team activities that occur between 18:00 and 23:00. Ongoing measurement and energy metering directed the project team to focus on the largest energy consumers.

The third lesson learned is that the sizing of the renewable energy production is always challenging. The RSF design team was forced to add additional parking space, shown in Fig. 11.4D, and cover it with extra PV panels to reach the energy neutrality. We recommend adding 20% extra PV panels in NZEB projects as a contingency plan to ensure meeting the energy neutrality target (Reeder, 2016; Eley et al., 2017). PV systems never put out quite as much power as expected. The uncertainty of weather, equipment performance, or service interruption is common in NZEB and, therefore, should be taken into account early on during the design phase.

4 CASE 3: ZERO ENERGY FACTORY, FLEURUS, BELGIUM

4.1 Project Description

Headquarters for the Kumpen company were required to host office spaces and an industrial storage hall in Fleurus, in Belgium. The project net surface area is 529 m^2 hosting 43 persons. As a construction company, the owner's approach was to build a regional operating center that is energy efficient and highlights the advantages of high-performance buildings. The building complies with the PH Standard and is an energy neutral building. The project includes a storage hall for goods, a small buffer zone for goods processing including small equipment and tools for the workers. Also, the building hosts offices for staff responsible for the management of construction sites and for the study of new projects to be developed in Wallonia (see Fig. 11.5).

The project team grouped three key players during the design stage. The project architect was Marcel Barattucci from Startech Management Group, the MEP engineering consultant was Boydens and an independent certification expert was recruited for the EPBD and PH Standard certification. The three design team members collaborated following an integrative design process while using three building performance simulation programs. The certification expert used Therm software for thermal modeling to revise the thermal performance and cost of all construction details. A PHPP excel sheet was used to verify compliance with the PH standard. For dynamic simulation, TRNSYS was used to verify comfort conditions and avoid potential overheating risks. The

Case 3: Zero energy factory, Fleurus, Belgium

FIGURE 11.5 (A) The office building in the front and factory in the back, (B) the main building entry and the security airlock system, (C) the ultra-efficient LED tubes and the linear ventilation supply in the center and exhaust on the side of the office space, and (D) the external automated shading blinds for solar protection.

certification responsible had to master all three simulation programs, prepare the certification documents, and submit the project for EPBD and PH standard certification.

The design of the project relied on two main principles. The first principle was to reduce the heat losses. This principle was realized through high insulation, optimization of construction details, avoiding thermal bridges, and maintaining the envelope airtightness. The second principle was to avoid overheating. This was achieved by limiting internal heat gains, installing external solar shades, and activating thermal mass.

4.2 Performance Characteristics

The total consumption for the Kumpen Headquarters is equal to 38 kWh/m^2 per year serving 34 occupants. The project attained the target values according to the PH standard. The energy consumption breakdown is 19% for heating energy needs, 6% for cooling energy needs, 60% for appliances, mechanical ventilation and plug load energy needs, and 15% artificial lighting energy needs.

Climate

The project location (Lat: 50.46, Long: 4.53, Alt: 40 m) is classified as temperate climate. The Köppen-Geiger climate classification is Cfb. The average annual temperature is 9.6°C in Seraing. In a year, the average rainfall is 856 mm. According to ASHRAE Classification the site location is cold humid (Climatic Zone 5A) with $3000 <$ Heating Degree Days $18°C \leq 4000$. Temperatures are relatively mild during the whole year with the average low at 1°C in winter and only 23°C in summer and winds tend to be slightly stronger in winter.

Envelope

The building geometry is relatively compact (1.83) and is raised on two floor levels. As shown in Fig. 11.5A, a separation wall between the "Storage Hall" and "Offices" was extended outside the building for architectural façade articulation. The offices had to be placed on the West side and perpendicularly to the access road. The choice of facade cladding materials also makes it possible to distinguish these two functions. The facades of the storage hall are made of solid wood panels. The West orientation is protected by external movable shades to limit the summer solar access during occupation period. An insulated brick wall footing is on the ground floor, while ETICS has been placed on the facades of the upper floor. The airtightness is 0.48 h^{-1} at 50 Pa. A slight overhang reinforces the distinction between the two floors. The entrance of the building is marked by a complementary SAS as an entrance, which makes it possible to limit the entry of cold air in the protected volume.

The interior layout of the offices was designed as follows: All the service rooms and the stairwell are grouped together and located along the separation wall with the hall. This interior design makes it easier to adapt office and meeting space to meet changing business needs (see Fig. 11.6). The technical installations of heating, ventilation, cooling, etc., are placed in the storage hall.

HVAC and RES Systems

Careful thermal zoning took place to reduce the heated and cooled building volume by eliminating a large part the floor plan from HVAC systems. As shown in Fig. 11.6, four thermal zones are space conditioned on the ground floor and three thermal zones are space conditioned on the first floor. MVHR supplies air to each thermal zone with 50 m^3/person. The total supplied air volume is 2500 m^3/h for the whole building. Mechanical night cooling operates between 18 and 8 hours with air flow of 5000 m^3/h. During working hours, active cooling is achieved through TABS as shown in Fig. 11.7A. The supplied air

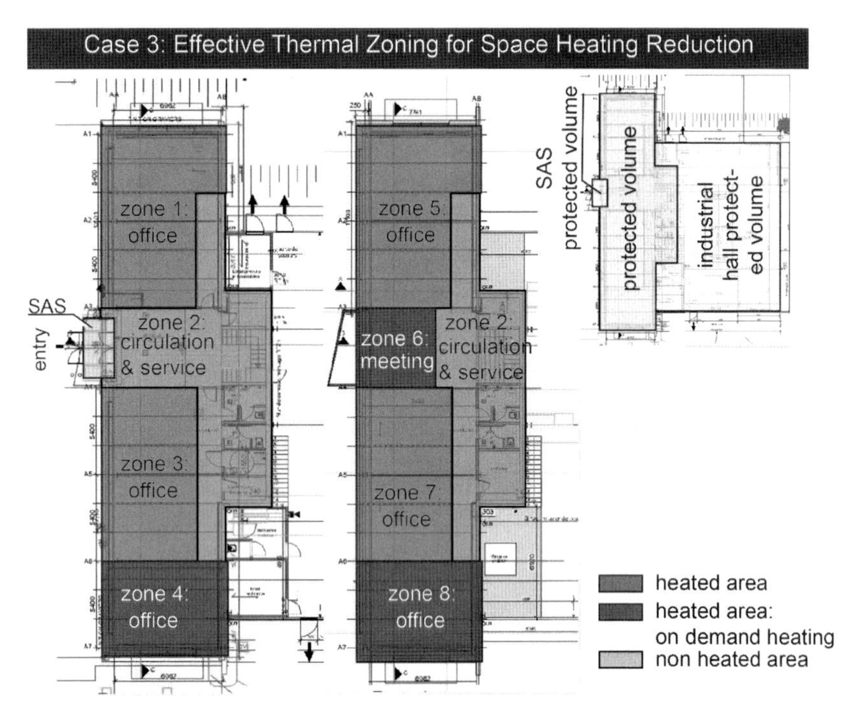

FIGURE 11.6 Thermal zoning or building compartmentalization for reducing and optimizing the volume of space heating (Marcel Barattucci).

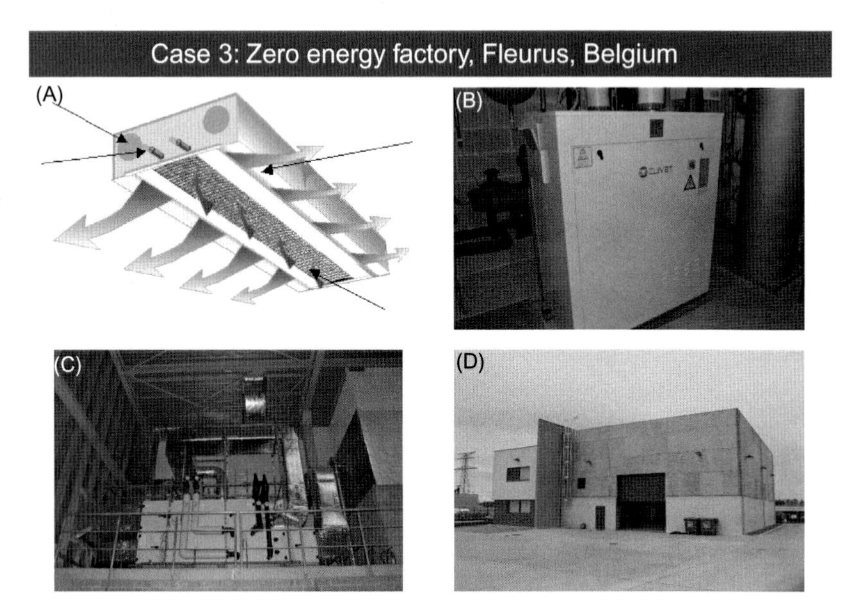

FIGURE 11.7 (A) The TABS is used as a cooling terminal for air conditioning (XPair, 2017), (B) internal unit of the water/water heat pump, (c) the air handling unit and heat exchanger unit, and (D) the storage hall on the right side is not part of the thermal heated zone.

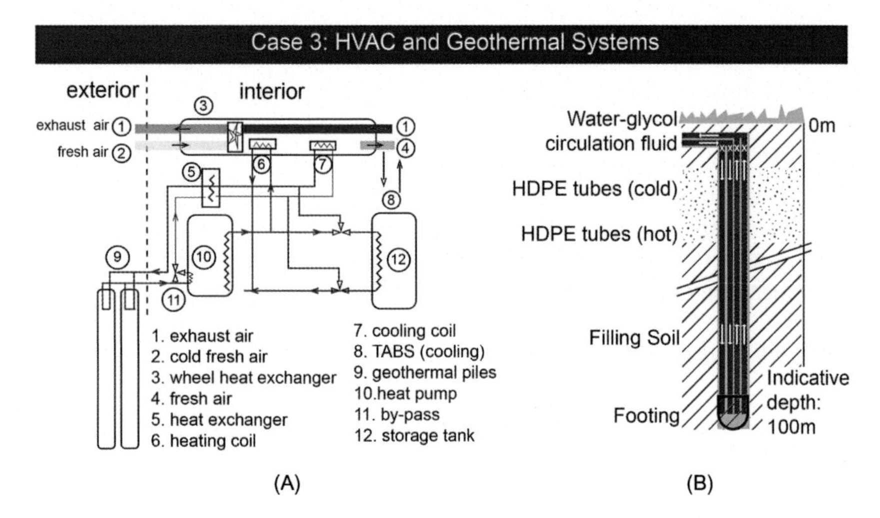

FIGURE 11.8 (A) HVAC scheme including the heat pump and water loops (Boydens), and (B) geothermal earth tubes (Boydens).

temperature in the TABS is 18°C and is achieved during the night by night cooling and during the day by a heat pump through chilled water loop of 16°C (sees Fig. 11.7A). A storage tank of 1000 L is coupled to the water loop for hot and cold water storage (Fig. 11.8A).

A heat pump provides the building heating and cooling needs. A water to water heat pump provides hot water during winter and cold water during summer (see Fig. 11.7B). As shown in Fig. 11.7C, an air handling and heat recovery unit provides fresh air to the office buildings. Ten geothermal probes are drilled in the ground at a depth of 100 m (Fig. 11.8B). During winter, the input water temperature is 10°C while the output temperature is 35°C. During summer, the input water temperature is 24°C and the output temperature is 16°C. The cost of the water/water heat pump system was higher than the air/water heat pump system (± 30.000€ against 8.000€) (Baratticci, 2016). A photovoltaic roof installation of 80 m^2 (40 panels) covers the annual energy needs. The peak power of photovoltaic is 10 kW peak DC with an expected annual energy yield of 8466 kWh/year mounted on top of the Storage Hall (Fig. 11.7D).

Plug Loads and Lighting

The design team reduced the internal heat gains through the use of high-performance lighting. To regulate the temperatures and reduce the operation hours of heating and cooling systems, a thermal mass was activated by the use of a concrete structure. The occupancy profile in the building allowed the use of thermal mass and couples it to night cooling (8:00–18:00). The lighting power distribution was 10 W/m^2 and the estimated occupant power density is 120 W/person for the 43 persons occupying the building. Personal computers had a power density of 80 W/workplace, a server of 230 W, and one printer of 2250 W.

Controls

The building depends on high-performance LED lighting system with automated controls based on presence sensors. Also, the exterior blinds are automated. Particular attention has been paid to limit summer overheating by using automated external blinds.

Feedback and Occupant Satisfaction

Less than 5% of overheating hours exceeded 25°C, based on data monitoring. The occupants moved into the building at the beginning of January 2014, a POE took place, and occupants were dissatisfied due to the low temperature. This was due to the calibration time required to activate the geothermal heat pump. By March 2014, the comfort conditions were improved and the level of satisfaction of the occupants reached 95%.

4.3 Lessons Learned

The first lesson learned is the importance of assigning a certification and modeling expert for NZEB projects. For a design team without previous experience of NZEB, the presence of an energy expert was the key to a successful PH certification and NZEB performance. The energy expert was responsible for performance quality control and to assemble all the documentation from the design and construction team. In the case of the Kumpen NZEB project, the energy expert used building energy modeling (PHPP calculation, Therm, EBPD tool, and TRNSYS) to assure the performance during design. More importantly, the energy expert in the case of Kumpen was part of the commissioning team, which resulted in ensuring the quality of implementation and communication with all the project stakeholders to achieve the energy targets.

The second lesson learned is the influence of the structural system choice on NZEB cost. NZEB involve assembling a wide range of materials from different sources to provide the required structural and thermal design. The selection of prefabricated construction modules compared to traditional construction reduced the additional investment cost to reach the PH requirements. Savings in initial cost are significant and the Kumpen NZEB cost were 2495€/m^2. The importance of modular and prefabricated construction in NZEB is high as the total investment cost is relatively low because the share of labor cost is relatively small.

The third lesson learned is reducing the space heating volume. A thermal zone is a space condition block within a building having sufficiently similar space conditioning requirements. The conditions of the thermal zone could be maintained with a single thermal control device. Thermal zones need not be contiguous to be combined into a thermal block. In the case of the Kumpen NZEB, the architect was aware of the challenges in achieving an energy neutral building. Therefore, he focused with the PH certifier to include only essential spaces and functions that need to be part of the HVAC conditioned thermal zone. As shown in Fig. 11.6, the space heated volume was reduced to the minimum excluding the circulation and building access. Those spaces were operating in a free running mode to reduce the overall conditions space. As a consequence, the building could achieve its performance target much easier and maximize IEQ in the space conditioned volumes or thermal zones.

5 CASE 4: ZERO ENERGY SCHOOL, BILZEN, BELGIUM

5.1 Project Description

The regional Flemish Government decided to realize a number of pilot projects as part of the EU 2020 targets to reach nearly zero energy buildings. The aim was to build high-performance school buildings that were comfortable with extremely low heating cooling need and could be energy neutral. The pilot school passive schools were launched in 2007 by the Flemish government. The Flemish government wanted to raise awareness about healthy and energy-efficient schools and provide a new educational infrastructure to achieve its learning goals. Twenty-one schools were selected—the first eight pilot schools have now been completed. In 2013, the regional government decided to build 65,000 m^2 of passive schools and nearly zero energy schools spread across all provinces in Flanders.

In this context, the construction of the *T'Piepelke* school in Bilzen (Belgium) comprises three classrooms, six classrooms for primary school, a teacher's room, a conference room, a secretarial room, a large gymnasium space, and a small multipurpose room with two changing rooms, and two rooms for the neighborhood operation (Fig. 11.9A). The passive school consists of two parts: The elementary school and the community center with community functions. A bridge, which simultaneously creates the indoor play space, connects the two parts (Figure 11.9B). For the school, the innovative activity in grade classes is central: Classes are clustered with a mobile wall. The neighborhood houses a multifamily room which can be converted into a much-needed dance hall for the city's neighborhood. The project net surface area is 1410 m^2 hosting 240 students (see Fig. 11.9C).

Case 4: Zero Energy School, Bilzen, Belgium

FIGURE 11.9 (A) The North Façade of Bilzen School, (B) the entrance and the bridge, (C) an example of the classroom with mechanical ventilation ducts, and (D) the South façade with automated external blinds.

The project team grouped three key players during the design stage. The project architect was LAVA architecten, the MEP engineering consultant was IVV technieken, and JC engineering was recruited for the EPBD and PH Standard certification. The project was commissioned in 2014 with a total occupied surface area of 964 + 447 m^2. It also received PH certification.

5.2 Performance Characteristics

The total energy consumption of the school is equal to 40 kWh/m^2 per year serving 250 occupants (heating/cooling—11.6/9.6 kWh/m^2 per year). The project attained the target values according to the PH standard. The electricity consumption breakdown is 48% for lighting, 35% for mechanical ventilation, and 17% for appliances, plug loads, and pumps.

Climate

The project location (Lat: 50.87, Long: 5.51, Alt: 58 m) is classified as temperate climate. The Köppen-Geiger climate classification is Cfb. The average annual temperature is 9.7°C in Bilzen. In a year, the average rainfall is 806 mm. According to ASHRAE Classification the site location is cold humid (Climatic Zone 5A) with $3000 <$ Heating Degree Days

18°C ≤ 4000. Temperatures are relatively mild during the whole year with the average low of 1°C in winter and 23°C in summer and winds tend to be slightly stronger in winter.

Envelope

The two connected buildings are very compact volumes with a north–south orientation (see Fig. 11.9D). In winter, southern-oriented windows provide (partial) heating through passive solar gains. The south-east to south-west oriented windows avoid overheating in the summer. The openings are equipped with triple glazing and high-performance thermal interrupted wood-aluminum profiles (Fig. 11.10A).

HVAC and RES Systems

The new building has an energy efficient technical installation including, among other things, a high-performance CO_2-controlled VAV system. For ventilation, the school has been divided into two zones, each with its own ventilation system. The first area is that of the gymnasium, the second area is the one that includes the rest of the school. The ventilation of the gym is made to ensure an air exchange of 2000 m^3/h (see Fig. 11.11).

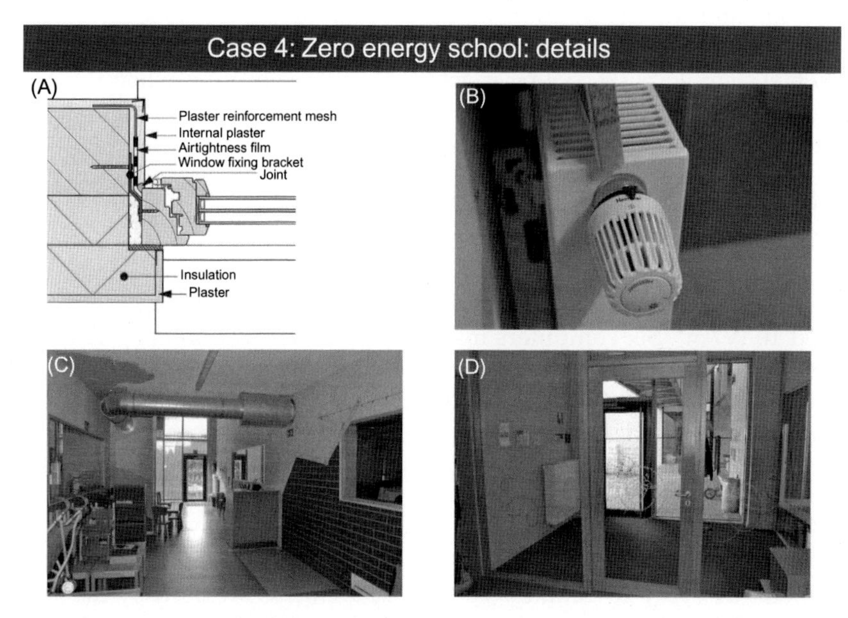

FIGURE 11.10 (A) The architect selected a hydronic heating system and chose radiator valves to allow the simple management of the heating system, (B) the ETICS is used to guarantee airtightness and low conductivity, (C) the main circulation corridor where acoustic insulation took place as a result of the PEO, and (D) the SAS is an important element to keep the building air-tight.

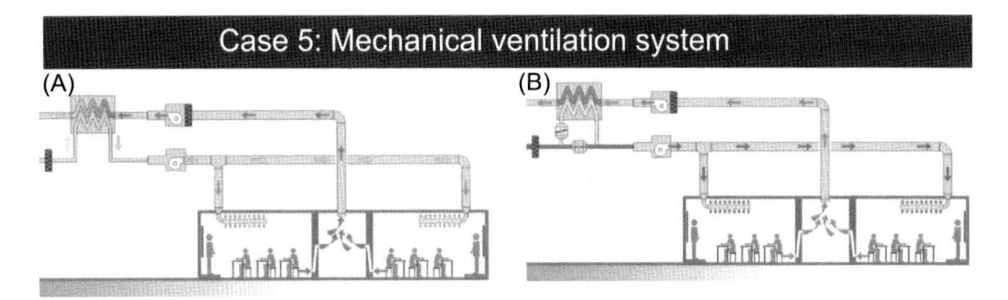

FIGURE 11.11 (A) The variable air volume delivery depends on presence sensors and thermostats, (B) the night cooling system's operation is coupled to the internal walls temperature and operated during the night. Source: *Illustration adapted from Abbeloos, G., 2011. Equipements Techniques d'une Ecole Passive (Master thesis). UCLouvain, Louvain-La-Neuve, Belgium (in French) (Abbeloos, 2011).*

Heat production is provided by a gas-fired condensing boiler with a power of 30 kW. The distribution system is based on a hydronic system. Heating takes place on low-temperature regime and gets distributed through an underground floor heating system for the ground floor classrooms. Radiators were placed in the upper classrooms and office spaces. As shown in Fig. 11.11A, mechanical ventilation is separated from heating. Mechanical ventilation is mainly responsible to deliver fresh air and air distribution in classrooms and office spaces. During extremely cold days, the school spaces can be heated through the ventilation system. Passive cooling is achieved via free cooling when the outside air is colder than inside (see Fig. 11.11B). The thermal mass of the building stores the coolness and emits it throughout the day.

Plug Loads and Lighting

Most classes have a large share of daylight due to the large window areas, which limit dependence on artificial lighting. The classes are equipped with energy-efficient LED luminaires. The reduced LPD targeted 10 W/m^2. Six computers installed in the school's offices and classrooms, as well as printers and photocopiers. All of these devices are equipped with on–off switches, to avoid using energy when the device is not in use. The aim was to reduce plug loads and improve controls for shedding loads during unoccupied periods.

Controls

Measurements are taken using temperature sensors that record the temperature every 20 minutes. These measurements are used by the BMS to adapt the heating and cooling demand of the HVAC system over time. Lights are turned on automatically in classrooms based on motion detection. Daylight-dependent control of lighting reduced energy consumption. In order to prevent glare, internal blinds were installed. However, after occupation automated outdoor protection screens were installed. On extremely windy days external screens are automatically set up, based on a weather station installed on the South façade, to prevent potential damage. Blinds can be manually lowered by occupants.

Feedback and Occupant Satisfaction

After 1 year of occupation, continuous overheating problems were experienced. Therefore, in 2014 the school administration decided to invest 13,000 euros and install external solar shading screens (see Fig. 11.9D) (VTM, 2016). However, summer comfort remained a serious challenge for this school. In 2016, a long heat wave hit Belgium resulting in dismissing pupils. For example, pupils had problems with concentration and had to put their lunch boxes in cooler boxes. Conventional schools were also closed during the heat wave.

An investigation by the Flemish Agency for Infrastructure in Education (AGION) provided evidence that summer remained a challenge in many passive schools in Flanders. The report indicated the overheating problem of the high-performance school in Flanders due to:

1. Lack of solar protection.
2. Undersized ventilation systems.
3. Missing or undersized air conditioning systems.

5.3 Lessons Learned

The case study of Bilzen School is extremely important because it represents 22 other schools that were built to attain the PH standard requirements and achieve energy neutrality in Flanders, Belgium. Several lessons learned can be shared from this case study. The first lesson learned is the underestimation of overheating risk in NZEB school buildings. The architects designed the school without any solar protection and relied mainly on internal shading. The second crucial mistake was done by the energy consultant who did not properly size the ventilation and cooling system. Even after installing automated external blinds, the displaced air and cooling load did not meet the high-intensity occupancy needs. Failing to underestimate the uncertainties associated with weather during the design phase resulted in a serious overheating problem that was related to the poor design of passive and active strategies. We recommend future design team to not undermine or overestimate the effect of passive design and prepare NZEB schools for worst case scenarios to face extended and extreme heat waves.

The second lesson learned is that the use of HVAC technologies must be simple in high-performance schools. In NZEB schools, many HVAC technologies are used per building resulting in a complex operation. The complexity of HVAC systems can lead to losing control of the different HVAC components and their management. As a consequence, the risk of attaining comfort can be easily increased. Lessons learned from practice indicate the importance of separating heat distribution from mechanical ventilation. In schools, the use radiators are effective because users are familiar with their use and heat can be provided fast. School teachers need to react to student comfort needs. The presence of a thermostat valve of a hydroid radiator in a classroom assures teachers and makes them feel they can control the indoor environment. The temperature increase that can be directly felt in association with radiator use is much more effective than waiting for the AHU to response to deliver conditioned air. In parallel, mechanical ventilation should be sized to deliver the minimum required fresh air individually from heating or cooling system. Therefore, the best way to achieve the targeted thermal comfort and air quality requirements is to separate heating and cooling from ventilation.

The third lesson learned is related to the school operation and facility management (FM). For small and middle sized NZEB schools without full time operation managers, we advise using manual controls of HVAC systems allowing users to engage and regulate their environment. We advise the use of manual controls in NZEB for buildings without FM staff or skills. In this case, the use of familiar comfort controls that can be easily operated following a shallow learning curve is recommended. Controls need to be clear to the occupant and have defaults to run the school.

Finally, the fourth lesson learned from the Bilzen school shows the importance of taking into account acoustic design solutions. The soft-landing and POE revealed a remarkable problem of noise in the school. Therefore, future school designs should significantly challenge acoustic conditions, requiring innovative acoustic design solutions. Good acoustic conditions in schools benefit the staff and pupils by improving learning attainment, behavior, concentration, and reducing teacher absenteeism due to loss of voice.

6 CASE 5: MULTIFAMILY DWELLING, BRÜTTEN, SWITZERLAND

6.1 Project Description

In the municipality of Brütten in Switzerland, a 1010 m^2 multifamily house dwelling capable of operating year-round without any external sources of energy has been built in 2016. The multifamily dwelling is a pilot project by Umwelt Arena Spreitenbach, which is also the developer, who has specialized in informing as broad a swath of the population as possible about sustainable construction and renewable energy. The municipality granted the permit for a nine-flat building between 80 and 145 m^2 in size (Hempel, 2017). The building block is not connected to the grid. All the electrical power comes from the sun and the heating comes from a geothermal system. The architecture is a symbiosis of design and technology with dark colored photovoltaic and rooms that let in the sunlight (see Fig. 11.12A and B). Architect René Schmid based his design concept on the three principles:

1. Generate maximum thermal energy with minimum electrical energy.
2. Efficient storage of electrical and thermal energy with minimal losses and maximum additional consumption.
3. Consume energy according to a limited budget.

The house has no eaves and the terraces are inserted into the volume to avoid thermal bridges. The monolithic character of the façade was meant to host a maximum of custom-designed thin-film solar modules (see Fig. 11.12A and B). High-powered PV panels are installed on the roof. The enclosures for the terraces and windows are clad in wood.

Case 5: Multifamily Dwelling, Brütten, Switzerland

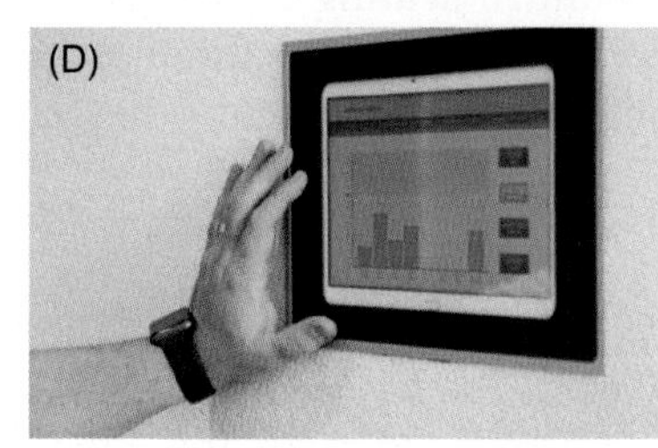

FIGURE 11.12 (A) 3D view for the autonomous multifamily dwelling in Brütten (photo credit: Umwelt Arena Schweiz 2016), (B) the main façade from the Unterdorfstrasse in Brütten (photo credit: Umwelt Arena Schweiz 2016), (C) the hydrogen tank ensures the long-term storage of energy (photo credit Hoval, Available from, http://www.hoval.com/press/com-9/the-worlds-first-energy-self-sufficient-multi-family-house (accessed March 2018), and (D) the control flush mounted display has an integrated room temperature controller and provide instantaneous feedback on the status of energy consumption and the daily energy budget (photo credit: Swisscom).

6.2 Performance Characteristics

The total consumption of Brütten Multifamily Dwelling is equal to 35 kWh/m^2 per year serving 25 occupants. The energy consumption breakdown is 14% for heating energy needs, 57% for appliance and plug load energy needs, 14% for mechanical ventilation, and 15% for artificial lighting energy needs.

Climate

Brütten is an ideal location as it lies 640 m above sea level and is above the fog line, as well as being on the route between Zürich and Winterthur. The project location (Lat: 47.47, Long: 8.67, Alt: 610 m) is classified as a temperate climate. The Köppen-Geiger climate classification is Cfb. The average annual temperature is 8.4°C in Brütten. In a year, the average rainfall is 1120 mm. According to ASHRAE Classification the site location is cold humid (Climatic Zone 5A) with 3000 $<$ Heating Degree Days 18°C $\leq$4000. Temperatures are relatively mild during the whole year with the average low of 0°C in winter and 17°C in summer.

Envelope

The building is highly insulated as part of ultra-energy efficiency concept. A cost-optimization took place to determine the quantity of thermal insulation while meeting the Minergie Standard requirements—similar to the PH requirements (Umwelt Arena, 2016). The insulated and airtight envelope is responsible for reducing the energy loads by 50%.

HVAC and RES Systems

Ensuring the comfort of the home was achieved through low-temperature heating and MVHR. Heat generation and distribution was optimized to meet the autonomous nature of the project. A water heat pump with a geothermal probe is used for heating. A heat exchanger is used for heat generation from outside air and warming up the controlled residential ventilation. This decreases the pressure lost within the system and electrical energy can be saved in the operation of the heat pumps (Umwelt Arena, 2016).

The short- to medium-term (1–3 days) storage of the electrical energy produced with the PV system is ensured with corresponding short-term storage (battery technology). For long-term storage, a gassy storage medium in the form of hydrogen is used (power to gas) (see Fig. 11.12C). Conventional water storage is available for the storage of short-term surpluses (heating water storage). Innovative storage solutions are used for long-term storage. In the loaded state, the heaters are used for heating. After discharge, the storage tanks can be used as a heat source for heat generation by the heat pump.

Nonreflective photovoltaic modules are used as facade elements. The roof is covered with photovoltaic modules. The solar energy is converted into an electric current by the solar cells and temporarily stored, 2–3 days, in daily and medium-term storage batteries. For long-term storage, there is a conversion process from electric current to hydrogen. The hydrogen is temporarily stored and can be converted back into electric and thermal energy when required. The photovoltaic energy can be converted into hydrogen for long-term storage of up to 25 days (Umwelt Arena, 2016). This is often necessary in December and January to bridge the low sunlight gap. The heat, which is generated during the hydrogen production, is lifted on a high temperature level by the heat pump in the summer and is stored in thermal long-term storages or used to produce domestic hot water (DHW). Solar energy is used to run the heat pump. The heat pump is responsible for producing DHW and heat to operate the underfloor heating system, as well as to charge the short- and long-term hot water storages. A heat exchanger is coupled to the shower to gain heat from the drained shower water. The elevator works with an energy recovery system (Hempel, 2017).

Two large water tanks store the heat for use during winter. In summer, 250,000 L of water are heated to 65°C, and in autumn and winter the heat is slowly circulated in the building hot water loop by the heat pump (Umwelt Arena, 2016). The storage and BMS includes a power-to-hydrogen gas plant that transforms surplus solar-generated electricity in summer into hydrogen which is likewise then stored for winter. If needed, a fuel cell then converts the hydrogen back into electricity. Solar power is conserved using a large lithium-iron-phosphate battery which can provide power for 3–4 days at a time (see Fig. 11.13).

Plug Loads and Lighting

The lighting in the building is made exclusively by the latest LED technology. Energy consumption is kept to the minimum with the use of LED lights and A + + + household appliances.

Controls

The building operates smart appliances easily accessed using a PC, smartphone, or tablet. Artificial lighting operation is based on preset schedules that are also used for operating heating systems and blinds. Control systems can be modified according to occupant needs.

Demand-dependent control regulates the energy flow in the building. The BMS detects the energy demand, evaluates it, and guarantees optimum control or regulation (Umwelt Arena, 2016). This allows energy consumption and energy supply to be coordinated optimally at any time, and users are informed about their energy consumption. If electric generation is low, the control system selects the warmest energy source

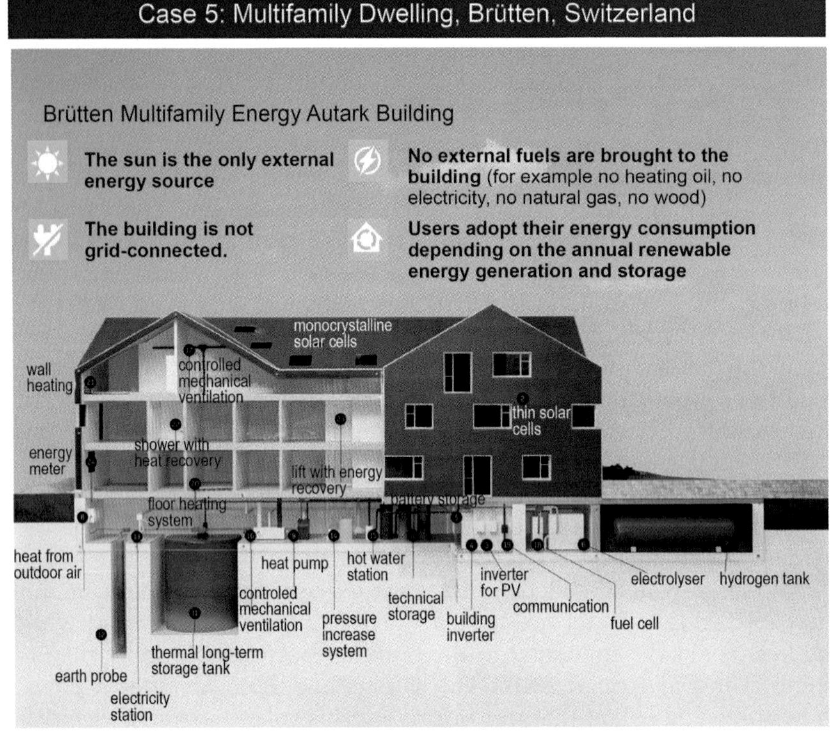

FIGURE 11.13 Schematic cross section explaining the difference energy systems and building services for the first energy autarchy family housing in the world (photo credit: Umwelt Arena Schweiz 2016). *Available from <http://www.umweltarena.ch/uber-uns/energieautarkes-mfh-brutten/> (accessed January 2018).*

to achieve a maximum degree of efficiency on the heating pump (COP). This reduces the electrical power used by the heat pump, as well as the total electrical energy consumption.

Feedback and Occupant Satisfaction

Special focus has been on user behavior. An energy monitoring and BMS records each tenant's energy consumption and provides real-time feedback according to a daily energy consumption budget. The building has a built-in information system helping the residents optimize their energy usage of warm water, lighting and heating, PCs, and other electrical devices. In each apartment, a tablet is installed on the wall which indicates how much percent of the daily, weekly, and monthly energy consumption is used in the home (see Fig. 11.12D). Tenants do not pay direct energy costs for electricity and heat. However, each tenant receives an electricity budget with a bonus/penalty scheme. During summer, residents reported their ability to consume only 50%–60% of their daily electricity budget. However, during winter the percentage is likely to be higher and when tenants consume more than 100%, they have to pay for it. For example, a look at the tablet on the wall in one of the apartment shows that on Monday only 30% was consumed, but on Tuesday about 75% is consumed. Residents recognized that on Monday they ate out, so they did not cook and were out of the house for a long time. On Tuesday, on the other hand, they used the washing machine and dryer. This shows the change of behavior in relation to the use household appliances (Raymann, 2016).

Also, residents reported high residential comfort and satisfaction after 1 year of occupancy. After 1 year, all residents respected the energy budget as planned and reported a change and adaptation of their behavior. Tenants indicated that they lived comfortably and consumed 2200 kWh in their first year of operation, compared to about 4400 kWh that would be consumed in a similar $4^1/_2$-room apartment. However, many occupants did not appreciate not being able to tilt the windows (Raymann, 2016).

Moreover, CO_2-neutral mobility was provided through electric vehicle charging points without additional energy requirements. The residents of the solar-powered multifamily house are provided with two eco-friendly vehicles: An electric vehicle and a bio/natural gas car. The electricity for the electric car is produced with the in-house PV system, as much biogas is available for the bio/natural gas car as can be obtained from the biological waste of all the inhabitants (compogas process).

6.3 Learned Lessons

The Multifamily dwelling project in Brütten is considered a state-of-the-art zero energy building. Several ideas and concepts are successfully applied in this project, which is strongly inspired by the 2000-Watt society concept. The 2000-Watt society is an environmental vision, first introduced in 1998 by the Swiss Federal Institute of Technology in Zürich, which pictures the average. industrial country citizens reducing their overall average primary energy usage to no more than 2000 W (48 kWh/day) by the year 2050, without lowering their standard of living (see Chapter 2: Evolution of Definitions and Approaches). In this context, two key lessons learned can be summarized from this project.

The first lesson learned is the importance of providing direct feedback to users to ensure that they are engaged in achieving the performance goal. By displaying the electricity and water consumption, occupants were motivated to live slightly more economically. For example, the POE study reveals that occupants would hang up the laundry, instead of using the dryer.

The second lesson learned is the importance of setting an energy budget for NZEB. The electricity budget forced the tenants to rethink their lifestyles as they only have as much energy as the building is able to produce and store. The energy budget made occupants aware about energy consumption and the energy efficiency of their household appliances. This includes being aware about energy intensive appliance such as a dryer and clothing iron. Occupants reported their increase of awareness and they started to pay more attention to their energy consumption by using energy-efficient equipment. The TV has the energy label A + + +, and the artificial lighting is provided by LED lamps. Occupants became more aware of their energy consumption since they moved in the multifamily dwelling in Brütten (Raymann, 2016). These tenants are reducing their energy footprints to almost zero and helped develop a sustainable lifestyle. At the same time, they changed their lifestyle without restrictions.

7 DISCUSSION

In this chapter, five NZEB case studies were presented to identify characteristics that promote good indoor environmental quality (IEQ) and achieve energy neutrality. This chapter builds upon several other studies identified in Section 1 and looks at different KPI and HVAC technologies presented in Fig. 11.14. We identified several common errors and mistakes that design teams went through in the five case studies. Section 8 (Lesson Learned #11) highlights those common learned lessons.

Based on Fig. 11.14 we created Table 11.1. The table allows us to compare the five case studies and quantify their KPIs. The projects represent

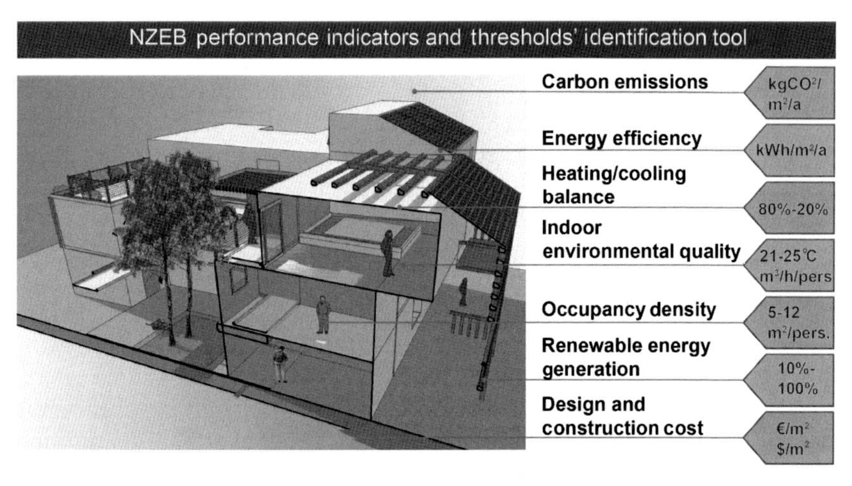

FIGURE 11.14 The different performance indicators that were used to generate Table 11.1 (see Chapter 3: NZEB Performance Indicators and Thresholds for details).

TABLE 11.1 Comparison of the Key Performance Indicators of the Five Case Studies

	Carbon Emissions ($kgCO_2/m^2$ per year)	Energy Efficiency (kWh/m^2 per year)	Heating/Cooling Balance in %	IEQ Satisfaction (%)	Occupancy Density ($person/m^2$)	RES Generation	Cost (€/m^2)
Office Govt.	100	45	85/15	65	15	60% On-grid	2805
Office Govt.	200	80	40/60	95	12	100% On-grid	2000
Industry	50r	35	100/0	90	12	100% On-grid	2495
School	35	21.6/19.6	60/40	75	6	85% On-grid	1950
Housing	30	25r	95/5	95	30	100% Off-grid	2600

five state-of-the-art NZEB that varied in function and context. One of the key take-away messages from our comparison was that building performance assessment is not enough. The indicators listed in Table 11.1 are not enough to evaluate the performance of NZEB. The best way to assess these buildings is to perform POE through daily life observations and monitoring of the building's performance and occupant's experience together.

The second important message is that without a systematic approach to close the energy performance gap, improve building performance, and occupant experience, the AEC industry cannot change for the better. In order to make NZEB mainstream, we need a continuous and systematic approach to improve the project delivery process, technology integration, occupant engagement, and experience (see Chapter 12: Raodmap for NZEB Implementation). For each of the presented case studies the author evaluated the NZEB project through interviews and POE and discussed with the designers, builders, and operators how they identified problems and resolved them. The case studies revealed that many design firms and operation managers analyzed their buildings and implemented actions to improve their performance. By examining the monitoring outcome or getting feedback from users, corrective actions were taken. However, what was missing is the last most important step in any process of defining lessons learned. There was the missing step of reviewing the problem and problem-solving process to avoid similar situations in future. As shown below, improving NZEB performance and occupancy experience requires three major steps:

1. Evaluating the problem—clarifying the nature of a problem and recognizing it.
2. Managing and resolving the problem—identifying steps and implanting action to solve the problem.
3. Avoid similar situations in future—review the process and apply preventive measures.

We believe that rather than focusing on step 3, most projects are trapped in their own errors when it comes to NZEB design, construction, and operation. We would like to highlight the importance of step 3 and that a systematic approach of POE for NZEB is missing.

To sum up, we need a dual assessment approach to evaluate NZEB. The dual approach is a mixed building related and occupant related KPI. At the same time, there is a need to focus more on cases studies documentation to distill lessons learned and allow new design teams to avoid common problems related to NZEB design, construction, and operation. There is no need to repeat the same pitfalls in every new

NZEB. The documentation of NZEB and their performance using long-term monitoring data is the first step toward knowledge dissemination and experience sharing.

8 LESSON LEARNED #11

In this chapter, five NZEB case studies were presented in detail including the lessons learned from them. The buildings represent five different typologies and a diversity of use. They do also represent the differences in NZEB performance thresholds, standards, cost, and processes. Those lessons learned are summarized here and organized into five groupings:

8.1 Integrate Building Modeling and Perform Design Reviews

As NZEB have been recently introduced worldwide, most designers and consultants have limited experience. Thermal comfort is the priority in NZEB. The case studies revealed that there is an overestimation of passive cooling strategies. Natural ventilation is limited in NZEB if high IEQ is required. Dynamic building modeling can produce accurate estimates of energy consumption and provide good estimates of thermal comfort, air quality, and overheating risk. The key lesson is to take all of the simulation estimates with a grain of salt and to apply some rudimentary overheating risk analysis to any claims of the passive cooling effectiveness in NZEB, absence of any static simulation result and use only dynamic simulations.

Design reviews can bring a better quality to the early design stages and make sure to size SASs, envelope thermal resistance, glazing to wall ratio, shading, and other common issues related to the building geometry and envelope. Another key lesson is to perform rigorous design reviews and apply some design checklists to make sure that heat gains are minimized. Make sure that key design elements of NZEB, such as SAS, shading devices, thermal bridge free walls, and PV friendly roofs are integrated in the design and sized properly. Design reviews can make sure that the envelope is airtight, the window-to-wall ratio is fit to climate, and that the potential HVAC system choice is a perfect match to the building fabric and function. This includes fire safety compliance and good acoustic conditions. Design teams should explicitly take into account the need for fire safety planning in NZEB. The design of a NZEB and the materials used in its construction can increase its inherent fire load and dramatically change the way fire propagates. Particular reviews for commercial and public NZEB such as schools, hospitals, and high-rise buildings should assess the higher risk in case of fire. The same applies to assuring good acoustics in buildings with high occupancy density.

8.2 Assure Continuous Monitoring and Performance Follow Up

Real-time monitoring and users feedback is vital in NZEB and can be only achieved through qualified operators. The energy trending is valuable for ensuring performance. Ongoing measurement and monitoring is the only way to achieve the NZEB performance target in the long term. ASHRAE high-performance building magazine reviewed 90 case studies and found an average of four HVAC technologies per building (Maor, 2016). This reflects the complexity of technology and the need for intensive technology examination. Design and operation teams must ensure tried and tested products and systems performance as expected. BMS subcontractors should be obliged, through contractual commitment, to adapt the BMS system to people's behavior and need in balance with the performances targets. Without continuous follow up NZEB, lose their value and cannot reduce their overall building energy use.

8.3 Define Performance Targets for the Occupancy Stage and Set an Energy Budget

A building performance-based contract is the only way to bring users and building operators together during the occupancy stage. By setting an energy budget or a limit for energy use intensity, the building operation team will be forced to invest in soft-landings and POE. Building operation teams should also invest in educating occupants to increase their awareness. The case of the multifamily dwelling in Brütten is an interesting case in setting an energy budget for tenants. This incentivized energy-conserving measures and changed occupant behavior. The idea is potentially successful, however it involves educating occupants and raising their awareness about how their lifestyle influences energy consumption.

8.4 Plan Soft-Landing Before Commissioning

The measurement and verification systems need to be commissioned prior to commissioning all the building systems. Plan soft-landing before the commissioning phase and make sure to include it in the RFP and project contract. The key lesson is to perform soft-landings and to plan ahead for any claims from occupants. Add to the technological complexity the changing occupant behavior. Complaints about discomfort can be avoided by measuring the comfort conditions and IAQ. A close monitoring of mechanized shading, comfort, or automated devices will likely assist in improving the visual comfort (including glare), thermal comfort (including overheating and draft), and air quality (including humidity and fresh air).

8.5 Anticipate Add-Up Plug Loads and Systems Operational Deficiency

Take into account add on plug loads and lighting. Plug loads and lighting are the fastest growing sources of energy use in NZEB. In offices, they account for 30%–60% of office and school electricity use. In parallel, HAVC systems and RES do not perform as expected in reality. The monitoring results in the five case studies indicate a discrepancy between the nominal power of PV panels and their real performance, the same was identified regarding the HVAC systems COP. The key lesson is to take all the system performance estimates with a grain of salt and to apply some rudimentary check-ups to existing systems or equipment. Add a contingency plan by oversizing the RES up to 20% (Eley et al., 2017) and back-up systems to meet the heating loads and, more importantly, cooling loads during summer.

8.6 Design Flexible Building Management Systems With Open Communication Standards

Lessons learned from case studies indicate the importance of selecting a flexible BMS. This requires open systems protocols and a scalable platform to allow the building operation team to change and modify the operation of the building HVAC systems and controls. Depending on the occupant behavior and usage changes, BMS requires to be flexible and adaptable to new operational situations. Building automation and control of appliances, HVAC and RES systems, storage systems, and electric vehicles offer great energy savings and flexibility in response to the grid distribution system and price signals, and to balance renewable energy. Therefore, BMS should be based on open communication standards in compliance with IT compatibility. Also, BMS installation and commissioning should be quick and easy. BMS components should be tested by independent certification authorities (meeting international BACnet standards and European recommendations) to offer measurable reliability from the automation to the management level of the building systems.

References

Abbeloos, G., 2011. Equipements Techniques d'une Ecole Passive (Master thesis). UCLouvain, Louvain-La-Neuve, Belgium (in French).

Athienitis, A., O'Brien, W. (Eds.), 2015. Modeling, Design, and Optimization of Net-Zero Energy Buildings. John Wiley & Sons, Berlin, Germany.

Attia, S., 2018. Regenerative and Positive Impact Architecture: Learning from Case Studies. Springer International Publishing, London, UK, ISBN: 978-3-319-66717-1.

Ayoub, J., Aelenei, L., Aelenei, D., Scognamiglio, A. (Eds.), 2017). *Solution Sets for Net Zero Energy Buildings: Feedback From 30 Buildings Worldwide*. John Wiley & Sons, Berlin, Germany.

Baratticci, M., November 22, 2016. Quel intérêt pour les entreprises d'investir dans des bâtiments zéro énergie? Exemple pratique: le nouveau siège de Kumpen Wallonie a Fleurus, I'ICEDO.

Busch, B., 2010. Climatisation: la bonne solution?, passive.be.

DOE, 2012. The Design-Build Process for the Research Support Facility, DOE/GO-102012-3293, June 2012.

Douin, J.H., 2017. Comportement thermique de la Cité administrative de Seraing (Master thesis). Liege University, Belgium (in French).

EIA. (2014). Colorado State Energy Profile: Colorado Quick Facts. Retrieved from http://www.eia.gov/state/print.cfm?sid = CO (accessed: 25.01.18)

Eley, C., Gupta, S., Torcellini, P., Mchugh, J., Liu, B., Higgins, C., et al., 2017. A Conversation on Zero Net Energy Buildings (No. PNNL-SA-128289). Pacific Northwest National Laboratory (PNNL), Richland, WA.

EN, I., 2000. 13829-2000 Thermal Performance of Buildings—Determination of Air Permeability of Buildings-Fan Pressurization Method. British Standard Institute, Britain.

Heaps, C., 2015. Army Net Zero: Lessons Learned in Net Zero Energy; NREL (National Renewable Energy Laboratory) (No. NREL/BR-7A40-62946). NREL (National Renewable EnergyLaboratory (NREL), Golden, CO.

Hempel, L.O., 2017. Autonomous House, p12-17. Available from <https://www.busch-jaeger.de/uploads/tx_bjeprospekte/ABB_Puls_0117_englisch_web.pdf> (accessed September 2017).

IEA, Task 40/Annex 52, 2008. Towards Net Zero Energy SolarBuildings, IEA SHC Task 40 and ECBCS Annex 52 <http://task40.iea-shc.org/> (accessed September 2017).

IFE (2005) Engineering a Sustainable World: Design Process and Engineering Innovations for the Center for Health and Healing at the Oregon Health and Science University, River Campus.

Kurnitski, J. (Ed.), 2013. Cost Optimal and nearly Zero-Energy Buildings (nZEB): Definitions, Calculation Principles and Case Studies. Springer Science & Business Media, London, UK.

Lesage, P., 2015. La ventilation naturelle intensive des bâtiments et leur confort en été – Etude de systèmes inspirés par la stratégie de refroidissement des termitières (Master thesis). Liege University, Belgium (in French).

Maor, I., 2016. Evaluation of factors impacting eui from high performing building case studies. High Perform. Build, ASHRAE, Atlanta, USA.

McDonell, G., 2003. Displacement ventilation. Canadian Arch 48, 32–33.

Neo Construct & IDES Engineering, 2009. La Cité Administrative de Seraing – Naissance du plus grand bâtiment public passif Wallon, Dossier de Conception Energétique.

Pless, S., Torcellini, P., 2012. Controlling Capital Costs in High Performance Office Buildings: A Review of Best Practices for Overcoming Cost Barriers. Preprint. NREL CP-5500-55264. National Renewable Energy Laboratory, Golden, CO, <http://www.nrel.gov/docs/fy12osti/55264.Pdf>. Retrieved from <http://www.ashburnerfrancis.com/SiteAssets/news/Controlling%20Capital%20Costs%20in

%20High%20Performance%20Office%20Buildings%20A%20Review%20of%20Best%20Practices%20for%20Overcoming%20Cost%20Barriers.pdf>.

Raymann, F., 2016. Wir leben komfortabel ohne Stromanschluss. Available from <https://www.swisscom.ch/de/storys/morgen/wohnen-im-energieautarken-haus.html> (accessed September 2017).

Rebours, C., 2016. Analyse du confort thermique et des mouvements d'air d'un bâtiment de bureaux passif – Etude du cas de la Cité Administrative de Seraing (Master thesis). Liege University, Belgium (in French).

Reeder, L., 2016. Net Zero Energy Buildings: Case Studies and Lessons Learned. Routledge, New York.

Torcellini, P., 2010. Monitoring Results of the Research Support Facility, Presentation, Golden, CO.

Torcellini, P., Pless, S., Lobato, C., Hootman, T., 2010. Main street net-zero energy buildings: the zero energy method in concept and practice. ASME 2010 4th International Conference on Energy Sustainability. American Society of Mechanical Engineers, pp. 1009–1017, Retrieved from http://proceedings.asmedigitalcollection.asme.org/proceeding.aspx?articleid = 1607404.

Umwelt Arena, 2016. Erstes energieautarkes mehrfamilienhaus in Brütten. Available from <http://www.umweltarena.ch/uber-uns/energieautarkes-mfh-brutten/> (accessed September 2017).

Voss, K., Musall, E., 2013. Net Zero Energy Buildings: International Projects of Carbon Neutrality in Buildings. EnOB, 2013, Berlin, Germany.

VTM, 2016. Passiefscholen kreunen onder hitte. Available from <https://nieuws.vtm.be/binnenland/206605-passiefscholen-kreunen-onder-hitte> (accessed August 2017).

Xpair, 2017. Poutre froide: climatisation modulaire. Available from <https://conseils.xpair.com>.

Further Reading

Andersen, R.V., Toftum, J., Andersen, K.K., Olesen, B.W., 2009. Survey of occupant behavior and control of indoor environment in Danish dwellings. Energy Build. 41 (1), 11–16.

ASHRAE, 2002a. Guideline 14-2014 (Supersedes ASHRAE Guideline 14-2002), Measurement of Energy, Demand, and Water Savings. Available from <http://www.techstreet.com/products/1888937>.

ASHRAE, 2002b. Guideline 14-2002, Measurement of Energy and Demand Savings. Available from <https://gaia.lbl.gov/people/ryin/public/Ashrae_guideline14-2002_Measurement%20of%20Energy%20and%20Demand%20Saving%20.pdf>.

Attia, S., Eleftheriou, P., Xeni, F., Morlot, R., Ménézo, C., Kostopoulos, V., et al., 2017. Overview and future challenges of nearly zero energy buildings (nZEB) design in Southern Europe. Energy Build. 155, 439–458. ISSN: 0378-7788, <https://doi.org/10.1016/j.enbuild.2017.09.043>.

Construction 21, 2017a. <https://www.construction21.org/belgique/case-studies/be/cite-administrative-passive---ville-de-seraing.html>.

Construction 21, 2017b. <https://www.construction21.org/belgique/case-studies/be/construction-du-nouveau-siege-regional-de-kumpen-en-wallonie-a-fleurus-un-batiment-passif-et-zero-energie.html>.

PMP, 2014. Passive Architecture: Strategies, Experiences & Viewpoints in Belgium Architecture Passive be Passive Platform Maison Passive, Brussels, Belgium.

CHAPTER

12

Roadmap for NZEB Implementation

ABBREVIATIONS

AEC Architectural, Engineering and Construction
ASC active systems controls
BIM building information modeling
BMS building managements systems
EPBD energy performance of buildings directive
EPC energy performance certificate
EUI energy use intensity
HVAC heating, ventilation, and air conditioning
IAQ indoor air quality
ICE information and communications technology
IEQ indoor environmental quality
KPI key performance indicators
NAHB national association of housing builders
NAR national association of REALTORS
NREL national renewable energy lab
nZEB nearly Zero Energy Buildings
NZEB Net Zero Energy Buildings
M&V measurement and verification
POE postoccupancy evaluation
RET renewable energy technologies
RFP request for purpose
SME small and middle enterprise
US United States

1 INTRODUCTION

In this final chapter, we present concrete guidance to businesses, governments, and nongovernmental organization, to reach the Net Zero Energy target on a national, regional, and provincial (state) scale and on

Net Zero Energy Buildings (NZEB)
DOI: https://doi.org/10.1016/B978-0-12-812461-1.00012-5

a project scale. After portraying NZEB definitions and frameworks in Chapter 2, Evolution of Definitions and Approaches and Chapter 3, NZEB Performance Indicators and Thresholds and discussing the different performance indicators in Chapters 5 to 9 we now provide detailed roadmaps for industrial (developed) and nonindustrial (developing) countries following two major approaches:

1. High-tech approach.
2. Low-tech approach.

The roadmaps provide a strategical plan with potential policies to implement NZEB on national or regional scale. The strategical roadmaps describe the key milestones that industrial countries reached to realize and mass-produce high-tech NZEB. We present the historical progress of regulations and technological advancement that lead to current status-quo regarding high-performance buildings in industrial counties. In parallel with this, we present an initial proposition for a strategical roadmap to achieve NZEB in nonindustrial countries. Then, we follow these strategical roadmaps with operational action plans in Section 4.

The key contribution of an operational action plan is that it is focused on the process of NZEB project delivery. We elaborate on the systematic planning and design steps that need to be taken into account in any NZEB project. The third aspect presented in this chapter is related to design tools and modeling of NZEB. We elaborate on the optimal use of building performance modeling and building information modeling (BIM) for the design, construction, and operation of NZEB. As shown in Fig. 12.1, the three major components of this chapter provide guidance on the approach, process, and tools necessary to achieve NZEB on a large scale and on a single project scale.

The chapter aims to provide strategic, organizational, and tactical aspects of NZEB. We present two high-quality roadmaps that can serve NZEB planning, construction, and operation following a high-tech or low-tech approach. The roadmaps are meant to help nations or regions to phase in NZEB on large scale. This is followed and intertwined with detailed guidance and advice on the process planning and design tools needed at project level. The formulation of roadmaps and developing action plans together with modeling tools and approaches prevent owners and building professionals from risks and make them become proficient in NZEB design, construction, and operation.

2 OPPORTUNITIES AND CHALLENGES OF SCALING NZEB

Achieving NZEB status is an ambitious goal. However, this goal is increasingly gaining momentum and becoming achievable across

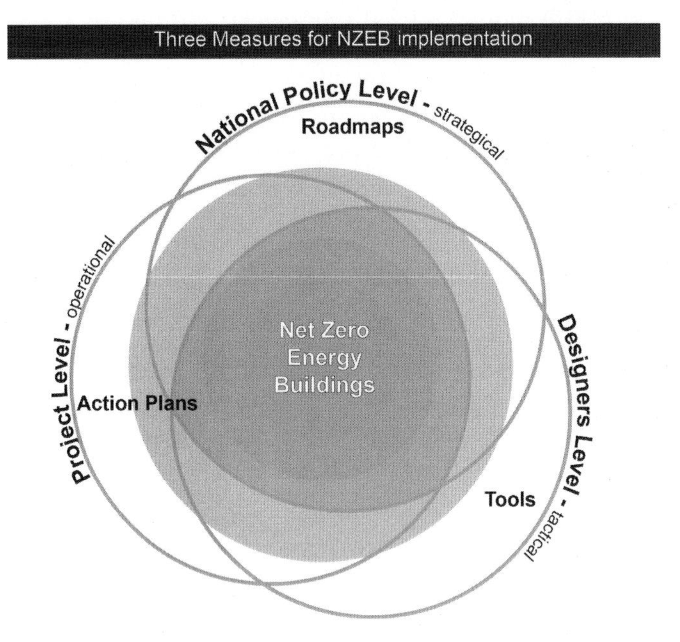

FIGURE 12.1 Three measures to accelerate the implementation of NZEB.

geographic markets and regions. To better understand the potential and risks associated with NZEB upscaling, in this section we present the opportunities and challenges to develop and create NZEB on a large scale.

2.1 Opportunities for Scaling-Up NZEB

In order to curb global greenhouse gas emissions and reduce carbon emissions, several provincial and municipal authorities and administrations will be responsible to specify new energy and carbon-relevant construction standards for new buildings. To make NZEB mainstream, actions needs to be taken on the four levels of governance:

- National Level
- Provincial/State Level
- Municipal Level
- District or Neighborhood Level

In the coming 30 years, governmental authorities will play a major role to promote NZEB and set a policy of concerted regulations that can accelerate the market uptake by setting minimum performance thresholds and performance indicators. As discussed in Chapter 2, Evolution of Definitions and Approaches and Chapter 3, NZEB Performance

Thresholds, NZEB require a legal framework and design approach that is based on simple design target metrics. At the same time, businesses and corporations will disrupt the market with products and technologies that will enable NZEB to be affordable, smarter, more connected, and grid resilient. The accelerating pace of affordable technological innovations (including photovoltaics, heat pumps, batteries, integrated HVAC systems, controls, lighting technologies, and automation systems) will ensure manufacturers and businesses play a major role in increasing NZEB market uptake.

Combining the energy demand side and the energy supply side in one facility, namely a NZEB, is a very cost-effective concept. The NZEB concept empowers users and makes them seek maximum energy efficiency while keeping an eye on their operational energy cost. There are three important opportunities for leading cities, municipalities and provinces/states, and innovation businesses and corporations to invest in upscaling NZEB:

1. Increasing demand of high-performance buildings by future owners

   Potential building owners are driving the market demand for ultra-energy-efficient homes. Building consumers are looking for better comfort, safety, accessibility, and affordability in their future homes (Lalit and Petersen, 2017). According to the US National Association of Housing Builders (NAHB), energy-saving features are on the top of building buyers' priority lists. The demand for high-performance building is driven primarily by Millennials, who seek to cut their operation cost (NAHB, 2016). According to the report of the National Association of REALTORS (NAR), young buyers are seeking affordable and high-performance NZEB. Similarly, in Europe, the market already signals a transformative change in consumer needs and market paradigms.
2. Achieving the ambitions of cities, municipalities, and provinces/states to reduce carbon emissions

   As discussed in Chapter 1, Introduction to NZEB and Market Accelerators, buildings consume between 30% and 40% primary energy in industrial countries and are responsible for at least 30% of carbon emissions. Cities and provinces/states are forced to meet the Paris Agreement targets so that the temperature rise remains below 2°C, and ideally below 1.5°C. Cities and municipalities are committed to meet the 2030 UN agenda for Sustainable Development and the Sendai Framework for Disaster Risk Reduction. Already, several cities recognize the challenge and therefore require ambitious performance thresholds and strict building codes for new constructions. For instance, EU building codes require all new buildings to be nearly zero energy efficient (Attia et al., 2017a). The revised EPBD requires provisions to promote elector-mobility to reinforce the use of building automation and control as well as the introduction of a "smartness indicator" to measure a building's capacity to use ICE and electronic systems. The council of European Municipalities and Regions and the United States have started to embrace climate and sustainability goals. Therefore, reaching the aggressive carbon-reduction goals are likely to be achieved by making the new building stock Net Zero Energy.

3. Renovating the aging building stock

   Almost 75% of the European Building Stock expected to be available in 2050 already exists today, Deep energy retrofits combined with renewable energy supply can accelerate the development of NZEB retrofits while also increasing comfort, safety, and affordability. Scaling up NZEB is a viable solution to turn cities into smart and carbon neutral cities. Encouraging deep renovation is an effective measure to upscale NZEB.

2.2 Market Barriers of Upscaling NZEB

Despite several opportunities mentioned in the previous section of the potential of upscaling NZEB, there are still hurdles and roadblocks ahead of NZEB mass scale development. The energy performance gap is causing a valuation uncertainty, the incremental up-front cost of poorly planned and designed NZEB projects, and nonperformance-based contracting are considered as implementation barriers (as discussed in Chapter 3: NZEB Performance Indicators and Thresholds) to upscaling NZEB. There are three major barriers that impede NZEB upscaling:

1. Economic: (Cost Optimal)

   As part of developing building policies for NZEB, it is crucial to set performance thresholds based on a cost-optimal methodology while taking into account the lifetime costs of the building (Atanasiu, 2013). In order to determine the cost-optimal energy performance level (energy performance level leads to the lowest cost during the estimated economic lifecycle) for NZEB, a comparative framework methodology needs to be established to evaluate the economic and environmental feasibility of all the possible designs (all the possible combinations of compatible energy efficiency and energy supply measures) for every country. The methodology should be used for determining cost-effective NZEB. The calculation of energy costs should be fed with the results of the energy performance calculations. This enables the effect of improved framework conditions (e.g., the introduction of economic incentives such as soft loans, feed in tariff, investment grants, etc.) to be assessed, shifting the economic optimum towards medium- or long-term targets.

Therefore, it is recommended to include ambitious (as well as highly ambitious) measures among selected packages in order to identify the remaining performance and financial gaps and to use these results accordingly to shape further policies and market support programs. However, cost-optimality is not considered in many countries and makes it difficult for private owners to estimate the economic benefits of NZEB.

2. Institutional: (Policy, local governance, codes)

   National and local authorities lack local governance support and national strategies to create an infrastructure for NZEB implementation. One of the main barriers to NZEB market uptake is weak institutional infrastructure. By institutional infrastructure we mean local authority officials dealing with building permits and energy performance certification. In many cases, local authorities are not in contact with local research centers to deepen their understanding of NZEB and its implementation requirements. As a consequence, little effort has been done to provide local guidelines for NZEB procurement. The strong climate variation in some countries demands flexible and regional climate performance requirements for NZEB. For NZEB definitions, descriptive parameters should be clearly identified, but their mandatory target values should be carefully tuned to the individual climates. In this context, many local authorities are not prepared to lead this transition.

3. Technical: (performance gap, overheating risks, simulation)

   One of the most common barriers for NZEB is technical and related to the energy performance gap. Failing to introduce minimum energy performance requirements for buildings, building elements, and technical building systems directly influences NZEB performance. Design and construction practice has generally failed in the past decades to merge new construction materials and HVAC system components with passive design measures. The potential and limits of passive strategies are often either overlooked or overestimated, with a general lack of objective assessment and optimized design. Poor design, construction, documentation, and management are widespread in the standard construction practices of past decades and are still not frequent today. As a result, many NZEB are frequently experiencing summer conditions which are out of the comfort ranges, whatever comfort model is chosen (Attia et al., 2017a).

   A second common technical barrier facing NZEB implementation is methodological. The methodology used to implement NZEB is partly related to inappropriate use of rules-of-thumb or calculation-based design approaches with little feedback from performance monitoring. Most building professionals and researchers rely on steady state simulation tools to address the design and construction of NZEB and have no links to laboratory or field measurements or real performance monitoring.

4. Procedural: (Delivering process, performance-based contracting)

   Project acquisition and delivery methods of NZEB are not conventional. One of the barriers of NZEB implementation is the lack of adopting performance-based contracting for NZEB. The design-bid-build project contracting and delivery is not the best project delivery process of NZEB. The aggressive energy efficiency targets and the intention to close the energy performance gap and provide high quality building performance requires performance-driven contracting methods to allow the design and construction team to achieve the high-performance targets. As discussed in Chapter 9, Construction Quality and Cost, the type of project delivery can influence the project cost significantly and reduce the energy performance gap for NZEB.

5. Industrial: (infrastructure, entities selling integrated solutions and services)

   The construction industry is not prepared with experience and products to deliver and supply the expected market demand. The sustainability business is fundamentally based on local industrial infrastructure of systems, product, and the national energy mix. Many countries are not well positioned to cater for high-tech buildings requiring innovative products, systems, and solutions aligned with the national energy mix and carbon emission reduction targets. There is a need for the development of a national and local bottom up environment of human and industrial infrastructure to carry out the transformation of building practices.

Based on the previous analysis, the construction industry needs to confront these major five barriers. Fig. 12.2 summarizes this section by listing the opportunities and barriers of NZEB upscaling. In the discussion section (Section 6), we elaborate on the potential impact of NZEB on the building sector and how to use the opportunities discussed in this section.

3 ROAD MAP FOR NZEB IMPLEMENTATION (STRATEGICAL FOR NATIONS)

A roadmap is a detailed plan to guide progress towards a target. This plan can be illustrated in a graphical format to provide an overview of a policy or goal and the associated deliverables presented on a

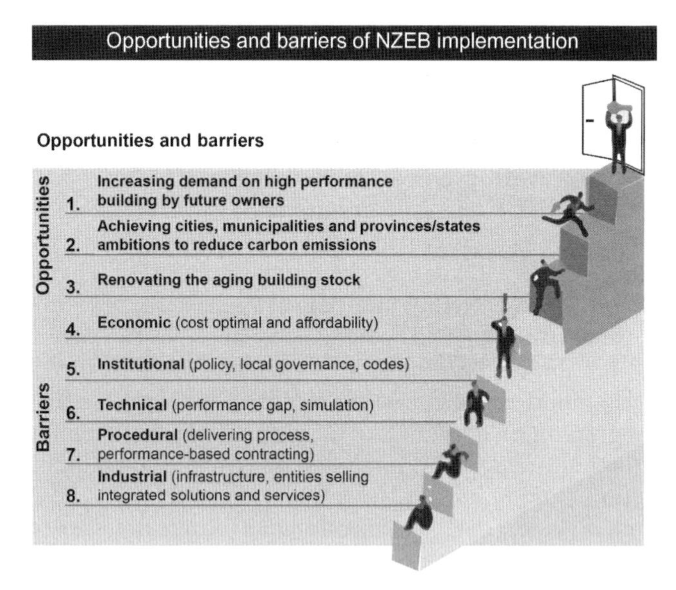

FIGURE 12.2 Opportunities and barriers of NZEB implementation.

timeline. A road map should be simple and free of minutiae. Therefore, in this section we present two different strategical roadmaps that help decision makers and inform important institutional and business changes. Based on case studies presented in Chapter 11, NZEB Case Studies and Learned Lessons, we identified the most important policies, milestones, and steps that need to be taken on a national or regional or provincial/state level. The roadmaps are presented representing two major approaches to achieve NZEB on a strategical level.

3.1 NZEB Roadmap for Industrial Countries

The International Energy Agency (IEA) Task 40 members complied and discussed the early definitions of NZEB comprising 20 industrial countries in 2008. The United States discussed the definitions within the Energy Independence and Security Act of 2007 and the European Union discussed these definitions within the recast of the EPBD adopted in May 2010. As shown in Fig. 12.3, the roadmap for high-tech NZEB is fundamentally based on the track record of building energy efficiency regulations that evolved since the energy crises in 1972 under the leadership of IEA, ASHRAE, and EPBD. The proliferation of the Passive House standard and the advancement of building service technologies are the foundations for introducing the NZEB target in the industrial North.

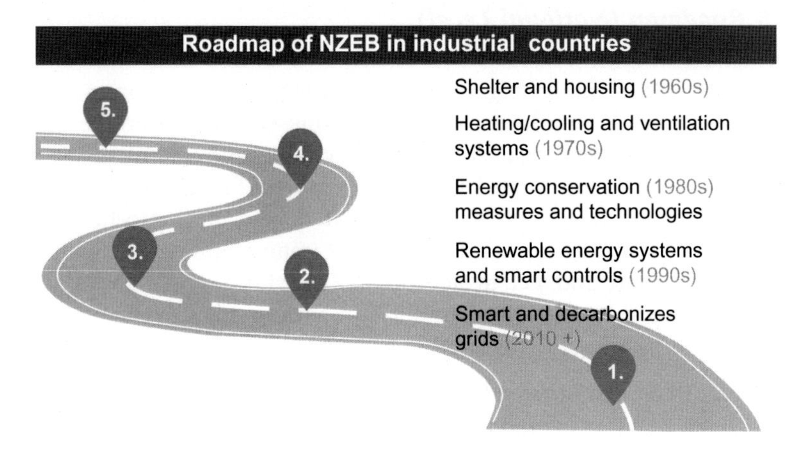

FIGURE 12.3 NZEB Roadmap in industrial countries.

Historical Context

The common practices of code compliances for building energy efficiency since the 1980s require applying energy conservation measures in the building practice. This was coupled to comfort and air-quality standards and technologies. According to Fig. 12.3, the key milestone for NZEB in industrial countries was to set a performance target for energy efficiency and renewable energy. However, this milestone is already succeeding three previous milestones that have been achieved. The first milestone was the post-World War II economic expansion that provided shelter and housing to all industrial countries including the United States, Western Europe, and East Asian countries. The Post-War housing boom resulted in the creation of affordable housing and businesses as authorities managed to provide a new form of living. The second milestone was in 1970 when most buildings in industrial countries were equipped with heating or cooling systems that avoid the use of firewood. The third key milestone was achieved in 1990 when most buildings incorporated energy conservation measures and technologies. Integrating renewable energy systems and smart controls is the fourth milestone. The fifth milestone that industrial countries are looking to achieve is to achieve smart NZEB and decarbonized grids. Industrial countries' roadmap towards NZEB is based on a long-term strategy that started in the after WWII to provide affordable and comfortable housing for most of its citizens. Until the 1980s, energy efficiency and energy neutrality was a dream. However, as a reaction to the Arab Oil embargo in the 1970s, the energy efficiency paradigm and energy neutrality paradigm became dominant (Attia, 2018). Since then, the plan to reach the energy neutrality goal of buildings, identify funding sources, and developing projects started.

NZEB Roadmap (National Level)

Industrial counties have the capacity to manage and control their building stock and plan its growth. To implement NZEB on a national level a road map is required to inform businesses and building owners to develop their projects. Authorities, construction industries, and energy industries should lead this process of NZEB uptake increase to improve the building stocks' energy efficiency and produce green energy. Many nations across Europe and many States in the United States should plan and act to achieve NZEB status and should follow the pragmatic implementation strategies mentioned below:

1. Milestone 1: Definition
   Develop a definition of NZEB including a set of performance indicators and performance metrics. The definition should quantify the performance threshold, e.g., EUI identified in Chapter 3, NZEB Performance Indicators and Thresholds. The performance threshold should set the greenhouse gas emissions limits and energy expectations. See Chapter 3, NZEB Performance Indicators and Thresholds.
2. Milestone 2: Comfort Model
   Establish a national level comfort model with a climate severity index to represent the climatic context on a local level. The comfort model should be based on the socioeconomic context to estimate the heating and cooling expectations and should include an adaptive comfort model and address fuel poverty (Attia and Carlucci, 2015a). See Chapter 5, Occupants Well-Being and Indoor Environmental Quality.
3. Milestone 3: Prototypes and Design Scenarios
   Develop optimal design prototypes for different building typologies and functions. By exploring different design scenarios experts should develop a series of prototypes and design strategies sets (Ayoub et al., 2017). The prototypes should integrate energy efficiency measures, HVAC systems, and RES adapted to different climatic and geographic regions context.
4. Milestone 4: Cost-Optimal ZEB and Incentives Program
   Establish cost-optimal NZEB prototypes every 5 years based on Milestone 3. Based on the cost-optimal approach, the investment gap and challenges to increase NZEB market uptake in new construction and renovations can be identified. This will inform and facilitate the decision-making process of governments, investment institutions, and individuals. This will also inform national governments to better provide subsidies and incentive programs for different design options, including energy conservation measures, renewable energy technologies (RET), and active systems controls (ASC) (Hamdy et al., 2017; Kurnitski et al., 2011).

5. Milestone 5: Project Database and Observatory
   Create a national database of NZEB projects documenting their performance as-designed and as-operated. The database should provide monitored data and lessons learned from real case studies. A national observatory will help to create a database of monitored NZEB.
6. Milestone 6: Technologies Set
   In relation to Milestone 8, Industrial Infrastructure, develop a set of technologies that will serve NZEB nationally. HVAC and control technologies adapted to the national NZEB prototypes and design scenarios. For example, the reliance on heat pumps, pellet boilers, district heating, or cooling should be supported and coordinated on a national level. The development of a technologies solution set should be industrial in coordination with local industrial and service providers in relation to building services and energy systems. See Chapter 7, Energy Systems and Loads Operation.
7. Milestone 7: Modeling Tools
   Develop a monitoring-based modeling tool that can be used by building professionals. The modeling tool should allow dynamic simulation in relation to national energy comfort standards. The tool should generally include prototypes, local schedules, local weather files, HVAC systems, and material libraries.
8. Milestone 8: Certification and Performance-based Contracting
   Develop certification schemes for NZEB. A certificate should expire after 5 years of the building construction date. The certification should be coupled to national standards, EUI, and indoor environmental quality (IEQ) performance thresholds. The certification should assure energy performance and performance-based contracting. See Chapter 4, Integrative Project Delivery and Team Roles.
9. Milestone 9: Industrial Infrastructure
   Create and support the industrial infrastructure behind NZEB. This should be based on involving industrial stakeholders dealing with products, manufacturing, and supply. This includes capacity building and convocational training for building professionals dealing with the design and construction processes. See Chapter 9, Construction Quality and Cost.

As shown in Fig. 12.4, the roadmap builds on previous experiences and knowledge presented in the previous chapters. The strategical approach presented in this roadmap facilitates the work of building professionals when they design NZEB on a project level, as shown in the Section 4.

NZEB roadmap on a national level

Milestones

1. Definition
2. Comfort model
3. Prototypes and design scenarios
4. Cost-optimal n/N ZEB and incentives programs
5. Project database and observatory
6. Technologies set
7. Modeling tools
8. Certification and performance-based contracting
9. Industrial infrastructure

FIGURE 12.4 Milestones of NZEB Roadmap (National Level).

3.2 NZEB Roadmap for Nonindustrial Countries

Energy transition and resource efficiency are as important in industrialized countries as in nonindustrial countries. The goal of reducing carbon dioxide emissions, increasing energy efficiency, and increasing renewable energy share became a reality for cities and buildings worldwide. The idea of setting an ultra-high-performance target for buildings coupled to an energy-use intensity index and carbon emissions intensity index is a sound idea to achieve the abovementioned goals. However, the way of reaching those goals is totally different in industrials countries and in nonindustrial countries. Nonindustrial countries face different types of challenges, according to their societal reality, that require different paths and approaches. In this section, we explore the historical context of nonindustrial countries with building energy efficiency and comfort, and provide a roadmap to reach NZEB status.

3.2.1 Historical Context

Typically, on the one hand NZEB are high-tech equipped facilities that are the fruit of an interconnected web of high ecological standards in construction and technology. They evolved over years of research and development and practical implementation to become energy independent and carbon neutral. Industrial countries addressed energy efficiency and renewable energy after meeting housing needs and achieving sufficient comfort conditions in their built environment. On the other hand, most nonindustrial countries are looking to close the housing shortage gap and slowly address comfort and indoor air quality issues. In this context, importing the NZEB concept as characterized in industrial countries is challenging. The evolution of comfort and indoor environmental quality in both sides of the world, North and South, are not the same. In fact, indoor environmental quality depends on socioeconomic status and progress. Therefore, the NZEB concept requires integration work to suit the needs of nonindustrial countries and reflect their socioeconomic progress. This can be achieved by integrating low-tech sustainable solutions and technologies that are economically justifiable and architecturally interesting. The use of passive and low-tech architectural solutions that meet the social needs of users and the ecological challenges facing humans in nonindustrial countries requires the development of a low-tech NZEB.

We can identify three themes that lead to approaching NZEB in a different way and finally lead to achieving low-tech NZEB. The first theme is related to the identification of the comfort level. In industrial countries, the comfort concept is well-established and the building services industry reached almost 100% penetration in buildings. In most industrial countries, there is a variety of comfort models too. With the incremental improvement of building energy efficiency in the past 30 years, the awareness about comfort and the advancement of comfort models resulted into narrow temperature ranges for thermal comfort (Attia and Carlucci, 2015a). The high-tech definition of NZEB is mainly based on stringent comfort models. However, the low-tech NZEB definition DOEs do not need to be based on those stringent comfort models. There is a serious opportunity to develop new definitions and concepts for NZEB in nonindustrial countries that integrate more tolerant adaptive comfort models, reflecting the socioeconomic status. As a direct result, the total energy consumption will decrease and the energy neutrality balance can be more easily achieved. Also, we recognize that an over focus on energy performance can lead to health and comfort problems. Especially in nonindustrial countries, where wood burning stoves are commonly used resulting into carbon monoxide, nitric oxide, nitrogen dioxide, and suspended particles, including benzo(a)pyrene (A Schueftan et al., 2016).

The second theme is related to the selection and use of active systems. As shown in Table 12.1, the high-tech approach is based on mechanical ventilation with heat recovery, which requires advance technologies. Those systems and building services installations are mostly imported and coupled to sophisticated controls. It is not necessary to follow the Passive House approach or the Active House approach or similar high-tech approaches (see Chapter 2: Evolution of Definitions and Approaches) using advances HVAC systems. Depending on the

TABLE 12.1 Comparison Between the High-Tech and Low-Tech Approaches for NZEB

	High-Tech Approach	**Low-Tech Approach**
EE Target	15–25 kWh/m^{2}y	min. 15 kWh/m^{2}y
RES Target	15 kWh/m^{2}y	30–45 kWh/m^{2}y
Envelope	Max. Insulation and air tightness static and adaptive models	Max. bioclimatic and passive design solutions
Thermal Comfort	Static and adaptive models	Adaptive models
Air Quality	1000 ppm	1000 ppm
Behavior	Conscious and based on rigid operation schedules	Conscious and adaptive
Systems	Mechanical ventilation with heat recovery, ultra-efficient HVAC systems.	Hybrid ventilation, with individual heating and cooling unit
Controls	BMS (see Fig. 10.7)	None or manual (see Fig. 10.7)
Monitoring	Real-time and full monitoring using smart meters.	Monthly and manual energy consumption readings
Operation	Full time dedicated expert or facility manager (see Fig. 10.7)	By users (see Fig. 10.7)
Cost	Cost-Optimality calculation every 5 years coupled to incentives	Cost-Optimality calculation every 5 years coupled to incentives

local climate and the technological infrastructure of countries in the South, a set of low-tech building service products and solutions can cater towards a low-tech NZEB. The use of ceiling fans or mobile stoves (gas canister) are two examples to avoid costly hydronic central heating systems. Another serious opportunity emerges here to develop new building service technologies that are adapted to the socioeconomic status and local energy market in nonindustrial countries.

The third theme is related to the influence of appliances and lighting on the total energy consumption. In high-tech NZEB, appliances dominate the energy use and the energy breakdown indicates a small contribution of heating or cooling loads to the total consumption (see Fig. 10.2). The ultra-low energy needs for heating or cooling due to the envelope insulation and airtightness magnify the influence of occupants' behavior and appliances. As a consequence, plug-loads, and lighting have a significant share of the total energy consumption. The energy performance gap in most NZEB is associated with the influence of occupants' behavior on the consumption patterns of lighting and plug-loads, besides space heating or cooling. In this sense, the challenge faced by industrialized countries is closely linked to the reduction of these burdens and in trying to educate users on more conscious use. In the case of nonindustrial countries, the low plug-loads and lighting can be a way to achieve net zero balance faster. In nonindustrial countries, the dominant energy use is influenced by energy needs for heating or cooling. This is followed by lighting and plug-loads. Users in nonindustrial countries do not have the conditions to have many of the technological devices associated with plug-loads, and in general users are more concerned about energy consumption (Attia et al., 2017b).

3.2.2 NZEB Roadmap (National Level)

Based on Table 12.1, the low-tech approach for NZEB is a practical approach. Success in nonindustrial countries should be measured by moving towards nearly zero energy buildings (nZEB) where practical and affordable measures can be applied. As shown in Fig. 12.5, the roadmap towards NZEB in nonindustrial countries can only be achieved in the long-term. In the short- and mid-term, nonindustrial countries should develop common national or regional visions integrating the industry, government, professionals, local community, and health institutions. The first step of achieving affordable, healthy, and energy-efficient buildings should focus on comfort and indoor air quality. The over focus on energy performance and energy efficiency in nonindustrial countries is not necessary because it can lead to health and comfort problems in highly polluted urban agglomerations with extreme climatic conditions. In parallel, improving indoor air quality and developing new comfort models should be the first milestone. Integrating low-tech strategies, solutions, and technologies that can be delivered by the local or regional industry including SMEs can be considered as the second key milestone. On the mid-term, low-tech nZEB that integrate RET and ASC can be achieved. From this moment on, nonindustrial countries can explore and plan their movement towards high-tech NZEB towards the fourth milestone in the NZEB national and strategical roadmap.

4 ACTION PLAN FOR NZEB PLANNING AND DESIGN (OPERATIONAL FOR PROJECTS)

In this section, we present an action plan to identify the most important operational and tactical changes that should take place on the project level. The previous section presented strategical roadmaps that need to be adapted on a national level in relation to policy, industrial infrastructure, and businesses. However, in this section we present detailed

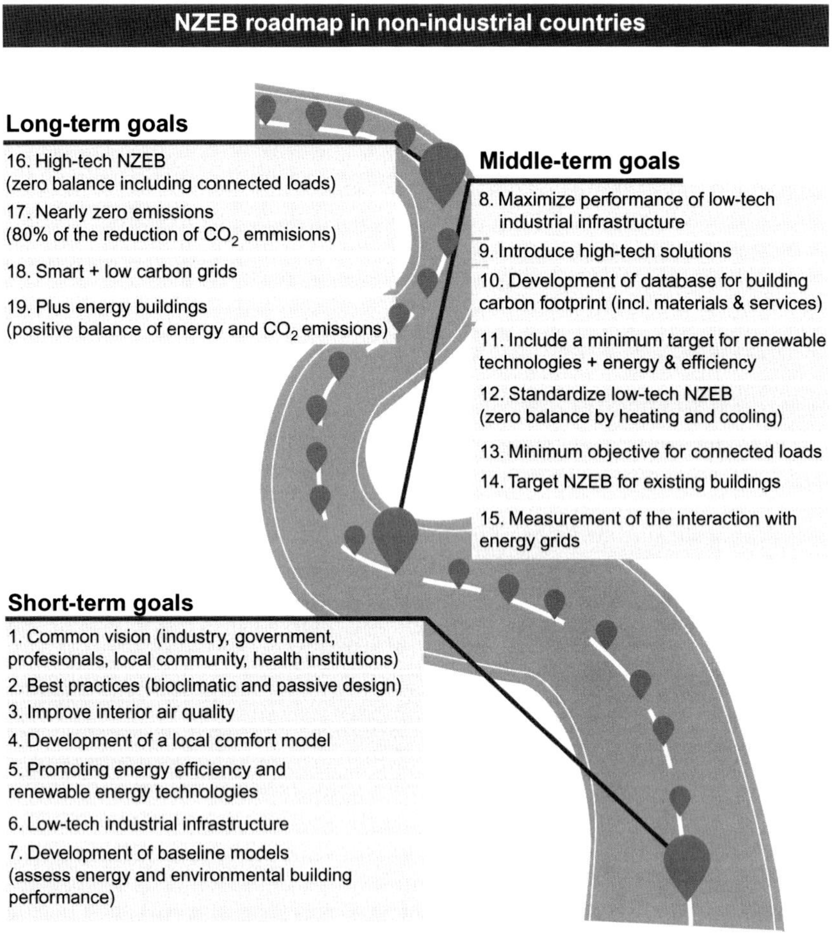

FIGURE 12.5 NZEB Roadmap for nonindustrial countries.

aspects of NZEB planning, design, construction, and operation that are useful for building teams designers, builders and operators. The action plan presents a high-level of informed decision-making that can guide the NZEB implementation process. The secret of achieving robust and resilient NZEB is based on proper planning, adapted design, and neat construction execution. The following action plan presents a detailed guidance and is illustrated in Table 12.2.

Step 1: Planning Stage

- Design a simple and cost-effective building.
- Use cross-discipline design and decision-making and create a multidisciplinary team.

TABLE 12.2 Action Plan and Checklist for NZEB

1. Planning Stage	Design a simple and cost-effective building	☐
	Use cross-discipline design and decision-making and create a multidisciplinary team	☐
	Engage experienced key subcontractors early in the design process	☐
	Incorporate early analysis and modeling tools to support the design	☐
	Incorporate a continuous value engineering process	☐
	Establish performance target thresholds and KPIs and set a green building certification scheme	☐
	Prepare a certification team for code, standards and green certification compliance	☐
	Plan the whole project delivery process and prepare the RFP	☐
2. Request for Purpose Stage	Interact with architectural and engineering team	☐
	Prepare the performance-based contracting requirements with measurable performance targets	☐
	Set-up the certification and standards for compliance	☐
	Engage in a Design and Building project delivering process	☐
3. Design Stage	Integrate the RFP requirements into design	☐
	Building modeling must be included in the design process	☐
	Review the design team submittals	☐
	Investigate the potential of electrification of HVAC systems	☐
	Avoid the overheating risk and assess the limitation of natural ventilation	☐
	Reduce the space conditioned volume	☐
	Take into account that add-on plug loads that build up during operation and foresee a contingency plan during the design stage.	☐
	Select energy efficiency features and optimize the envelope design	☐
	Address the construction nodes and make sure to use heat transfer modeling	☐
	HVAC systems and building controls do not perform as expected in many NZEB. Therefore, a contingency plan is required to be integrated during early design strategies	☐
	Downsize or eliminate heating or cooling systems equipment if possible and rely on tested technologies, flexible and open-box BMS	☐

(*Continued*)

TABLE 12.2 (Continued)

4. Construction Stage	Review the RFP requirements	☐
	Maximize the use of prefabricated modular construction elements	☐
	Investigate the cost reduction by optimizing the structural system	☐
	Make sure construction nodes are implemented properly and check the envelope airtightness	☐
	Plan the onsite inspections and conduct measurements and verification on-site for construction quality and system integration	☐
	The measurement and verification systems need to be commissioned prior to commissioning all the building systems and controls	☐
	Plan the soft-landing process and involve the building operation team in the commissioning process	☐
5. Occupancy Stage	Real time monitoring and user feedback is vital. The energy trending is valuable for ensuring performance.	☐
	Perform soft-landing and plan ahead because it takes time	☐
	Manage the on-going measurement and monitoring and facilitate resources for these	☐
	Educate and train occupants to take action to minimize wasted energy	☐
	Perform postoccupancy evaluation and make comfort a priority before energy efficiency	☐
	NZEB are based on layered approaches and integrated systems. They require continuous calibration and monitoring. Implement low-cost measures to find and fix wasted energy	☐
	Track successful NZEB lessons learned and best practices and share them through professional documentation	☐

- Engage experienced key subcontractors early in the design process.
- Incorporate early analysis and modeling tools to support the design.
- Incorporate a continuous value engineering process.
- Establish performance target thresholds and KPIs and set a green building certification scheme.
- Prepare a certification team for code, standards, and green certification compliance.
- Plan the whole project delivery process and prepare the RFP.

Step 2: Request for Purpose Stage
- Interact with architectural and engineering teams.
- Prepare the performance-based contracting requirements with measurable performance targets.
- Set-up the certification and standards for compliance.
- Engage in a Design and Building project delivering process.

Step 3: Design stage
- Integrate the RFP requirements into the design.
- Building modeling must be included in the design process.
- Review the design team submittals.
- Investigate the potential of electrification of HVAC systems.
- Avoid the overheating risk and assess the limitation of natural ventilation.
- Reduce the space conditioned volume.
- Take into account that add-on plug loads build up during operation and foresee a contingency plan during the design stage.
- Select energy efficiency features and optimize the envelope design.
- Address the construction nodes and make sure to use heat transfer modelling.
- HVAC systems and building controls do not perform as expected in many NZEB. Therefore, a contingency plan is required to be integrated during early design strategies.
- Downsize or eliminate heating or cooling systems equipment if possible and rely on tested technologies and flexible open-box BMS.

Step 4: Construction Stage
- Review the RFP requirements.
- Maximize the use of prefabricated modular construction elements.
- Investigate the cost reduction by optimizing the structural system.
- Make sure construction nodes are implemented properly and check the envelope airtightness.
- Plan the onsite inspections and conduct measurements and verification on-site for construction quality and system integration.
- The measurement and verification systems need to be commissioned prior to commissioning all the building systems and controls.
- Plan the soft-landing process and involve the building operation team in the commissioning process.

Step 5: Occupancy stage
- Real time monitoring and user feedback is vital. The energy trending is valuable for ensuring performance.
- Perform soft-landing and plan ahead because it takes time.
- Manage the on-going measurement and monitoring and facilitate resources for these.
- Educate and train occupants to take action to minimize wasted energy.

- Perform postoccupancy evaluation and make comfort a priority before energy efficiency.
- NZEB are based on layered approaches and integrated systems. They require continuous calibration and monitoring. Implement low-cost measures to find and fix wasted energy.
- Track successful NZEB lessons learned and best practices and share them through professional documentation.

5 MODELING AS A TOOL FOR NZEB DESIGN (TACTICAL FOR DESIGNERS)

Advanced modeling techniques are the main way to lead designs towards the optimal design, construction, and operation of NZEB. As mentioned in Chapter 4, Integrative Project Delivery and Team Roles, the design of NZEB relies on tools that must be used during an integrated design process to achieve high degree accuracy in performance prediction and building construction. Appropriate modeling of building integrated energy systems and optimal control strategies can only be tested and verified through building performance simulation. Based on systematic mapping of different case studies we identified two main tools that must be used during the development of NZEB (Attia, 2016; Attia and Bashandy, 2016, 2018). In this section, we present building performance modeling and BIM as tactical tools that should be used by designers.

5.1 Building Performance Modeling

The design and modeling of Net Zero Energy Buildings is a challenging and complex problem of increasing importance. Informing the uncertainty of designers during early design stages for decision-making is very important (Attia, 2012; Attia et al., 2012; Athienitis and O'Brien, 2015). The role of building performance simulation and modeling tools in supporting and providing information for design decision-making is essential for NZEB. Modeling tools can help to:

1. Develop and evaluate building energy efficiency.
2. Set building performance levels based on cost optimality.
3. Evaluate impact and direction for ECM, RET, and ASC

By building modeling we mean dynamic performance simulation methodologies and optimization. Appropriate modeling is essential to assess thermal, electrical and daylighting performance of NZEB. Modeling tools enable design teams to accurately estimate and balance the relationship between meeting building loads, finding fit-to-purpose

HVAC systems, and the supply and demand side of energy. The interaction between different parameters and building variables requires integrated and complex modeling packages to assess the interactions between the design features. Using building modeling during early design stages is effective to meet the performance goals (Attia et al., 2013a). There are five areas where NZEB modeling should take place during NZEB design:

1. Cost and Budget Modeling
   Cost and budget modeling for NZEB should take place during early design stages to avoid extra costs while investigating potential savings for NZEB facilities. Lifecycle cost analysis should be performance-based on cost models that deal with design features, energy performance, cost analysis, and capital investment (Hamdy et al., 2017). An analysis of performance and cost parameters can provide a wider perspective to integrate cost-effective measures and perform early value engineering (DOE, 2010).
2. Comfort and Overheating Modeling
   Comfort modeling should take place as early as possible to select the fit-to-function comfort model while avoiding overheating risks. Using dynamic thermal modeling can inform the design team on potential risks (Attia and Carlucci, 2015a). Designers and modelers must make sure to use the correct climate data and seek model calibration after construction.
3. Energy Performance Modeling
   Energy performance modeling allow assessing the site conditions, building size, and building mass during early conceptual design phases. Architectural design features such as orientation, glazing area, envelope composition, and thermal massing can be explored. The integration of natural ventilation and solar shading as part of the architectural design can lead to significant energy savings and should be extensively explored through parametric analysis. Active systems including HVAC systems and RET should be also modeled. ASHRAE high-performance building magazine reviewed 90 case studies of high-performance buildings and found four HVAC solutions on average per building (Maor, 2016). Energy performance modeling can help in the selection of HVAC solutions to achieve IEQ and energy efficiency. Modeling can help design teams to reduce cooling and heating loads while optimizing the system efficiency. The energy performance modeling will help to select and size RET and assess different control models, occupancy, and load profiles.
4. Daylighting Modeling
   Daylighting models are influential to reduce energy consumption and determine optimal glazing areas, orientation, and shading.

Lighting is typically the largest energy end-use in office buildings. Advanced daylighting modeling tools should be used to optimize the design of natural lighting characteristics and louvers. Daylighting modeling has to be coupled to energy modeling to decide on lighting control concepts in relation to lighting and occupancy sensors.

5. Plug and Process Loads Modeling

   Advanced plug and process loads modeling is another important task that should be addressed during early design stages. To reach high levels of energy efficiency, plug and miscellaneous loads should be properly estimated. They can help the design team to set an energy budget to limit the plug and process loads. Reducing plug loads should be achieved with the help of energy simulation tools.

Building performance modeling (BPM) can be supportive when integrated early on in the design process. Simulation in theory handles dynamic and iterative design investigations, which makes it effective for enabling new knowledge, analytical processes, materials and component data, standards, design details, etc., to be incorporated and made accessible to practicing professionals. Informed decision-making (or informed design choice) forms an essential basis for the design of NZEB. Providing knowledge prior to the decision-making can influence the decision attitude and lead to nearly optimal and optimal design solutions (Attia, 2012a). Automated mathematical building performance optimization (BPO) paired with building performance simulation (BPS) is a promising solution to use as a means to evaluating many different design options and obtain the optimal or near optimal solution for a given objective or combination of objectives (e.g., lowest lifecycle cost, lowest capital cost, highest thermal comfort) while achieving fixed objectives (e.g., net zero-energy) (Attia et al., 2013b, 2015b). Addressing all building design parameters in a holistic approach allows the optimization of geometry, envelope, comfort, systems, and renewables. The often-ill-defined design problems of NZEB can be defined as problems with explicit multiobjective criteria. This will establish fully integrated NZEB designs where the builder designers can act to influence the direction of the optimization (Attia et al., 2013a).

5.2 Building Information Modeling

The digital transformation of the construction industry will advance the project delivering process of NZEB and even Net Zero Energy Communities. The manufacturing of NZEB is already influenced by data-driven automation, inventory management, production controls, and digitalization strategies. In the near future, the physical work of building construction and operation will be simulated and managed through virtual operational software. BIM can streamline and standardize the construction process of NZEB. BIM allows better coordination to manage drawings on the construction site while enabling tracking and updating information continually in real-time and as-built. This brings the design and construction team members closer together resulting in optimizing the project delivery process and reducing costs. BIM allows expanding NZEB during design and on-site while improving the efficiency, safety, and cost. Therefore, we consider BIM as a powerful tool to upscale NZEB design and facilitate their implantation. With the interoperability of different building-related software tools, NZEB will be created based on an integrated workflow that takes into account the lifecycle of buildings.

6 DISCUSSION

NZEB are growing at a global level and the construction industry must be prepared for this market growth. Governments, developers, and service providers require roadmaps and action plans across the world. In this final chapter, we presented several roadmaps, action plans, and tools to facilitate and increase the market uptake of NZEB. In this discussion section, we would like to highlight the top key measures that can amplify the contribution and value of NZEB in the near future.

6.1 Implementing NZEB at a District Scale

Moving from single scale implementation to district scale implementation can accelerate the market uptake and reduce the project's cost. We learned from case studies that operating buildings collectively on a cluster level, community level, or district level can be more efficient and cost effective. Achieving NZEB status on a district-level and coupling them to micro grids is promising and maximizes optimal management interactions. Implementing NZEB on a district scale creates a stronger connection between building energy users and renewable energy system capacities on a larger scale. The increase of demand-supply and the direct interaction between different buildings on the same project site can lead to a locally centralized energy generation. This allows for leverage of synergies such as thermal load electric supply aggregation. The use of central heating or cooling systems becomes cost-effective and boosts energy efficiency, while aggregating buildings in a district can facilitate loads management and make the micro grids more robust and reliable. As a consequence, developers and tenants can reap the benefits of those synergies, reduce their carbon footprint, save money and reduce the upfront investment.

6.2 Promoting Cost-Neutrality

Using informed design is not only effective to scale NZEB on a district scale, but also to become cost-effective and agile. As shown in Chapter 9, Construction Quality and Cost, NZEB can be cost-neutral compared with conventional buildings. A NZEB can be built at no cost increment if it is well designed. The first goal for design teams is to set the performance indicator thresholds and fix the building budget. Owners should find design and contractor teams that will deliver the NZEB while maintaining a cost-neutral budget. Following an integrated design process that empowers contractors leads to more informed and better equipped building professionals to control and reduce any NZEB project budget. Based on Fig. 9.3 the proper selection of project delivering types can lead to a decrease the building cost up to 30% (see Chapter 9: Section 2.2). Modularity and off-site prefabrication can reduce the upfront cost significantly. Also clustering NZEB together on a district scale will likely yield extra cost-reductions. It is crucial to manage the initial cost of NZEB and capitalize the benefits of eliminating operational costs.

6.3 Amplifying NZEB Value

The creation of livable, healthy, comfortable, resilient, and environmentally sustainable buildings can motivate and mobilize future tenants and local building industries. Next to the benefits of producing energy on-site, well-being and indoor environmental quality is important to many owners. NZEB can provide superior comfort, daylighting, and better indoor air quality. The creation of a significant market demand for NZEB based on quality and human factor values can drive the market. Living in a NZEB can bring important valuable health and productivity benefits for individuals and businesses (see Chapter 5: Occupants Well-Being and Indoor Environmental Quality). Amplifying the value of volatile fossil fuel price fluctuation immunity, more stable financial conditions for households and businesses, and a healthier and more sustainable life can make families and businesses willing to seek NZEB solutions. This can be used as a decisive marketing advantage towards the upscaling of NZEB.

7 LESSON LEARNED # 12

NZEB are growing at a global level and the construction industry must be prepared for this market growth. To guide the incremental transition to developing and operating NZEB, guidance is needed on strategic, organizational, and tactical levels. Therefore, in this chapter we presented roadmaps, actions plans, and modeling tools as key parts to solve the NZEB design and upscaling challenge. The roadmaps identified the critical requirements, NZEB performance targets, technology levels, and milestones for meeting those targets. The roadmaps, action plan, and modeling tools identify precise objectives and help focus resources on critical technologies that are needed to meet NZEB objectives. However, the roadmaps must be a perceived need by all those concerned. The process needs input and participation from different relevant groups, which bring different perspectives and timelines to milestones. The roadmap needs broad participation (industry, tenants and suppliers, as well as governments and authorities). In nonindustrial countries, the roadmap development process should be need-driven rather than solution-driven. There must be a clear specification of the milestones and these must be coupled to action plans.

Three measures can amplify the contribution and value of NZEB in the near future:

- Implementing NZEB at a district scale
- Promoting cost-neutrality
- Amplifying NZEB Value

These measures are helpful to promote NZEB if coupled to the presented roadmaps, which is conducive to addressing complex problems such as the acceleration of NZEB market uptake. Owners of NZEB tend to be motivated if the three abovementioned measures are addressed. The emphasis on building a network of all stakeholders, developing a common understanding of NZEB challenges and solutions, and encouraging technological, operational, and societal innovation can increase the private sector adoption of NZEB.

References

Atanasiu, B., 2013. Implementing the cost-optimal methodology in EU countries: Lessons Learned from three case studies. The Buildings Performance Institute Europe (BPIE), March 2013.

Athienitis, A., O'Brien, W. (Eds.), 2015. Modeling, Design, and Optimization of Net-Zero Energy Buildings. John Wiley & Sons.

Attia, S., 2012. A Tool for Design Decision Making-Zero Energy Residential Buildings in Hot Humid Climates, PhD Thesis. UCL, Diffusion universitaire CIACO, Louvain La Neuve, ISBN 978-2-87558-059-7.

Attia, S., 2016. Towards regenerative and positive impact architecture: a comparison of two net zero energy buildings. Sustain. Cities Soc. 26, 393–406. ISSN 2210-6707, https://doi.org/10.1016/j.scs.2016.04.017.

Attia, S., 2018. Regenerative and Positive Impact Architecture: Learning from Case Studies. Springer International Publishing, London, ISBN: 978-3-319-66717-1.

Attia, S., Bashandy, H., 2016. In: Belis, Bos, Louter (Eds.), Evaluation of Adaptive Facades: The Case Study of AGC Headquarter in Belgium. ChallengingGlass 5 – Conference on Architectural and Structural Applications of Glass. Ghent University, Belgium, ISBN 978-90-825-2680-6.

Attia, S., Carlucci, S., 2015a. Impact of different thermal comfort models on zero energy residential buildings in hot climateEnergy Buildings 102, 117–1281 September 2015, ISSN 0378-7788 . Available from: http://dx.doi.org/10.1016/j.enbuild.2015.05.017.

Attia, S., Gratia, E., De Herde, A., Hensen, J., 2012. Simulation-based decision support tool for early stages of zero-energy building design. Energy Building 49, 2–15. Available from: https://doi.org/10.1016/j.enbuild.2012.01.28. June 2012, ISSN 0378-7788.

Attia, S., Gratia, E., De Herde, A., 2013a. Achieving informed decision-making for net zero energy buildings design using building performance simulation tools, Int. J. Building Simulation, vol. 6–1. Tsinghua-Springer Press, pp. 3–21, https://doi.org/10.1007/s12273-013-0105-z.

Attia, S., Hamdy, M., O'Brien, L., Carlucci, S., 2013b. Assessing gaps and needs for integrating building performance optimization tools in net zero energy buildings designs. Energy Building. Available from: https://doi.org/10.1016/j.enbuild.2013.01.16. Available online 28 January2013, ISSN 0378-7788.

Attia, S., Hamdy, M., Carlucci, S., Pagliano, L., Bucking, S., Hasan, A., 2015b. Building performance optimization of net zero-energy buildings. In: Athienitis, A., O'Brien, W. (Eds.), Modeling, Design, and Optimization of Net-Zero Energy Buildings. Wilhelm Ernst & Sohn, Berlin, Germany, https://doi.org/10.1002/9783433604625.ch05.

Attia, S., Eleftheriou, P., Xeni, F., Morlot, R., Ménézo, C., Kostopoulos, V., et al., 2017a. Overview of challenges of residential nearly Zero Energy Buildings (nZEB) in Southern Europe. Energy Buildings .

Attia, S., Hamdy, M., Ezzeldin, S., 2017b. Twenty-year tracking of lighting savings and power density in the residential sector. Energy Buildings 154, 113–126. 24 August 2017, ISSN 0378-7788, 24 August 2017, https://doi.org/10.1016/j.enbuild.2017.08.041.

Ayoub, J., Aelenei, L., Aelenei, D., Scognamiglio, A. (Eds.), 2017. Solution Sets for Net Zero Energy Buildings: Feedback from 30 Buildings Worldwide. John Wiley & Sons.

DOE, 2010. NZEB: Cost Analysis and Cost Modeling: Net-Zero Energy Buildings Expert Roundtable III, U.S. Department of Energy | USA.gov.

Hamdy, M., Siren, K., Attia, S., 2017. Impact of financial assumptions on the cost optimality towards nearly zero energy buildings- a case study. Energy Buildings 153, 421–438. 24 August 2017, ISSN 0378-7788, https://doi.org/10.1016/j.enbuild.2017.08.018.

Kurnitski, J., Saari, A., Kalamees, T., Vuolle, M., Niemelä, J., Tark, T., 2011. Cost optimal and nearly zero (nZEB) energy performance calculations for residential buildings with REHVA definition for nZEB national implementation. Energy Buildings 43 (11), 3279–3288.

Lalit, R., Petersen, A., 2017. R-PACE: A Game-Changer for Net-Zero Energy Homes. Rocky Mountain Institute (RMI), Boulder, CO.

Maor, I., 2016 Evaluation of Factors Impacting EUI from High Performing Building Case Studies, High Performance Buildings. High Performing Buildings, Fall 2016, pp. 38–39.

NAHB, The Millennial Home Buyer. National Association of Housing Builders, Survey, US. Available from: https://www.nahb.org/en/research/design/the-millennial-home-buyer.aspx (accessed January 2018).

Schueftan, A., Sommerhoff, J., González, A.D., 2016. Firewood demand and energy policy in south-Central Chile. Energy Sustain. Dev. 33, 26–35.

Further Reading

Andersen, R.V., Toftum, J., Andersen, K.K., Olesen, B.W., 2009. Survey of occupant behaviour and control of indoor environment in Danish dwellings. Energy Buildings 41 (1), 11–16.

ASHRAE, 2002a. Guideline 14-2014 (Supersedes ASHRAE Guideline 14- 2002), Measurement of Energy, Demand, and Water Savings. Available from http://www.techstreet.com/products/1888937.

ASHRAE, 2002b Guideline 14-2002, Measurement of Energy and Demand Savings. Available from https://gaia.lbl.gov/people/ryin/public/Ashrae_guideline14-2002_Measurement percent20of percent20Energy percent20and percent20Demand percent20Saving percent20.pdf

Heaps, C., 2015. Army Net Zero: Lessons Learned in Net Zero Energy; NREL (National Renewable Energy Laboratory) (No. NREL/BR-7A40-62946). NREL (National Renewable Energy Laboratory (NREL), Golden, CO.

Index

Note: Page numbers followed by "*f*" and "*t*" refer to figures and tables, respectively.

D

E

F

G

J

K

L

T

U

Printed in the United States
By Bookmasters